Environmental Biochemistry

Environmental Biochemistry

Neelima Rajvaidya
Dilip Kumar Markandey

A.P.H. PUBLISHING CORPORATION
ANSARI ROAD DARYA GANJ
NEW DELHI-110 002

Published by
S.B. Nangia
A P H Publishing Corporation
4435-36/7, Ansari Road, Daryaganj
New Delhi 110002
Ph.: 23274050
E-mail : aphbooks@gmail.com

2026

Rs.4495/-

Printed at
Balaji Offset
Navin Shahdara, Delhi 110032

CONTENTS

Chapter -1

UNDERSTANDING BIOCHEMICAL ASPECTS IN RELATION TO ENVIRONMENTAL SCIENCE AND TECHNOLOGY

There are few terms, which has to be define for understanding the concept of environmental biochemistry in Environmental science and technology :

Biology : study of life, study of living systems.

Ecology : relationship of a group of organisms with their environment, physical, chemical and other forms of life. Other forms of life in the system compete for life-giving substances and may in fact prey on each other for food (predation).

Life : has may definations, may be considered animate versus inanimate.

Organisms : living entities, an object which is alive. Organisms may be unicellular (bacteria) or multicellular (plants, animals etc.)

Protoplasm : living material in cell, heterogeneous suspended or dissolved solids in an aqueous solution.

INTRODUCTION

In relation to the subject of water pollution, the background is that of sanitary engineering. Biological concerns began with application of engineering technique to public health problems. Some micro-organisms are *etiological* agents, that is, they are responsible for causing a disease. In the 1850's the engineers' prima concern with water was the problem of delivery of water and its drainage or removal. At this stage the concern was more the field of a hydrologist than a sanitary engineer. Later interest developed in the control of disease which lead to water and wastewater treatment. The recent concern has begun a new field of environmental engineering. Their interest is to protect the health and *well-being* of the population as well as keeping an eye on air, land and water resources. In subsequent decades, the concerns of environmental technologists were as :

1. Water quality management
2. Air resources engineering
3. Radiation protection
4. Industrial hygiene

Now, solid waste have also been added to the list of concerns. These are implications of biology in all of these concerns.

Of current interest, detection, isolation and identification of viruses are also included in Environmental studies. The techniques for isolating, culturing, or identifying viruses etc. are now well described. Generally speaking, the concern of the environmental technologists were/are to minimize or eliminate stress on the environment. In earlier engineering work with water quality, engineers stuck strictly to physical processes. They removed suspended solids by sedimentation or filtration for dissolved solids. At this point chemists contributed their quantitative terms. Not using biology yet, engineers looked for way to measure or describe pollution. Based on chemical data, the isochlor method was used in New England. In other words, chlorides in the water were measured and any increase on this base information would indicate pollution. However, Chlorides don't disappear in the streams, that may have purified themselves or microorganisms. Chemical Nitrogen was used as an indicator of pollution and showed up on health hazard curves of the following type (Fig. 1.1)

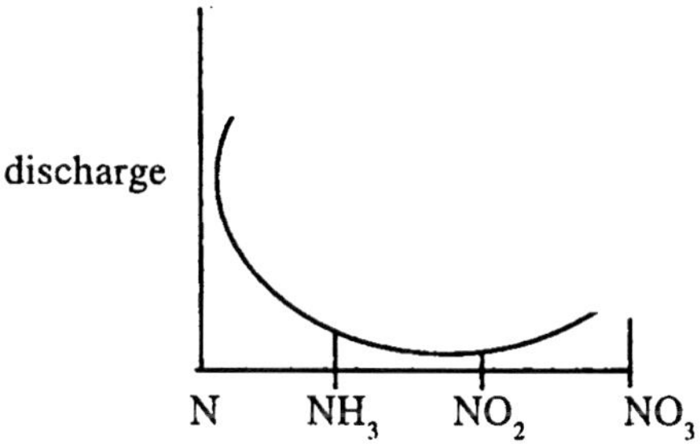

Fig. 1.1 : Chemical nitrogen as a indicator of pollution

The biological degradation of proteins in nature as well as in treatment system follows these steps and different organisms metabolize product at each step :

Amino acid ⟶ ammonia ⟶ nitrites ⟶ nitrates

There are also different organisms which will reverse these processes, which will be discussed in detail elsewhere in this volume. Chemical data is expressed in terms of concentration. When biological factors were recognized the chemists joined the biologists, biological data is based on counting and identifying organism. The biologists' taxonomy caused a lack of communication with the chemists. The engineers didn't understand any of them. Epidemics of waterborne diseased demanded an increased concern for biological aspects. The Broad street Pump, Soho, London 1855, an outbreak of cholera was traced to a single contaminated pump by John Snow.

An influx of Russian immigrants put a stress on the water supply. Hamburg of Germany drew and discharged its water directly from the Elbe River and had an outbreak of cholera. Altuna had no epidemic although it also used the Elbe River across from Hamburg. The difference is traced to the slow sand filter that Altuna used for the water supply.

By 1870-1900, etiological agents had been identified as a group of *Pathogenic Bacteria (Pathogens*=desease producing organism), in the following manner :

- *Salmonella typhosus* causes Typhoid.

- *Shigella* causes Bacillary disentery. and,
- *Salmonella paratyphy* causes Paratyphoid.

Biology became the concern of water quality engineers, technologists and scientists and to identify the disease of water they counted the pathogen. Many pathogens are hard to grow in culture, moreover, even if pathogens are present, there will be very few to count, making preventive counts a time-consuming job. So they looked for an indicator or type of organisms. This proved to be the coliform bacteria, which are present in the intestinal tract. Using this as an indicator, it is easier to test for the bacterial quality of water, if coliforms are present, then pathogens can be present, too.

Coliforms are good indicators also because they are easy to isolate and easy to identify. Coliform have greater resistance to chlorination than pathogens and therefore, enough chlorine to kill the coliforms, will be enough to kill the pathogens. Techniques for getting a coliform count have been refined to get quicker results. Cultures are grown on membrane filters.

Today, interest has shifted to the viruses. Enteric (intestinal) viruses are more resistant to chlorination than the coliform test. The coliform test may no longer be workable in this case so the search is on for a new test indicator organism for chlorination levels. All these organisms are the interest of the environmental microbiologists and biochemists, especially the viruses. In water quality and treatment processes, microorganisms are of increasing concern. Some microorganisms cause taste and/or oder problems. The technologists needs to know about the organism in order to design controls. Some organisms are able to clog pipes, for example, the slime growth in toilet tanks. Other organisms can cause corrosion in pipes.

Organisms need not be microscopic to be of concern to water quality scientists and technologists. Englebrecgt sites a case in 1960 where nematodes or round worms were making their way into the water supply. The nematodes range from 1-1.5 mm. in size, big enough to be visible. Not only were they repulsive to see floating around in the drinking water, they proved to be a health concern because their food material is bacteria. There was concern that bacteria, possibly pathogenic, could be smuggled through treatment by staying alive in the alimentary tract of the nematode. The nematodes could get through the sand filters because of their movement. It proved that pre-chlorination controlled the wiggling of the worms enough that they could be trapped in the sand filters. Chlorination does't kill the worms, it just stuns them enough that they won't wiggle through the filters.

The focus of biological concern for the technologists is the decomposition of organic matter. The activity of these organism is important to the design and running of activated sludge and trickling filter, both based on biological processes. Sanitary microbiology encompasses knowledge of may fields like bacteriology, virology, physiology, mycology, protozoology and cytology. An important biological aspect of microbiology is habitat. Microbial ecology is a special problem orientation for the subdivisions of microbiology like Pathogenic microbiology, agricultural, industrial-and geo-microbiology etc.

PROTOPLASM

Cell growth means cell synthesizing its own cellular material by process metabolism. *Protoplasm* means living material.

In general, protoplasm is a complex mixture. It differs slightly from one organism to another and there may be find differences in the same organism from time to time. However, there is a remarkable similarity of constituents in the protoplasm of all organisms. Various chemical compounds in protoplasm are found suspended, dissolved or ionized, inorganic etc. forms. There are 100,000's of organic compounds in protoplasm. Most are macromolecules or polymers.

Carbohydrates are found in cellular material, some, as simple sugar, but usually as complexed as polysaccharides. Carbohydrates may be stored for reserve food in the form of cellulose or glycogen. Nucleic acid is the smallest in amount, but are immensely important because they transfer hededitary properties. Table 1.1 gives an average composition of various cells with respect to carbohydrates, fats, protein, nucleic acid etc.

Table 1.1 : Constituents of bacterial, animal and plant cells

Biochemicals	*Bacteria*	*Animal*	*Plant*
Carbohydrates	2%	1-2%	20% (cellulose)
Fats, lipids	2%	13%	1-2%
Proteins	10-12%	15%	2%
Nucleic acid	....	less than 1%	....

A CELL

A cell is generally about 1μ in size. Other organic compounds are used by the organisms to nurish themselves, to take in energy.

Plants derive their energy from solar radiation in the process of photosynthesis. Animals presumable derive their energy from chemical activity. This energy is not used directly by the cells. By a series of reactions, the energy is put into storage, than it goes to the cells.

The mechanism of the energy transfer cycle is brought about by ADP/ATP. *ADP* = adenosine diphosphate. *ATP* = adenosine triphosphate.

The parent compund of ADP/ATP is adenine :

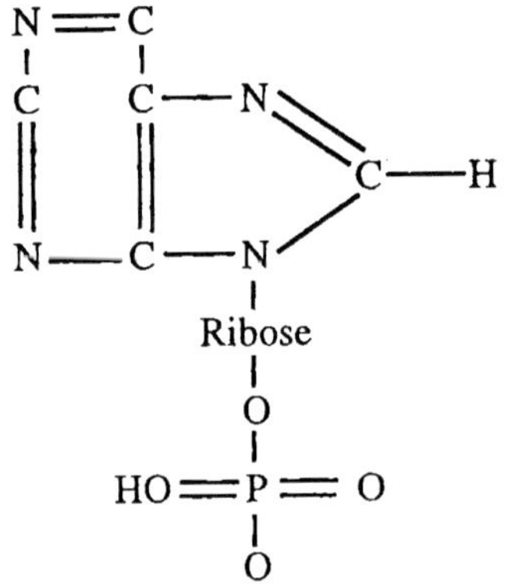

```
N ══ C
|    |
C    C ── N
‖    ‖      ╲╲
‖    ‖        C ── H
‖    ‖      ╱
N ── C ── N
          |
        Ribose
          |
          O
          |
   HO ══ P ══ O
          |
          O
```

One cell measures about 1μ and is contains 2/3rd to 3/4rd water as an important constituent. The chemical components of a living cell are : Carbon, Nitrogen, Hydrogen, Oxygen, Phosphorus, Sulfur, Chlorine, Molybdenum, Copper, Calcium, Magnesium. Generally 70% Carbon and 10% Nitrogen and traces of the other elements are present in enzyme systems, but are necessary for proper functioning of the biological system.

When the waste is deficient in some nutriets, sometimes it is necessary to add these trace nutrients to waste to enhance organismic metabolism. These elements exist in the forms of elements, compounds or in ionic solution. There are many complex organic compounds arranged in specific ways to carry on life functions.

A chair has cell fibres arranged in a specific way, but it is inanimate, not living. A living cell has chemical cellular activity. Compounds are elements in living cells are not static. They are dynamic, constantly interacting, combining and separating.

There are three more important terms, to be described in relation to environmental Biochemistry :

Metabolism : the sum total of cellular chemical activity. Metabolism can be measured by respiration.

Catabolism : The breaking down of food to release energy for cell activity. The reaction is of the following type :

$$C_6H_{12}O_6 + O_2 \xrightarrow{\text{bacteria}} CO_2 + H_2O$$

This reaction does not take place in one stage; there are many steps to get these and products

Anabolism : intermediate products of above reaction may be used for cell growth.

An organism metabolizes glucose and oxygen in a series or reactional to get the end products as carbon dioxide and water. These reactions are facilitated by enzymes which act as mediators of reactions. The resultant energy of these reactions is captured by the ADP, which takes on a phosphate and becomes ATP. If the cell uses the energy held in reserve by the ATP, the ATP loses a phosphate and becomes ADP again. In the basic metabolic reaction, as organic matter is oxidized, something is reduced. The final hydrogen acceptor is oxygen which forms the water and product

Cell is the basic unit of life. The simplest form of life is one cell. What can be defined as life is a matter of conjecture. Life at this simplest level may defined as particular chemical composition of fine, delicate structures, which carry out the following functions :

1) can nourish self, take in food and use it.

2) has a mechanism to derive energy from food.

3) can synthesize, can undergo measurable growth or repair damage, which known as *Biosynthesis* i.e. makes own growth.

4) exhibits *homeostasis* i.e. capacity to control internal environment, control to maintain stable conditions; maintain conditions necessary to life.

5) can reproduce (or replicate in one-celled organisms).

6) an adapt or evolve to long-term changes in the environment by genetic mutation.

In a unicellular organism, all these functions are carried out in one cell. The chemical components undergo biochemical and biophysical reactions. In this respect, a unicellular organism is more complex than a multicellular organism. (Fig. 1.2).

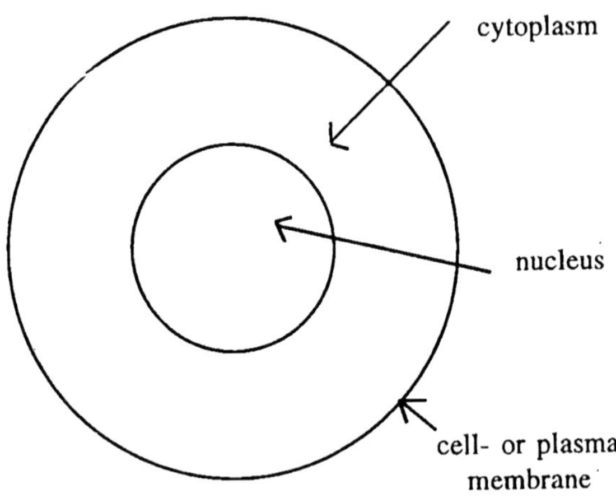

Fig. 1.2 : The cell (diagramatic)

Cytoplasm is the basic inner material of a cell. *Nucleus* is relatively large body within the cytoplasm; it carries the DNA and RNA, which control hereditary factors. *Plasma or cell membranes* is made up of carbohydrates, proteins and lipids; controls what gets in or gets out of the cell, especially in unicellular organisms, which have not alimentary tract.

Organelles are small, fine structures in the cytoplasm which carry out certain chemical reactions or other specialized activities. e.g. Examples mitochondria, microsomes, vacuoles. Unicellular organisms carry out all life functions within themselves. They are the simplest organism in that there is little specialization, but they may be considered most complex in that they carry out all functions in one cell. Multicellular organisms are distinguished by their pattern of organization. The way the cells are attached together gives the organism its characteristic external shape and size. As the number of cells increases the structural complexity of the organism also increases. Cells of higher plants and animals are differentiated to perform specific function.

In multicellular organism, we can see the followings :

Tissue

Many similar cells attached together to perform a given function. e.g. epithelial tissue (skin), connective tissue, muscle etc.

Organs

Aggregation of different cells or tissues to perform specific functions. e.g. heart, kidney, eye etc.

Organ Systems

Organs which work together, each performing a specific step of a life function. Ex : digestive system=esophogus, stomach, liver, panceras, etc.

Characteristics of living organisms

Here are discuss only those characteristics, which are the result of metabolism not shown by non-living things. These are following :

1) ***Growth*** : than more increase in size, a living organism takes in

substances, that it needs or wants; it organizes the substances, synthesizes its own cell material and reproduces.

Combining biosynthesis and energy releasing chemical reaction of a unicellular organism :

Organic Matter + O_2 + Bacteria ⟶

Bacteria+Bacteria' + $CO_2 + H_2O$

Biosynthesis chemical reaction release energy

(This is NOT a single-step process)

B.O.D. measures the difference between the distribution of new growth and end products i.e. $CO_2 + H_2O$.

2) ***Responsiveness*** : a living organism reacts to changes in the environment: the type of response is determined more by the organism than by the stimulus.

3) ***Reproduce*** : not all material is used for function and growth of an organism. Offspring of an organism resemble the parent but often have slight differences. This slight difference is the basis of evolution.

In the life process the organism carries out a variety of processes. A cell carries out the processes as the first cell.

There are two broad classifications of functions :

1) Metabolism - the sum total of chemical reactions in the cell.

2) Self-perpetuation.

Basal Metabolism Rate is a measure of the oxygen taken up and the carbon dioxide produced. Not all energy in bio-oxidation is used by the cell, some is lost as heat. In metabolism, cell uses large molecules for energy production by breaking them down to derive what they want by the process, called *catabolism.* Cells can build up through synthesis of their own complex compounds from simpler ones is the *anabolism.* Both anabolism and catabolism go on simultaneously but independently and continuously.

The process of *metabolsim* presented in fig. 1.3, where nutrition-provides the raw material for cell activity and production and utilization of energy is performed by respiration and synthesis.

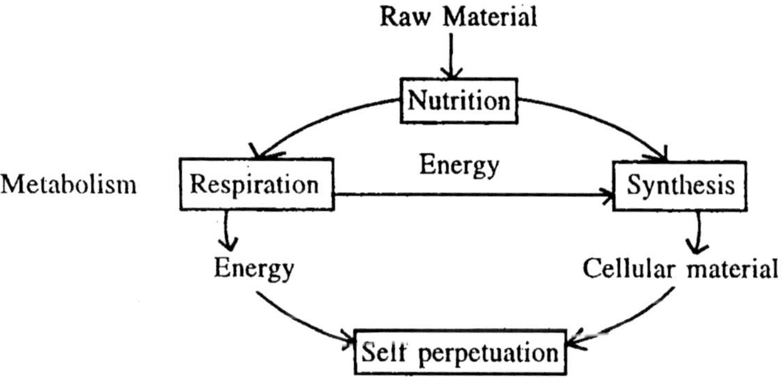

Fig. 1.3 : *Self-perpetuation* - sets apart living cells

ACCLIMATION OF ORGANISMS

The Three following activities help the organism to cope with disruptive or destructive elements :

1) *Steady state* - organism receives stimulus and steady state mechanisms adjust reactions to self-preservation.

2) adaptation and,

3) reproduction.

Steady state

Adjustment of respiration to existing conditions e.g. once usual food is gone, the immediate response is to turn to something else. It is this mechanism, which makes the activated sludge process work. Steady state also includes the various mechanisms of an organism, may use to real predators.

Homeostasis

Allows permits an organism to live as long as it inherently can.

Adaption

Adjustments to long-term environmental changes, usually by mutation of hereditary properties.

Reproduction (Replication in unicellular) Reproduction in an orderly series of events, not spontaneous generation. This is self-preserving of the species in that the organism outwits the environment by reproducing before death. Reproduction may be sexual or asexual.

Growth

Refers to an increase in size of a cell or in reproductive terms refers to an increase in the population of cells. Growth may refer to mass or the number of cells.

Chapter -2

ENVIRONMENTAL BIOCHEMISTRY IN RELATION TO CHEMISTRY

In general, Environmental Science and Technology touches on-

1. Engineering and Technology.
2. Organic and biochemistry. and,
3. Biology and Microbiology.

An environmental design Scientist and/or Technologist, must identify and control the source of pollution. In water quality control, he needs to develop systems to :

1. Survey and,
2. Monitor Water.

CHEMISTRY FOR THE ENVIRONMENTAL SCIENTISTS AND TECHNOLOGISTS

A familiarity with chemistry, in general and organic chemistry in particular, is necessary for Environmental Scientsist and Technologists in order to characterize waters and organic compounds, so that they may design controls. It is also essential, because there is recent concern about trace organics and carcinogens in water supplies.

Organic chemistry and biology involved in the Environmental Scientists and Technologists interest in the recycle of wastes. In fact,

in many treatment systems, Environmental Scientists and Technologists use microbes to degrade wastes in aerobic and/or anaerobic environment for biomass and/or energy production.

Organic chemistry and biological functions are involved in the Environmental Scientists and Technologists concern with receiving streams and/or aquatic resources. Those who live in the downstream below industries of paper production, may pick up mercury from industrial waste. Mercury by itself would just stagnate on the bottom of these streams, but often it forms a reaction with biological products e.g. anaerobic bacteria disgesting sludge on the bottom of the stream give off methane as a waste product of their metabolism.

$$Hg + CH_4 \longrightarrow CH_3 - Hg$$

(methyl mercury, a very toxic substance)

It is believed that refractory materials like plastic and other synthetic organics etc. are resistant to biological degradation. Some research investigations show that these things are also subject to degradation but the rate of this can be accomplished is extremely low.

ORGANIC CHEMISTRY

At the time of evolution and development of chemical science, the name organic at first, pertained to the plants and animals. Originally, organic chemistry was concerned with the chemistry of living material, but to-day, it includes products synthesized from organic compounds.

About 2 century ago, Lavoisier discovered that the combustion of living material released CO_2 and H_2O. He also discovered that different organic compounds of the same weight used different amounts of oxygen, when burned to achieve their material or mass balance. About a century ago, other scientists discovered a reaction which turned an inorganic compound to an organic one-

$$NH_4OCN \longrightarrow CO - (NH_2)_2 \text{ Urea}$$

This was the first synthesizing reaction. Organic compounds may contain Phosphorus, Nitrogen, Sulfur, halogens, but it is known as the chemistry of Carbon. Organic Chemistry is basically a non-mathematic science. Originally, organic chemistry divided into the following series :

Aliphatic : - e.g. fats, oils, grease, tallow etc.

Aromatic :- e.g. Organic compounds, which had an odour.

Organic compounds have the following characteristics :

1. They are combustible (Inorganic salts are not combustible, but they fuse instead).
2. They can be found as a gas, liquid or low melting solids (Inorganic compounds are mostly solid, a few gas).
3. They are generally less soluble in water (Those unusually soluble, sugar, alcohol, acids, have a functional hydroxyl group - OH).
4. They are generally non-ionic (except certain acids and proteins).
5. They are food for bacteria (though few bacteria feed on inorganic matter e.g. Nitrosomonas).
6. They have high molecular weights and large, complex molecules.
7. They have a different type of bonding.

There is a pivotal carbon atom which shares electrons. e.g., carbon tetra-chloride.

```
       Cl
       |
Cl --- C --- Cl
       |
       Cl
```

The manner in which the elements in molecules are arranged is important and the possibilities are infinite.

The graphic form of CH_4 is :

```
     H
     |
H -- C -- H      methane
     |
     H
```

When an carbon attach to another atom carbon in a chain by single bond :

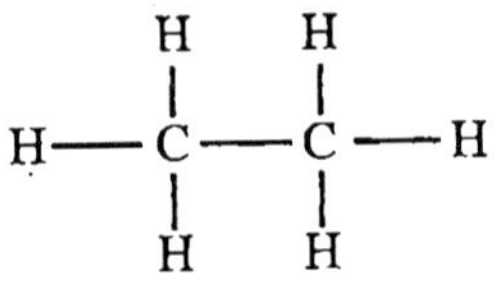

```
    H   H
    |   |
H — C — C — H
    |   |
    H   H
```

ethane C_2H_6

There can be double bonds between carbons :

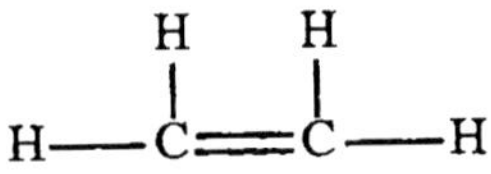

```
    H   H
    |   |
H — C = C — H
```

ethylene C_2H_4

There can be triple between carbons :

H-C = C-H acetylene C_2H_2

The structural formula of ethane for example-

$H_3C — CH_3$

There can be a branched carbon chain e.g. n-Butane and isobutane are both C_4H_{10} but have vastly different graphic formulae. This is an example of chain isomerism, which is described in the following sections.

ISOMERISM

It may be defined as "***the sample molecular formula with different graphic formulae is known as Isomerism***".

e.g. Butane is the first of that series to have isomers. In which, different isomers will have different functions e.g.

C_6H_6O :

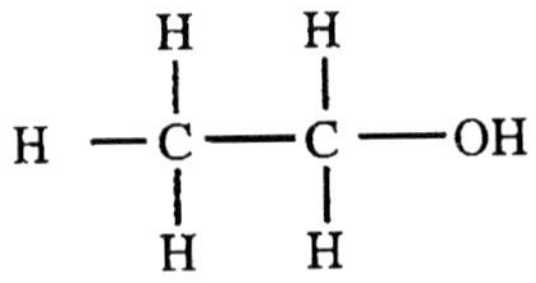

```
    H   H
    |   |
H — C — C — OH
    |   |
    H   H
```

ethanol — CH_3CH_2OH (B.P. 78°C)

```
     H       H
     |       |
H—— C —O—— C ——H
     |       |
     H       H
```

Dimethyl ether -CH_3-O CH_3 (B.P. —25°C)

Aliphatics

Straight or branched chain carbon compounds are knwon as aliphatics or aliphatic compounds. Details of these compounds were given elsewhere.

Aromatics

A ring structure of 6 carbon atoms with three double bonds in alternate positions. e.g. simple benzene (an organic solvent).

Heterocyclic compounds

A ring structure with at least one element other that carbon e.g. furaldehyde or furfural.

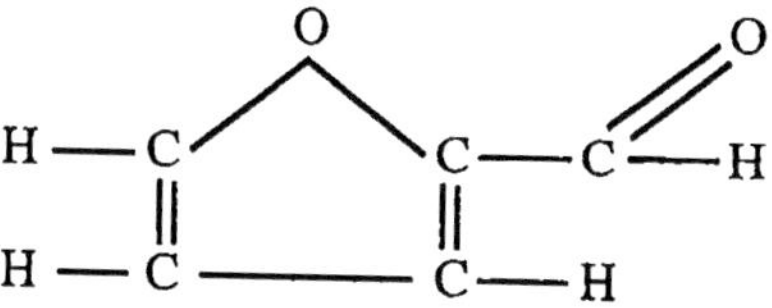

Furaldehyde or furfural is of significance to Environmental Scientists and Technologists as a waste product of cereal industry, coming from oat hulls and corn cobs.

ALIPHATIC HYDROCARBONS

Saturated

Single bond between carbons (alkanes typically natural gases and petroleum products, also known as *paraffin series* or *methane series.*

Unsaturated

Having at least one double bond (alkanes) and those having at least one tripe bond (alkynes).

Alkanes

Methane series - specific increments - $C_nH_{2n}+2$, homologous series. This will react readily with other elements, specially in synthesis.

They are in the industrial waste of the pterochemical industry and react easily with chlorine.

NOMENCLATURE :

Ethane reaction with chlorine

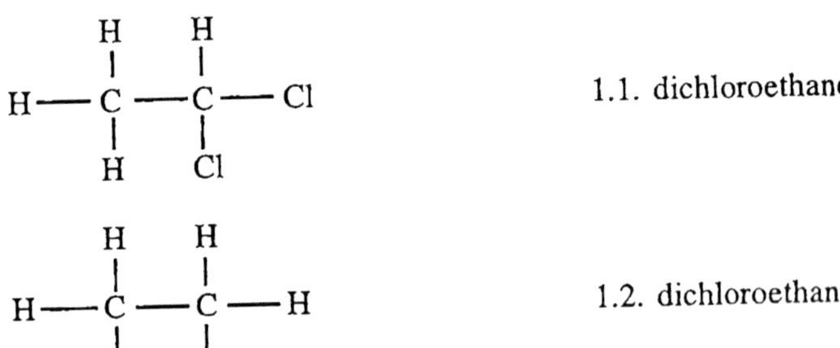

The different positions of the chlorine make these two different compounds and they react differently to biodegradation.

Methyl group - H_3C—R

Ethyl group - H_5C_2—R etc.

RULES OF THUMB IN I.U.C. NOMENCLATURE

- The name is derived from the longest carbon chain.
- Position are numbered to give the smallest possible numbers.

Aliphatics

Straight or branched chain Hydrocarbons.

Saturated

Inert single bond IUC Suffix

Alkanes

Paraffin Series Methane Series	**-ane**

Alkyls

Paraffinic Hydrocarbons - **yl**

which react

Chemical reactions of saturated hydrocarbons

Need more extreme conditions to react than unsaturated -

- Oxidation.
- Substitution of H with halogens.
- Pyrolysis or cracking. and,
- Biological Oxidation.

Unsaturated Hydrocarbons

React with ease, polybond. Following series are known as unsaturated hydrocarbons :

Alkanes

Ethylene series olifins	One double bond between two adjacent C atoms.	} **—ane**
Diolefins -	two double bonds in molecules.	
Polyenes -	more than 2 double bonds in molecule.	
Acetylene series -	one triple bond between adjacent C atoms.	

Chemical Reactions of unsaturated hydrocarbons

1. Odidation
2. Reducation
3. Addition
4. Polymerisation
5. Bacterial Oxidation

Table 2.1 : Types of Aliphatics and their chemical characteristics

Compound	*Derivation*	*R=alkyl radical General structural formula*	*IUC Suffix*	*Linkage or group*
Alcohol	Theoretical oxidation of hydrocarbons	R—OH e.g. CH_3OH	—al	Carbonol (C—OH) hydroxy group (OH)
Glycol	2 hydroxy groups on 1 molecule			
Glycerol (or glycerine)	3 hydroxy group on 1 molecule			
Aldehyde	oxidation of primary alcohol	$R-C(=O)-H$ e.g. CH_3CHO	—al	Carbonyl (C=O)
Ketone	oxidation of secondary alcohol	$R-C(=O)-R$	—one	Carbonyl (C=O)

1	*2*	*3*	*4*	*5*
Ether	dehydrating agent on alcohols	*R*=R—O—R		
Acids	highest oxidation state before CO_2 + H_2O	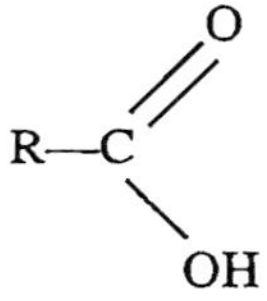 R—C(=O)—OH	—oic	Carbonyl (C(=O)—OH) 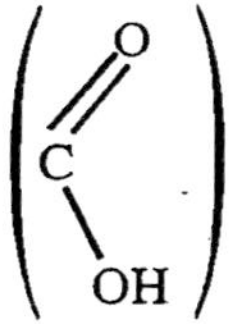
Esters	hydrolysis of acid + alcohol	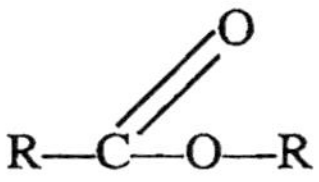 R—C(=O)—O—R	------	acetalte —C—C(=O)—O 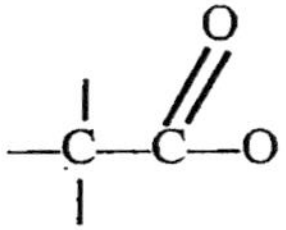

ALPHATIC HYUDROCARBONS

Table 2.1 gives an idea about various type of aliphatics.

Alkanes

The group of single bond carbon atoms, whose graphic structure is a chain, branched or not (-ane suffix).

Akyls

Radicals of the alkane group (-yl suffix) e.g. the detergent component ABs-alkyl benzene sufonate -used to cause problems in sewage treatment plants. detergents contained 70-80% polyphosphates ($Na_5P_3O_{10}$) to attack & loosen soil & about 30% ABs, a surface active ingradient which attached to the soil particles & fouled them away in the form of foam. It was found that ABs is biologically resistant to degradation & went through water treatment without losing its power to foam & froth whenever agitated. The graphic formula of ABs is -

O
||
C — C — (benzene ring) — S — O   Sulfonate group
||
O

alkyl radical etc.   Benzene

Detergents makes now use LAS (linear alkyl sufonate) deleting the benzene ring which was resistant.

Alkanes occur often in natural gases & petroleum Alkanes having 1-5 carbon atoms in a chain are gases, having 5-17 C are liquids & having more than 17 C are solids (fats).

Petroleum compounds yield countless synthetic compounds under processes of cracking etc. One produce of raw petroleum is gasoline. (Octane rating of gasoline = the mixture of n-haptane & isoocatane Isooctane 2, 2, 4 trimethyl pentane in IUC nomenclature).

N-haptane causes no knocking in the internal combustion engine. N-haptane is rated 0 in knocking & iso-octane is rated 100. The octane rating is the ratio of two.

UNSATURATED HYDROCARBONS

Alkanes : Have at least one double bond between two carbon atoms. It reacts to chlorine & produces oily compounds.

The graphic formula of

This diolefin is important to the rubber industry.

Ethylene :

$$H-\overset{\displaystyle H}{\overset{|}{C}}=\overset{\displaystyle H}{\overset{|}{C}}-H$$

1.3 Butylene :

$$H-\overset{\displaystyle H}{\overset{|}{C}}=\overset{\displaystyle H}{\overset{|}{C}}-\overset{\displaystyle H}{\overset{|}{C}}=\overset{\displaystyle H}{\overset{|}{C}}-H$$

The double bond of the oilefins is unstable & readily joins with halogens, especially chlorine, which is of significance to environmentalists. Chlorination of wastes as a disinfectant measure is hindered by the large chlorine demand of the olefins. Disinfectants in the form of HOCl or OCl^- react with the olefins to break the double bond but joins with the compound leaving not disinfectant capabilities.

CHLORINE DEMAND

The dosage of chlorine applied to give measurable residual chlorine after 15 minutes. Colour wastes from the canning of vegetables e.g. carotene, an unsaturated compound, cause colour problems as well as exerting a large chlorine demand.

Colour is difficult to remove. Compounds which contribute colour are generally biologically resistant. Colour removing agents such as activated carbon or azone can be used, but are very expensive methods, often too expensive for smaller firms. Black sulphide liquor wastes from wood processing industries comes from the resinous material & are extremely difficult to remove.

ALKYNES

Have at least one triple bond between carbon atoms. e.g.

acetylene $H - C \equiv H - H$

1. Butyne

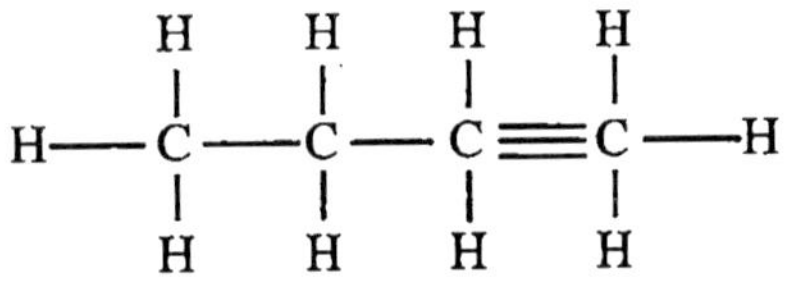

$$
\begin{array}{ccccccccc}
 & & H & & H & & H & & H & & \\
 & & | & & | & & | & & | & & \\
H & - & C & - & C & - & C & \equiv & C & - & H \\
 & & | & & | & & | & & | & & \\
 & & H & & H & & H & & H & &
\end{array}
$$

Alkynes & alkanes are compounds starting polymers. To discover the same in industrial wastes, look at a compound used in the industry because processes are not 100% efficient. Double or triple bonds are unstable & indicates that they are biodegradable & will react easily with chlorine.

Chapter -3

Microorganisms : General Characteristics and Properties with reference to Environment

Microorganism is a collective term encompassing many types of organisms, plants and animals, which are small in size and considered as individual living entities. Morphology, various types of microbes can simply be differentiated on the basic of by their size effect :

1) By seeing them and,

2) Their metabolism and chemistry.

Bacteria have a *large surface area to volume ratio.* Bacteria have the same volume as plant cells but 10X the surface area. This large surface area affects the exchange of material in and out in terms of nutrition. The greater surface for ingestion of food and discharging of waste leads to a greater rate of metabolism and a corresponding greater rate of reproduction. Some bacterial reproduce every 20 min.

Types of classification

Classification is a way of organizing the large compilations of observations and provides a system for identifying a large number of organisms.

I. ***By ecological system*** - where in the environment the organism is found. e.g. terrestial, aquatic, marine etc.

Aquatic Distribution	Nekton (free-swimming)	
	Plankton : floating, suspended	Phytoplankton / Zooplankton
	Benthic (bottom, in mud)	

II. *Classifying organisms as plants or animal.*

III. *Might develop a system of classifying organisms of special interest to ecological scientists and technologists.*

1) ***Pathogens*** - classify them in order to design control to eliminate (a) bacterial; (b) viruses; and (c) parasites, e.g. malaria parasite, which cause malaria.

2) ***Vector*** - organisms that can carry disease producing organism, e.g. mosquitos and flies.

3) ***Index Organisms*** - Indicate environmental quality, e.g. coliform bacteria groups.

Diversity index - measure of stress on a stream by a census of the community. Pollution is indicated, if there are few species but a large population (vice versa if NOT polluted). Another indication of pollution is the existence of pollution-tolerant organisms, e.g. sludge worms.

4) ***Decay*** - organisms which degrade organic matter, i.e., in activated sludge.

5) ***Nuisance*** - organisms which affect the aesthetic quality of water, small, taste etc. Sulfate reducers, present especially in warm water which give the water the smell and the taste of H_2S (rotten eggs). Presence of the nematodes is also important as water quality is concerned.

6) ***Conversion or resource recovery organisms*** - organisms which produce some useable product organic waste, e.g. methane forming bacterial or single cell protein by yeast from the biodegradable effluents.

This biological system of classification is *Taxonomy* or *Systematics*.

The Biology system of classification is Phylogenetic. It is a scientific description and organism are named according to specific (1) function, (2) structure, (3) development, and (4) evolutionary history. Accordingly,

Animals - mobile, seek out preformed food, variety of colours and textures.

Plants - stationary, manufacture own organic matter, most are green in colour (photosynthetic).

The Animal Kingdom is broadly divided into Vertebrates and Invertebrates, those with or without a centrally, protected nerve cord. An example of a microbial invertebrate is the protozoa.

Plants have many divisions, mosses, sea plants, etc. The plant like algae, fungi and bacteria etc. are of interest to enviromentalists. Bacteria are considered plants because it is believed that their cell wall was cellulose originally similar to plants and because they are essentially non-mobile, have a simple call structure and are *holophytic* in nutrient solution, that is, they manufacture their own food.

Bacteria	*Genus*	*Species*	*Remarks*
	Escherchia	*coli*	(occasionally written *E. Coli.*)
	Escherchia	*frundi*	
	Salmonella	*typhose*	(etiological agent or typhoid occasionally written S. typhose)

Species are distinguished by their activity.

In the early 20th century, a biologist named Hackle questioned the accuracy of having the two kingdoms divided into plant and Animal and proposed a third division for unicellular organisms, which did not quite fit either Animal or Plant exactly. He named this third division as Protista.

PROTISTA

Protista includes organisms which are green, manufacture their own food BUT move or organisms, which do move BUT do not eat other organisms. *Protista* have simple structures, do not have much variety among organisms and are mostly unicellula with a few simple multi-cellular organisms. It includes alges, bacteria, fungi and protozoa.

Protista is divided into *Lower Protista* and *Higher Protista. Lower Protista* are *procaryotic*, that is, there is no nuclear membrane so the nucleus is somewhat diffused and in direct contact with cytoplasm. This group includes bacteria and the blue-green algae. Fig. 3.1 shows the classification of living being.

HIGHER PROTISTA ARE EUCARYOTIC

That is having a distinct nucleus as do the other plants and animals. This group includes algae, protozoa and fungi. (The root word "Eu" means "true").

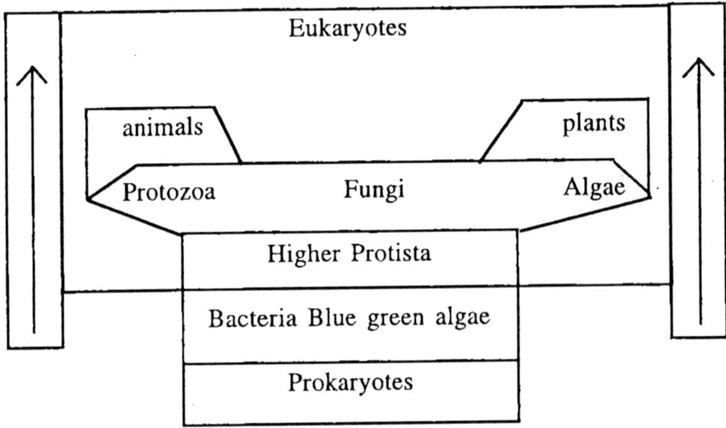

Fig. 3.1 : Classification of living being

In ecological sense, organisms can also be classified as *Produces or Consumers.*

Producers

Plants which are photosynthetic use the sun's energy plus CO_2 to produce their own food. This general reaction is:

$$CO_2 + H_2O \xrightarrow{\text{radiant energy}} (CH_2O)_x + O_2$$

By the fixation of CO_2 plus energy plants can make their own food (organic matter).

Consumers

The food produced by photosynthesizers, are used by other organisms, called consumers. Animals are very much dependent on producers as they can not synthesize their own food.

Plants derived their energy from the sun and therefore, need not be mobile. Animals, on the other hand, must seek out in search of their food.

Protista include both producers (algae) and consumers (protozoa).

In the ecological system, the *decomposers* are the intermediate organisms. The decomposers feed on organic wastes and dead organic matter returning the minerals to the carbon cycle or the geochemical cycle. The organic matter used by the decomposers is said to be *stabilized* or *mineralized* back to its component elements. There is not known living matter that cannot be decomposed. Decomposers are fungi and certain bacteria.

Eucaryotic cells

Cells having a distinct nucleus and nuclear membrane. Eucaryotic cells have distinct chromosomes in the nucleus which line up clearly in cell division (mitosis).

Procaryotic cells

Have no nuclear membrane and there is not distinguishable organization prior to division. Bacteria and Lower Protista, procaryotic in nature. Lower Protista like Blue-green algae-photosynthetic, unicellular, haivng general habitat in fresh water/aquatic environment.

Bacteria

Myxobacteria

non-photosynthetic, short rods, divide simple binary (transverse) fission:

- Asexual reproduction
- generation time-as low as 20 minutes.
- difficult to distinguish parent cell
- found in soil, water as well in air in various forms.

Spirochetes

unicellular, do not have a rigid cell wall,

- helical or curved rods
- found in tooth tartar among other place.
- *Typonema pallidum*-etiological agent of syphillus.

Eubacteria

unicellular, non-photosynthetic --includes most bacteria.

Bacteria are unicellular microscopic and have 1700 different species. All the known species are identified in Bergey's *Manual of Determinative Bacteriology* from pure cultures.

Classical definition of Bacteria

Extremely minute, rigid, unicellular, apparently sexless, free of chlorophyll and other photosynthetic pigments, multiply by binary fission with resulting off spring of relative equal size.

Bacterial come in three different forms : Rod-shaped, Round or Spirochetes.

BACTERIA (SINGULAR BACTERIUM)

Definition : small, rigid, unicellular, no photosynthetic material, sexless, reproduce by transverse fission into 2 equal cells.

Bacteria are classified and identified by the following information : 1) Morphology (form), 2) Staining characteristic, 3) cultural properties, 4) Biochemical activity.

MORPHOLOGY - SHAPE, ARRANGEMENT AND SIZE

1. Shape and arrangement

a) ***Spherical*** (for slightly elliptical) coccus, 1 cocci. ○

i) ***Diplococci*** - two separate cells but stuck together in pairs (diplococci)

e.g. *Diplococcus pneumoniae* causes bacterial pneumonia.

* Diplococcus is also a genus name.

ii) ***Streptococcus*** - engaged in chains.

*Also a genus name.

e.g. Hemolytic streptococci are the type which cause strap throat. It is cultured on blood agar (*Agar* - a semi-solid media which serves as nutritional base for culturing bacteria), the hemolytic streptococcus cause lysis of the blood cells for nourishment; this forms a clear ring around the strep colony in the culture ("Hemo" means blood).

iii) ***Staphylococcus*** - arranges in irregular bunches.

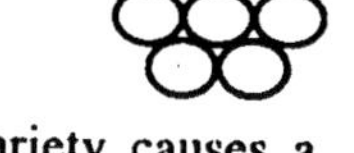

e.g. *Staphylococcus aureous* - certain variety causes a kind of food poisoning.

iv) ***Tetrads*** - arranged in fours, divides in two planes.

v) ***Sarcina*** - arranged in eights, divides in three planes.

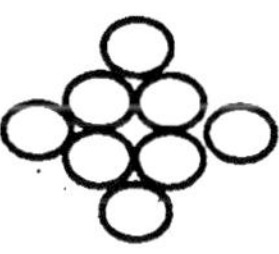

e.g. *Sarcina lutea* - an air contaminant.

- More cocci in the air, more resistant to desiccation (drying).

- More bacilli in dilute media, more surface to volume ratio.

b) ***Bacillus*** - rod-shaped, cylindrical, some squared at the ends some rounded at ends.

*Also a genus name

e.g. *Bacillus subtilus* - harmless, found in the soil or in hay or grass infusion, common in lab exercises - Bacillus aren't usually seen in special arrangements. Occasionally there are diplo-bacillus but unusual arrangements like these are not criteria for identification.

Occasionally Bacillus for raft or palisade arrangements.

e.g. 

e.g. *Mycobacterium tuberculosis* causes T.B., *Coryanbacterium diptherise* causes deptheria.

c. ***Spiral or Halical***

e.g. Vibrio comma ('sprochets) causes asiatic cholera

2. Size generally 1μm, size varies with species

a. cocci - less than 1μ range .5 to 1.5 μ

b. bacilli - 1/2μ wide and 3/4 to 3 or 4 μ long

c. spiral - less that 1μ wide and 1 to 2μ long

d. smallest - microplasma - cocci - .125μ

All Cells	***May be***
Cell wall	Pili or Fimbrae (hair)
cytoplasm	Flagella (tail)
cytoplasmic membrane	Capsule or slime layer
Inculsions	Endospores

Some cells are characterized by reaction to differential stain, or Gram stain. Cells are either Gram +ve or Gram –ve.

Pili or Fimbriae

— organelles of attachment.

— on some Gram-cells.

— basically protein, 30-40 angstroms wide, several 100/cell.

— certain bacteria which are obligate aerobes stay on the top surface of liquid or attach to glass. The cells attach to each other by pili.

— Formation of activated sludge depends on organisms flocculating. It is conjectured that organisms may flocculate because of pili.

— pili are NOT involved in mobility.

Flagella

— organelles of mobility.

— bacteria are generally not mobile but some are mobile in liquid.

— the flagella originate in the cytoplasm, it is fragile and very small making it difficult to see, some staining techniques enhance visibility.

— basically protein, molecular weight 40,000.

— commonly found with bacillary forms, few cocci are mobile.

— arrangement of flagella is used for identification :

  — Monotricate— one flagella at one end.

  — Amphitricate-one at each end.

  — Peritricate— flagella all around cell.

  — Lophotricate-tuft of flagella at one end.

  — Atricate - No flagella.

Capsule or Slime layer

— viscous material

  — compact=capsule

  — diffuse=slime layer

— influenced by physical or chemical properties of the environment.

— cells with slime layers look shiny moist inculture (mucus type) e.g. Pseudomonas aeruginosa etc.

— another guess at why organisms flocculate is that with this slime layer the cells stick together.

- It is formed by the cell and form sin accumulation around the outside of the cell.
- It is not known exactly what the slime layer comes from or what purposes it serves. A few conjectures are that it may be a resevoir of food material or it may be an accumulation of waste. Wherever it comes from it seems to serve as a protective coating. The relation of capsule or slime layer to protection was investigated by various experiments. It has been observed that *Diplococcus pneumoniae* in culture loses its ability to form capsule. Relation of capsule to virulence although bacterial lose the capsule in culture it appears that they can regain this capacity after several generations of being in living tissue. e.g. mice in the lab infected with *Diplococcus pneumoniae* grown in culture did not show symptoms of the disease. The bacteria had lost its ability to produce a capsule. However, after several progressions of infecting other mice with already infected tissue, the bacterial seemed to eventually regain its capsule-making capacity and its virulence. There is a difference between infection and disease.
- Infection may harbour organisms without clinical symptoms. Another study was one to see if anthrax-causing bacteria which may be present on animal hides could survive the tanning process and be found in that industry's waste. Anthrax is a *zoonosis*, that is, it is transferred from animals to man. It is caused by *Bacillus anthracis* and does not appear to be able to survive the tanning process.
- The capsule or slime layer is composed of complex polysaccharides and some polypeptides.

Cell wall

- Gives the cell rigidity and shape, keeps insides in and outside out.
- The composition varies with different organisms; generally it is composed of polysaccharides or primarily glucose and galactose units, some protein and some lipid.
- difference according to whether the organisms are Gram +ve or Gram -ve.

— Gram +ve has fewer amino acids.

— Gram –ve more lipids, wider range of amino acids, also indications of 2nd layer of cell wall in Gram–ve; the inner basically lipid; the outer protein.

— cell wall is porous and allows larger molecules to pass through than the plasma membrane.

Cytoplasmic Membrane

— semipermeable.

— composed of phospholipids, protein and polysaccharides.

— membrane selectively controls what comes in or goes out.

— when a needed protein molecule is too big to pass through the membrane, the organism uses *exocenzymes* to break the large molecule outside by hydrolysis so that it can pass through the plasma membrane.

Uptake of nutrients

— Normally concentrations of organic matter are higher in the cell than in the environment.

— (1) *passive uptake* - more organics matter in environment.

— (2) *active transport* - more common, cell manufactures permease enzymes to facilitate passage of a large molecule through the membrane (Fig. 3.2)

— *osmosis*- movement of water in and out of cell : tendency to equalize density.

— In hypertonic solution, — cell will dye due to exo-osmosis.

— In hypotonic solution-cell will rupture from too much water.

— Isotonic solution-for culture medium; balanced salt.

— *physiological saline* for cell suspension = 0.85% saline (NaCl).

— *cytoplasm* : fluid portion of cell.

— *Ribosome* : area associated with RNA for protein synthesis.

— Nuclear material includes DNA, the hereditary function.

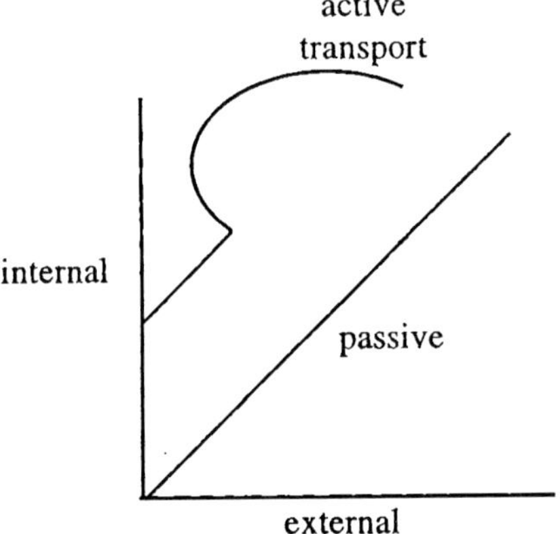

Fig. 3.2 : Active and Passive transport in microbial system

— *Inclusions* : deposits or small bodies of certain substances in the cytoplasm.

— The types of inclusions differ in different organisms.

— Voluting and Metachromatic granules stain under spècial techniques to become quite distinctive for identification.

— starch granules.

— sulfur deposits.

ENDOSPORES

Some bacteria cells have the capacity to transform cellular material to small ovals or speheres, which are highly resistant to heat and desiccation.

— the endospore is produced intracellularly.

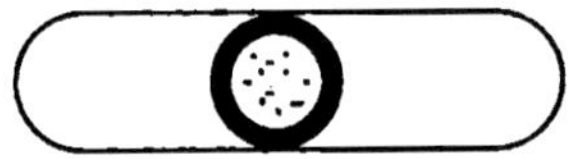

— Bacillary types usually produce endospores

— the genera *Bacillus* (aerobic) and *Clostridia* (obligate anaerobe) commonly form endospores.

— the endospore is not involved with reproduction, rather it is merely a survival tactic in adverse environmental conditions.

— species that form endospores : *Bacillus anthracis* and *B. subtilus.*

— species may be distinguished by where in the cell the endospore forms : end, middle or bulges.

— when the endospore forms the vegetative portion of the cell disappears but when favourable environmental conditions return, the cell will return to its normal state.

— the endospore in bacterial is *not* a reproductive feature as are the spores of yeasts and fungi.

— Species which form endospores : *Clostridia tetani* tetanus, *C. welchi (or perfringae)*, cause gangrene, *C. botulinum,* cause botulism Botulism is usually associated with improper canning.

STAINING PROPERTIES

Bacteria are hard to see unstained; they have the same refractive index as water. Bacterial can be observed with a wet mount with stain to see their motility, but without stain it is very difficult.

Generally, stain is used :

1) to render the microorganism visible.
2) to measure the size and shape.
3) to see internal and external structures.
4) to produce specific characteristic reactions.

In order to "fix" a smear to a slide it should first be air dried then immersed in a flame making the slide hot enough to coagulate the proteins BUT not schorched. When the smear is "fixed" it can be put in water without everything washing off.

PREPARING SLIDES

Slide may be prepared from either liquid or solid growth culture.

"Fix" smear to slide with heat, The heat is believed to coagulate the protein in the cells. In any event, once the slide is "fixed" it will not wash off.

Flood the slide with stain a predetermined length of time. Either a physical or chemical reaction between the stain and the cell occurs. The chemical reaction is more likely. The dye or stain colours the cell so that it is easier to see. Most of the cationic dyes are used for staining the bacterial smears, these are, generally, negatively charged.

Some common stains which are commonly used in environmental Microbiological studies are given in Table 3.1. Wash off excess stain and blot dry. There are special papers for this but a paper towel works well enough.

Table 3.1 : Commonly used stains

Stains	*Colour*	*Remarks*
Fucsin	red	greatest affinity for bacteria
Crystal Violet	blue/violet	greatest affinity for bacteria
Safranin	red	greatest affinity for bacteria
Methyl Blue	blue	greatest affinity for bacteria

Differential staining

Procedures using more than one dye.

Grams' stain procedure is described in Table 3.2

Table 3.2 : Grams' Staining for Gram +ve and Gram -ve bacteria

	Step	*Gram+ve*	*Gram-ve*
1.	Crystal Violet 1 min.; wash	violet/blue	violet/blue
2.	Iodine 1 min.; wash	violet	
3.	95% ethyl alcohol 30 sec.; wash	violet	colourless
4.	Safranin 20 sec.; wash+dry	violet	pink or red

— Iodine in step 2 is a *mordant,* that is, it enhances staining.

— The ethyl alcohol is a decolorizing step.

— Safranin is a counter-stain.

— Differential staining is important for characterizing bacteria.

By definition, *coliform bacteria* are Gram-ve, small, rod-shaped, and ferment lactose.

Characteristics of Gram -ve cells

— are believed to have lipids in their cell wall.

— are constantly Gram-ve; the state of being Gram-ve does not change.

Characteristics of Gram+ve cells

— have a magnesium RNA complex in the cell wall.

— being Gram +ve depends on the pH of the growth medium or the age of the cell (optimum is 18-24 hrs.)

— more susceptible to penicillin and sulfa drugs than Gram-ve.

— more susceptible to destruction by mechanical means.

— in the differential staining procedure, crystal violet and iodine react chemically, and theoretically form molecules too big to pass through cell wall.

Antibiotics (compounds originally detected as end products another organisms' metabolism) are effective against either Gram +ve or Gram -ve bacteria, but no two bacteria respond to the same antibiotic. Occasionally a doctor will prescribe a polyvalent antibiotic for an unknown aetiological agent.

Alexander Flaming discovered pencillin when a mold spoiled the culture he was working on. Spores in the air are a common lab contaminant. He noticed that the bacteria he was culturing was not present in a ring around the mold colony. The species of mold was *Penicillium notatum.* During WWII, pencillin was produced by large-scale fermentation. Now it can be synthesized.

Antibiotic agents are commonly found in soil. Antibiotic substances differ from chemical therapeutic drugs, such as sulfa.

Another differential staining procedure is the *Acid Fast Test.*

Procedure for *Acid Fast Test:*

1) First Stain with Carbon Fucsin
2) Heat
3) Ethyl talocohol (decolorizing)
4) Methylene Blue (counter stain)

— This test is important in detecting Tuberculosis bacteria.

— There appears to be many more acid fast bacteria is sewage than originally believed. Acid fast bacteria resist chlorination.

BACTERIAL NUTRITION

— Most information comes from laboratory study which does not account for conditions in nature which are affecting; predation, heterogeneous population, environmental stress, synergism. Synergism means two or more organisms or environmental factors join for a common end; neither can produce alone. Synergism may influence toxicity.

Autotrophs, all plants and some bacteria, by definition depend on only inorganic compounds for their nutrition. Generally, their source of Carbon is inorganic CO_2 or carbonate. Autotrophs require inorganic compounds in different degree:

1) Oligates are the organisms that inhabited by certain concentrations of organic materials.
2) synthesize organics from inorganic material deriving energy from the sun or simple oxidation reaction.

Examples of organisms using oxidation reactions for energy:

iron bacteria $Fe^{++} \longrightarrow Fe^{+++}$

nitrifixers $NH_3 \longrightarrow NO_2 \longrightarrow NO_3$

Heterotrophs, generally animals, require organic compounds for nutrients. They cannot synthesize organic matter and need their organic matter preformed. Heterotrophs use organic Carbon.

Autotrophs :

1) Photosynthetic
2) Chemosynthetic

Heterotrophs :

1) Saprophitic-use dead matter
2) Decomposers also use dead matter

Autotroph/ Heterotrophs is a rather arbitrary classification. There may be an intermediate stage depending on the environment.

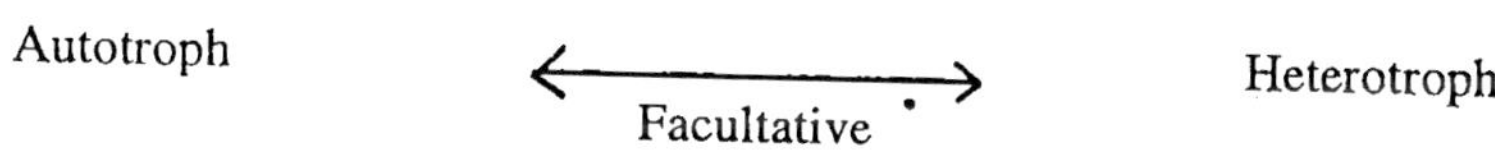

All organisms share basic chemical nutritional requirements.

NUTRITIONAL REQUIREMENTS

Seven basic Nutrition Requirements for all organisms:

1) **Source of Energy**

   a. ***Radiant energy***- photoptrophs-plants, algae+some bacteria

   b. ***Chemical energy***- chemotrophs-bacteria. Energy derived from oxidation differentiated by organic or inorganic.

      (i) Lithotrophic—inorganic ; and

      (ii) Organotrophic-organic

2. **Source of Carbon**

All organisms need C source. Some require additional Carbon compounds. Some need only organic compounds. Some need only organic compounds.

— Plants take in CO_2 during photosynthesis or carbonates from the soil.

— Some organisms use CO_2 as their sole Carbon source.

— CO_2 is used to build on to preformed carbon compounds through metabolism.

— CO_2 stimulates respiration. Man extracts a very small amount from the air which is difficult to measure since so much is produced in respiration.

3. **Nitrogen**

— There is a high amount of protein in cells.

— Occasionally it is necessary to add Nitrogen to waste lacking in this nutrient so that it can be biologically treated.

— Plants derive nitrogen from inorganic substances.

— Animals need their nitrogen in the organic form of proteins.

— Bacteria are versatile; different species can use nitrogen available in three sources:

1) Atmospheric

Nitrogen-fixing bacteria

Rhizobium=genus living in symbiotic relationship with legumes and forms nodules on the roots. Another important microbe is *Azobacter*=genus of free-living nitrogen-fixers some blue/green algae can use atmospheric nitrogen.

— Theoretically, one can stop the *eutrophication* of a body of water by removing Nitrogen and Phosphorus. But, nitrogen can still be present from run-off and nitrogen-fixers which is not as simple to control.

2) Inorganic Nitrogen

3) Organic Nitrogen—in the form of preformed amino acids

4) **Sulphur and Phosphorus**

— the amount necessary varies with the organism.

— both are essential cell components;

Sulphur ⟶ sulphides ⟶ amino acids

Phosphorus ⟶ ADP/ATP, phospho+lipids in nucleic acids

— Plants use inorganic sulphates.

— Bacteria use inorganic PO_4.

5) Several Metallic Ions

Na, Mg, Mn, Ca, K, Fe, Zn, Co, Cu etc. are also present in the cell. Metallic ions are co-factors with enzyme system. Different organisms need different elements in varying proportions. Metals are usually in very trace quantities. However, care must be taken in the compounds used in a culture, one must be aware that even a trace of a metal may be toxic. The vitamin Riboflavin contains Iron

6) Vitamins or "growth factors"

Vitamins are special nutrition needs for metabolism of energy and biosynthesis. Since vitamins are critical to the rate of metabolism and biosynthesis, they are known as the "growth factors". Necessary vitamins are provided in food or synthesized by the organism. Many animals, including Man, need their vitamins performed. Some organisms can synthesize all the vitamins they need, providing the necessary elements and compounds are available e.g. for an organism that needs Riboflavin (Fig. 3.3 and 3.4)

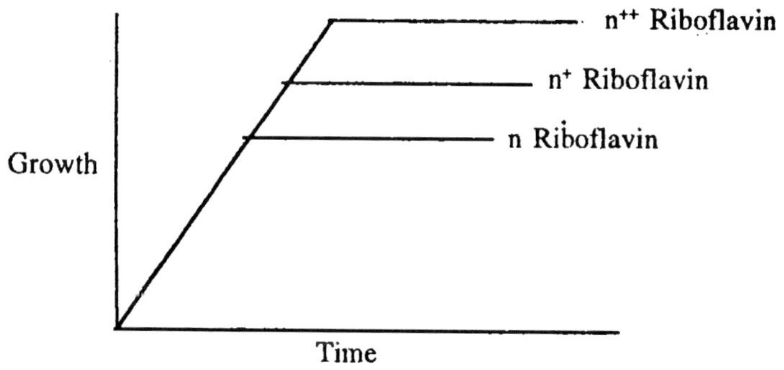

Fig. 3.3 : Requirement of growth factors in Microbes

The amount of growth is in direct relationship to amount of Riboflavin.

— In a pharmaceutical assay, the amount of a vitamin is measured in terms of growth units.

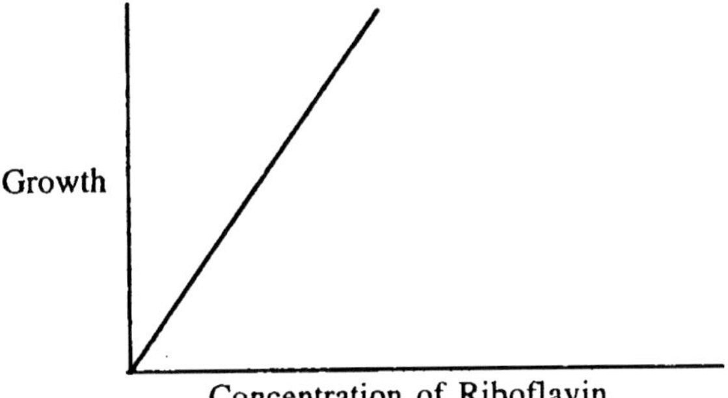

Fig. 3.4 : Requirement of growth factor in term of growth units.

7) **Water**

It may be argued that water is not technically a nutrient, but it is part of the nutritional requirements; it is necessary in nutrition. How much water is necessary is not clearly known and probably depends on the organism. A matter of current study is the amount of water necessary to most efficiently degrade solid waste biologically.

BACTERIOLOGICAL MEDIA

1) ***Synthetic or chemically defined media***

Synthetic or chemically defined media are those, where all chemical constituents and their concentrations are known. This type of media is made up of the purest constituents, reagent grade compounds and the purest of distilled waters. Once more complex compounds are added the media is no longer so precisely known, because, for example, components such as peptone or beef extract may differ slightly in composition from one batch to the next. An attempt was made to produce a media synthetically which would approach domestic sewage, thereby giving a constant for lab use. There are, however, many complex and variable components in sewage so making it synthetically with pure components is a close but poor second.

2) ***Nutrient Agar and Nutrient Broth***

Nutrient agar recipe : Beef extract 3g

Peptone 5g

Agar 15g

Water .... 1000 ml

— Nutrient broth is the same excluding the agar.

— Agar is only a solidifying agent and not a nutrient.

— a culture on the solid nutrient agar will show up in colonies.

— growth in the nutrient broth will increase the turbidity.

— Heterotrophs" "grow" (in population) when the sterile media is inoculated and incubated for a predetermined amount of time.

— Some organisms may need special nutrients added to the basic broth or agar. They are called fastidious organisms.

3) ***Special purpose media***

a) ***Enrichment*** additions to basic media to encourage growth of specific organisms e.g. yeast extract, blood cells (for Hemolytic Strep.) blood serum, extract from plants for plant associated bacteria or sterilized domestic sewage.

b) ***Selective*** addition of specific chemicals to inhibit growth of certain organisms and to allow the target organisms to flourish e.g. Crystal violet inhibits Gram + bacteria.

c) ***Differential*** incorporation of certain substances to encourage a certain kind of growth or to show special chemical reactions characteristics of the organism e.g. Sugar broth-different sugars are used with different tubes to show up different reactions to the sugars.

CLASSIFICATION BY NUTRITION

Based on energy source and external electron donor required for growth, microbes are also classified in floowing categories :

1) **Chemotrophs** : do not use solar energy; derive their energy from oxidation/ reduction reactions.

   a) ***Chemolithotrophs (Chemoautotrophs)*** These types of organisms can grow completely on inorganic material; use inorganic salts from oxidizable inorganic substrate for electron donor. e.g. Iron bacteria and nitrifiers.

b) ***Chemoorganotrophs (Chemoheterotrophs)*** : These types of organisms can do not use solar energy; energy derived from oxidation/ reduction reactions; oxidation/ reduction reactions involve organic compounds; derive CO_2 from organic compounds e.g. largest group of bacteria, including those in activated sludge, in soil and in air

2) **Photographs** : use radiant energy from the sun

a) ***Photolithotrophs*** : hydrogen is the external electron donor; uses inorganic CO_2 e.g. green plants green and purple sulphur bacteria

$$CO_2 + H_2O \rightarrow (CH_2O)_n + O_2$$

$$CO_2 + H_2S \rightarrow (CH_2O)_n + S$$

b) ***Photoorganotrophs*** : use radiant energy ; electron donor from organic matter, e.g. non-Sulphur purple bacteria.

GROWTH

Orderly increase of all components of organism which results in multiplication or replication being a consequence of growth, except some material in the cell, which is not essential to the protoplasm and is therefore not to be included in the definition of growth, e.g. storage materials, such as glycogen. In bacteriology, growth is measured in terms of change in the entire crop rather than change in size of individual cells. organic system to reproduce itself. It is important to know the difference between growth in size and growth in population/ density. *Growth measured in terms of density or turbidity is more important in environmental studies. Population growth can be measured as an increase in biomass (weight/unit volume) or as an increase in the number of cells (cells/unit volume). A unicellular organism dies when it has a irreversible loss of ability to reproduce itself; the cell may continue to metabolize however.

Problems of defining growth

If we define growth as cell number, how do we account for spores, or freeze-dried organisms which do not metabolize for a time but are still capable of reproducing. Similarly, if we define growth as reproduction how do we account e.g. drugs/ antibiotics create extreme stress, but are not lethal.

REPRODUCTION IN BACTERIA

1) ***Binary (transverse) fission*** - asexual type of reproduction (Fig. 3.5)

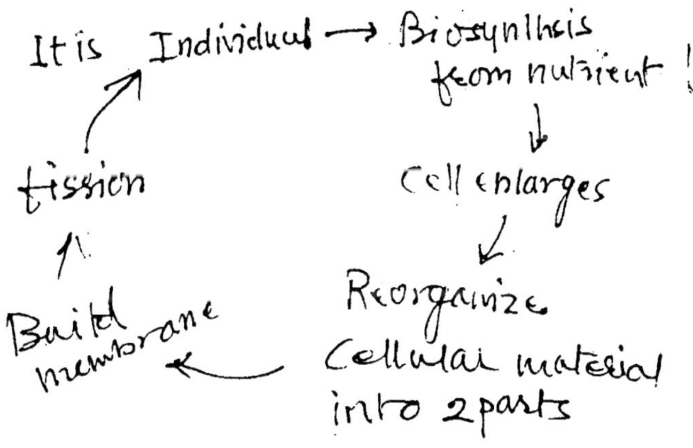

Fig. 3.5 Binary fision in microbes

It is most common in unicellular organisms.

2) ***Quasi-sexual reproduction***

Under certain conditions, some bacteria are identified as having 2 kinds of parent cells which is a kind of combination/ conjugation produce daughter cells with characteristics of both parents e.g. E. coli. Some molds show another type of quasi-sexual reproduction; two hyphae join and form a "zygote".

3) ***Budding***

Protoplasm flows through a break in the cell wall of the parent cell. This bud forms its own membrane and separates from the parent cell. Budding leaves scars in the cell wall of the parent cell; it can no longer bud at this point e.g. yeasts

4) ***Fragmentation***

Bits of a filament form their own membrane and break off of the parent cell and subsequent grows into an individual organism. e.g. Actionomycetes.

Measuring Growth (population/ mass rather than individual cells)

1) Cell count of population —up to millions/ml.

a) ***Total count*** - Microscopic estimation

— has sole advantage of being quick.

— put a known amount of the media on a known area under the microscope and count the number of organisms, usually the average of segments of the field.

Limitations

— cannot differentiate between living or dead cells.

— more interest in possible pathogenic organisms than number.

— extremely tedious method.

b) ***Viable Count***

— *Pour Plate* — mix 1 ml of sample with barely liquid agar medium (agar solidifies at about 15°C)

— incubate sample to form colonies.

— count colonies; if there were 100 organisms/ml in the sample and all were alive, there would be 100 colonies.

Limitations

— does not account for clumping of the cells in the sample.

— the time between taking the sample and making the test will produce a certain degree of error.

— this method is most practical for 30-300 colonies.

COLIFORM TEST

1) **Presumptive phase**

— Uses an *enrichment medium* of nutrient broth plus about 5% lactose.

— 1ml of the sample is put into each tube each containing a *durham* or inverted vial.

— gas in the inverted vial is positive presumptive evidence of coliform.

— negative samples are incubated another 24 hours to make sure. Total time of presumptive phase : 48 hours.

2) **Confirmed phase**

— Uses a *selective medium;* BGB (Brilliant Green + (ox) Bile) is added to the lactose broth; BGB suppresses non-coliform bacteria and inhibits certain other spore-formers.

— Incubates another 24 hours plus another 24 to make sure total time of confirmed phase : 48 hours.

3) **Completed phase**

— test out all coliform characteristics on samples of positive indications from first two phases, Gram-, rod-shaped, ferments lactose, etc.

Total time of completed phase: at least another 48 hours.

— completed phase is usually omitted unless positive coliform after first two phases.

Total time of coliform test : minimum 96 hours.

Membrane Filter Test

— time down to 12-18 hours.

— cellulose acetatemembrane traps anything larger than 54 0.45μ in size.

EMB media gives a metallic color to coliform.

MPN test is a technique for measuring growth.

— an actual count of coliform is gotten.

— the significant number derived from the samples can be checked against a table to get the most probable number of bacteria e.g.

Biomass Measurement

1) **By weight (most accurate) :**

— take the dry weight of the biomass.

— direct measure of organisms trapped in membrane filter.

— less direct is the weight of the Nitrogen in the biomass which is about 14% constant.

2) **Measure Turbidity :**

— incubate sample is tube for growth 24 hours

— calibrate the percentage of light transmittancy (fig. 3.5)

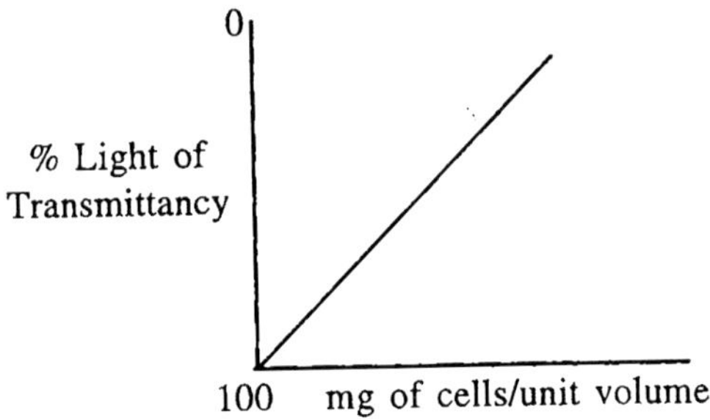

Fig. 3.5 : Turbidity Measurement

— needs homogenous cells

3) **Biomass measurement by Centrifugation :**

— measures the packed cells.

4) **Indirect measures :**

— measure the DNA or the ATP or the acid or CO_2 produced

Chapter -4

FUNGI (MOLDS), YEASTS AND VIRUSES — AN INTRODUCTION

SOME DEFINATIONS

Actinomyces : transition organism between bacteria and fungi-causes taste and odour problems

Fungi : also referred to as moulds, thallophytic plants (a division of plants which show no differentiation between stem, leaf or root), lack chlorophyll, are generally filamenous In the old classification, fungi were placed in the Plant Kingdom alongwith bacteria. In the new classification, fungi are in the higher Protista, along with blue-green algae, bacterial, etc. Fungi generally characterized by having filaments or *hyphae* (singular *hypha*). (Fig. 4.1).

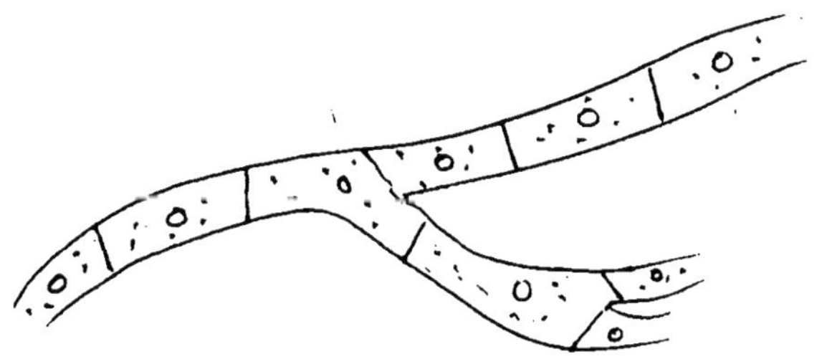

Fig. 4.1 : Fungal hyphae

Coenocytic : filaments are multi-nucleated, but have no division into individual cells (Fig. 4.2).

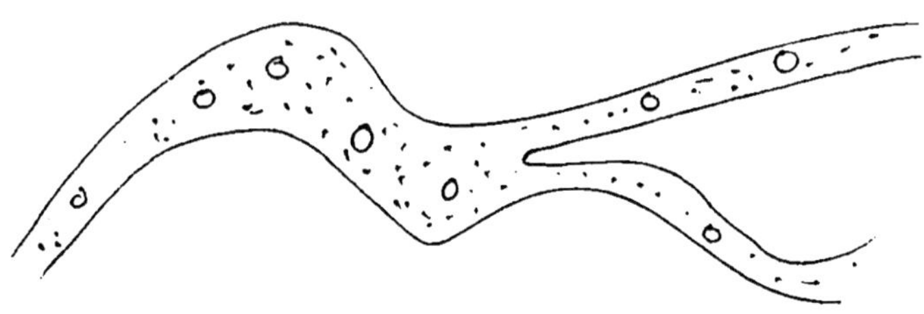

Fig. 4.2 : Coenocytic Mycelium

Mycelium : the mass of filaments. Molds are usually filamenous; yeasts are essentially not.

Fungi come in the forms of mushrooms, plant rusts, smut or puff balls, and mildew. Of significance are the molds, and yeasts, the range of unicellular fungi. Mushrooms are multi-cellular

Fungi are importance to man in many more ways. It has its importance in agriculture, as most of them are found in the soil. Similaryly, Yeasts raise bread. It is also important in fermentation. Some molds important in forming cheese *e.g. Penicillium rogueforti. Similarly, Penicillium notatum* is used in production of penicillin, an antibacterial agent.

Fungi are common lab contaminants. The spores in the air also are responsible for allergies. Fungi are classified either by morphology (form) or by their method of reproduction.

METHODS OF REPRODUCTION IN FUNGI

ASexual reproduction

1. Binary or transverse fission. (especially in unicellular fungi)
2. Budding (e.g. most yeast)
3. Spore formation (most common) (Fig. 4.3)

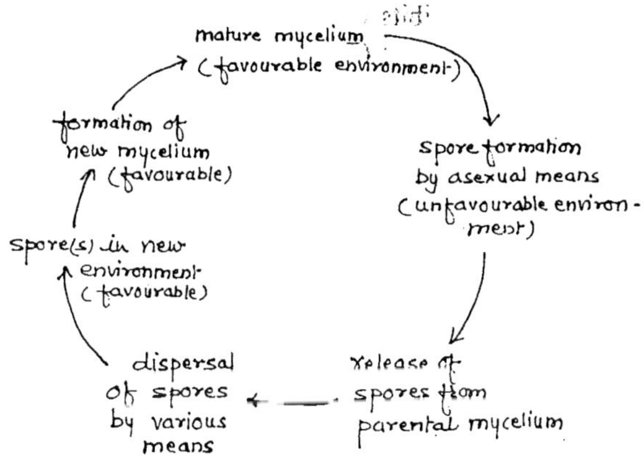

Fig. 4.3 : Asexual reproduction

Sexual conjugation

construction forms between two different hyphae producing a zygote or *zygospore* (a zygote has hereditary characteristics from both "parents")

Asexual spore formation

— Exogenous : spores form outside by parent fungi.

— organisms classified by method of spore formation and color of spores.

e.g. Penicillium, constriction of filaments form spores at the ends (Fig. 4.4).

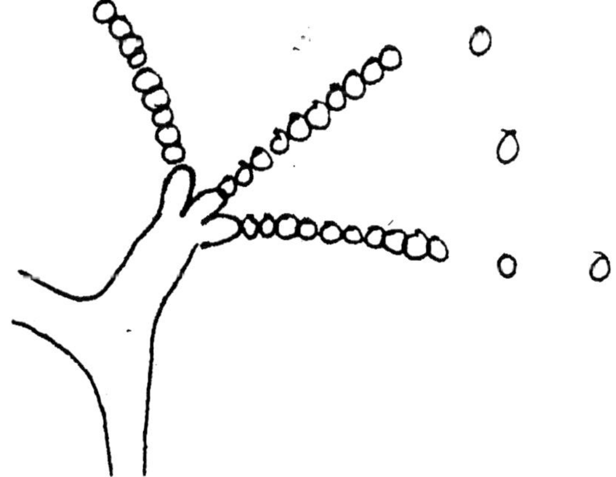

Fig. 4.4 : Exogenous spore formation in Penicillium sp.

Aspergillus sp. exihibits different spore forming mechanism; black or green-coloured spores (Fig. 4.5).

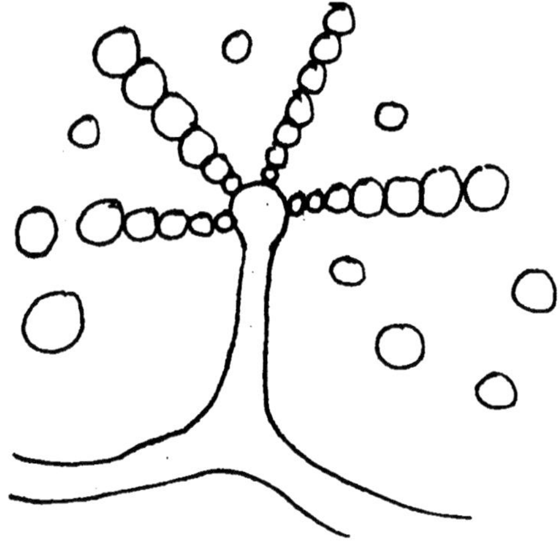

Fig. 4.5 : Exogenous spore formation in Aspergillus sp.

Endogenous spores

These are formed within the parent and released. In this type of mechanism, spores are released suddenly, propelling them into the air. e.g. *Mucor sp.* (Fig. 4.6).

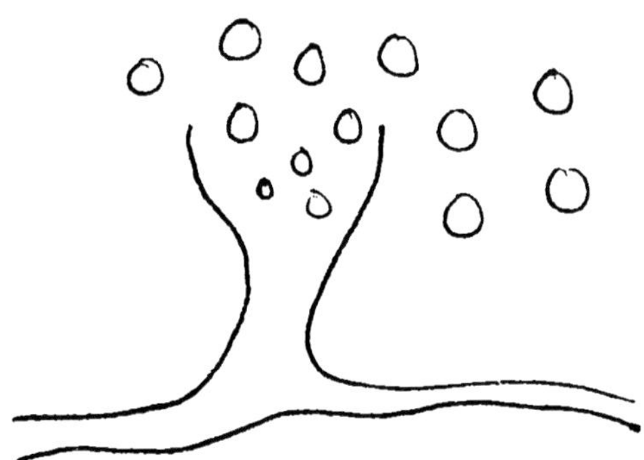

Fig. 4.6 : Endogenous spore formation in Mucor sp.

Molds

General Characteristics of molds are :

— saphrophytic; not photosynthetic.

— usually widespread in nature.

It can be cultured in the lab. It can tolerate to a wide temperature range. Mostly, these are aerobic in nature. Some tolerate or even prefer low pH, e.g. pH 3, 4, or 5 in solution. It can grow in a high concentration of sugar that would inhibit bacteria. Molds are important in : spoilage of foods, spoilage of cloth and leather as well as in agriculture.

YEASTS

General Characteristics of Yeasts are as follow

— have lost the mycelial habit; discrete cells rather than filaments.

— rather large 1-5μ; some up to 10μ.

— common among fruits, in orchards and on surface of grapes.

— found in air, on skin, in intestinal tracts of humans, insects and other animals.

— yeasts are being worked on as an indicator organism.

— yeasts are more resistant to chlorination, the buds being probably more resistant.

— yeasts are present in fecal matter Coliform bacteria are not good indicators. For viral forms, which are more resistant to calorination.

— reproduce primarily by budding.

— occasionally nuclear material divides and spores form .

— aerobic and/or anaerobic in nature.

— alcohol is the end product of anaerobic fermentation.

— under aerobic conditions, yeast cells reproduce.

— tolerate low pH, some very low. This tolerance to pH as low as 2,3, or 4 may make yeast a good test organism.

— are present in wastes.

— are high in protein, especially B-complex vitamins.

— have been used to produce single-cell protein.

— wastes which is high in organic matter may be sterilized and used to grow yeast for single-cell protein.

VIRUSES

History /

— Jenner in 1798 did first experiments on viral matter. He discovered a vaccine for smallpox.

— Louis Pasteur in 1884 developed a vaccine for rabies.

— Neither Jenner nor Pasteur knew the nature of the organisms they were working with.

— In the 1890's , Iwonowsky isolated the agent causing Tobacco mosaic.

— In 1989, Beijerinick was credited with naming the new agents "contagium vivum fluidum".

Definition : A virus is an infectious, *filaterable* particle that can be produced only with a specific host.

Composition = nucleic acid core surrounded by a protein coat (Fig. 4.7)

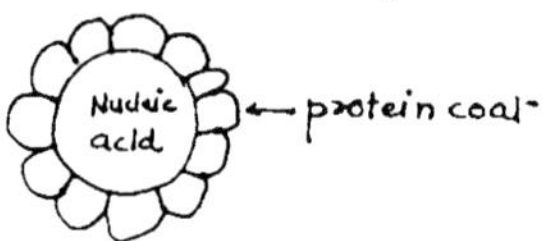

Fig. 4.7 : A Virus

Nucleic acid : assures the genetic continuity of the virus.

— offspring resemble parents

Protein coat : protective shell for nucleic acid.

— provides the specificity

Envelope : comes from the cytoplasmic membrane of cell.

— composed of lipids

Size : 15μ-400 μ

Scale of size : 1mm (Fig. 4.8).

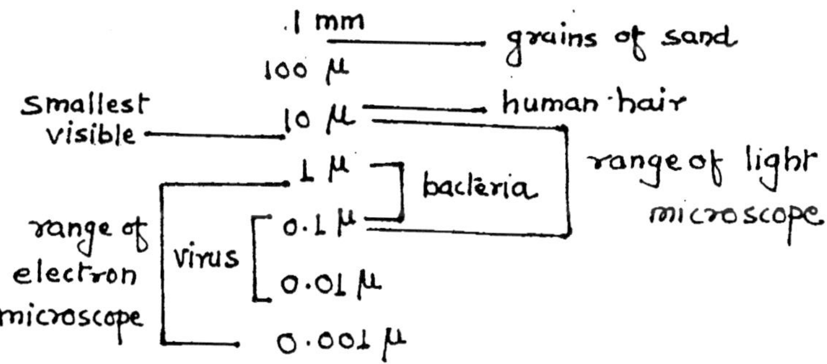

Fig. 4.8 : Chemical composition of a virus

Structure

1. Cubic e.g. polio or Herpes virus
2. Helical e.g. Tobacco mosaic virus (TMV)
3. Complex—combination of other forms
   e.g. bacterial viruses T_4, T_2

Classification

1. Type of host infected

A. **Animal virus, including man**

   e.g. pox, picorna

B. **Plant viruses**

   e.g. TMV

C. **Bacterial virus** (Bacterial phage)

   e.g. T_4, MS_2

2. *Type of tissue infected*

A. **Neurotrophic - nerve cells**

e.g. polio, encephalitis

B. **Viscerotrophic**

e.g. infectious hepatitus

intestinal infections

C. **Dermotrophic**

e.g. smallpox

D. **Pantrophic** (variety of host tissues)

3. Basis of coat
4. Basis of size

Replication

e.g. bacterial phage

Steps

1. **Adsorption**

— diffuses through host cell wall (Fig. 4.9).

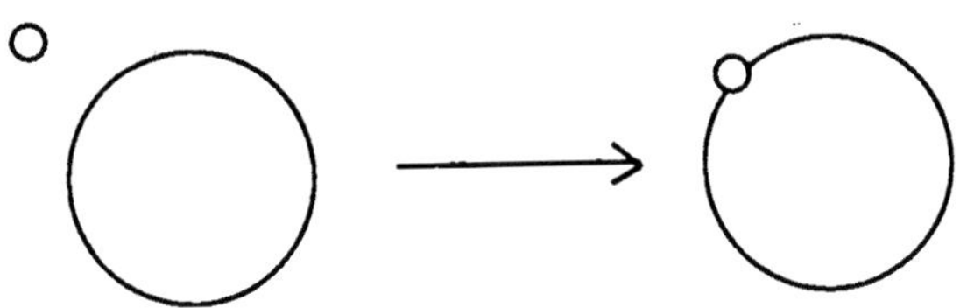

Fig. 4.9 : Replication in Viruses -Adorption

2. **Pentration**

— enzyme Lysozyme lyses host cell wall.

— virus injects its nucleic acid into cell.

— protein coat has no further function (Fig. 4.10).

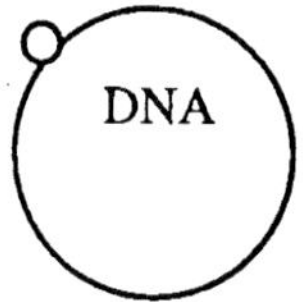

Fig. 4.10 : Replication in Viruses - Penetration

3. **Latent period**

— DNA of virus takes over function of DNA of host cell and directs host cell to synthesize for the virus (Fig. 4.11).

Fig. 4.11 : Replication in Viruses - Latent period

4. **Eclipse or synthesis period or maturation period**

Virus in host cell assembles parts, produces lyzozyme. (Fig. 4.12). Lastly, Host cell lyses and new virus are released.

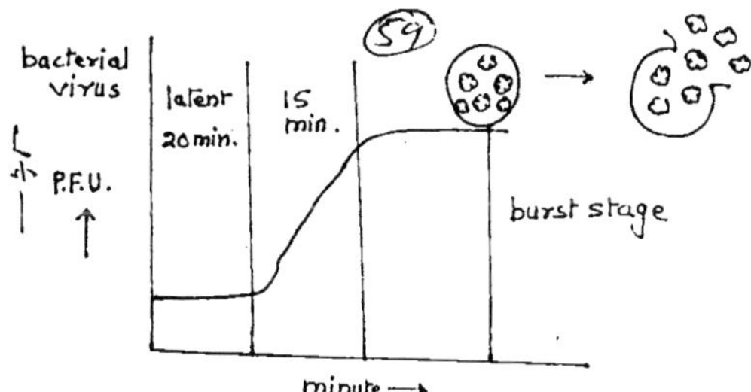

Fig. 4.12 : Replication viruses : Maturation & release phages

Animal virus replication

— entire virus particles enters cell, including protein.

Differences:

— *viropexis* = cell engulfs virus, protein coat removed inside cell.

— length of time of replicate in animal cell different.

— more animal viruses produced than bacteria phages.

Viruses of some significance

A. Entric viruses (fecal viruses)

1. **Enteroviruses**

These are naked in morphology i.e. no protein coat is occurred. The nucleic acid is single-started RNA. It vary from 15 to 20 μ in size and generally cubic in shape. Basically, it inhabit in alimentary tract. Some of the examples are :

A. polio viruses—polio myelitis

B. coxsackie — aseptic mengitis

C. coxsackie B— aseptic mengitis

D. Echo virus — aseptic mengitis

2. **Reoviruses**

It is 70μ size and its genetic material i.e. nucleic acid is double-strand RNA. It is cubic in shape and naked. It is the causative agent of respiratory diseases in children and infants.

3. **Adenoviruses**

It is 60-80μ in size. It's nucleic acid is double-stranded DNA. It is also cubic and naked. It is responsbile for causing upper respiratory infections.

4. **Infectious hepatitis**

— proven water transmitted.

— epidemics, New Delhi (1955-1956) 30-50,000 cases, 1,800 cases mostly in small water suppply system.

— not isolated.

Infective dose

1. $TCID_{50}$ enough to infect man.

TCID = Tissues Culture Infectious Dose for 50% or the smallest amount which produces 50% results in subjects

Concentration of entric viruses in water and sewage

Approximately 100 distinct enteric viruses isolated from water and sewage. Generally 200 virus units/gram of feces are present as common count. However, as infected man has 10^5—10^6 virus units/gram feces and coliform = 13 x 10^6/gram feces.

$$\frac{1}{65{,}000} = \frac{\text{enteric virus}}{\text{coliform bacteria}}$$

— raw sewage = 5,000 virus units/liter.

— polluted water = 10 virus units/liter.

Detection of viruses

1. sampling
   - A. grab method.
   - B. Gauze pad method (not quantitative, for qualitative estimation).
2. concentration
   - A. ultracentrifugation 100,000 (C).
   - B. ion exchange.
   - C. membrane filtration by adsorption (membrane 450).
   - D. insoluble polyelcetrolye.
3. identification-no single host.

Chapter -5

MICROORGANISMS IN AIR

Microbes can be present upto 35,000 ft. height, as suspended in atmosphere. They may spend entire life cycles in atmosphere. However, no aero-plankton-not known to reproduce in atmosphere. Study of biologically active entities, including cysts, spores, and pollen which must be small enough to become airborne and remain airborne for some time is called Aerobiology.

Importance

— airborne microbes and relation to animals and plants.

— public health problem when involving pathogenic organisms to man, other animals and plants.

The idea of air transmission of disease is ancient; believed by ancient Greeks, people in the Middle Ages and later.

Miasma theory=disease caused by bad air".

— precedes germ theory.

— before germ theory, disease was believed to be caused by "specific agents".

— In 1845, W.M. Bud blamed "specific agent" for spread of typhoic through water.

— Snow had same theory for cholera spread through water. Pater's microbiological studies of air disproved the theory of spontaneous generation. Pasteur did early research in aerobiology with his dust fall studies and his vacuum filtration devices for sampling air. Disease transmitted through the atmosphere which are studied most are human respiratory diseases. Only 1/3, 2/3 of these are caused by viruses.

Airborne organisms which may cause disease:

Adenovirus, bacteria like Streptococcus, Neisseria, Klebsiella etc. are some important airmicroflora responsbile for causing diseakes like diptheria, pneumonia, tuberculosis, whooping cough, mumps, influenza etc. in human beings. Additionally, Protozons cysts, Pinworm eggs, some tapeworm eggs, fungal spores can also be seen in suspended forms in air.

Sneezing

— 10,000-100,000 droplets at a force of 200 mph.

— no affects on organism from being airborne

— protected by the material expelled with.

— smaller droplets quickly dry.

— aerosols from of less than 50μ.

— sneezing produced droplet nuclei.

Coughing

— less violent: therefore, fever organisms.

Aerobiology is different for enclosed spaces (indoor environment).

Indoor air

Indoor environment contaminated by people by talking "body dust" as well as by other human activity. Small areas often have poor circulation. Indoor dust includes fragments of human skin which form "rafts" for carrying the bacteria. The rafts of skin are a pure culture for the airborne bacteria to surviving on. bed making stirs up lots of dust. It has also been known to stir up eggs of pinworms in bedding.

Outdoor air

Depending upon geographical and environmental variables, air microflora, in outdoor environment exhibits peculiar characteristics, which may affected by cooling towers' spray irrigation andwind action on water etc. large numbers of saporphyitic fungi, especially yeasts, have mechanisms for shooting spores into the atmosphere.

Pollen (male reproductive cell of flowering plants), which is also found in air may vary with season. In spring—tree pollen, in summer—timothy, grasses, and in late summer, autumn—weeds, ragweed can be possible to observe. These are causes allergic reactions in human being. Zobell found marine organism in the atmosphere at 0 ft. above surface of water 10.7/cu.ft. and at 1,600 ft. above surface of water 0.5/cu.ft.

Algae and Protozoa

Protozoa are not present in large amounts being airborne makes it possible for algae to inoculate unusual habitats. Chlorella is an algae found growing in cooling towers since it is not characteristic to the water, it has been brought there by air.

Fungi

Vast number of fungal spores even at 1 k.m. above the surface 2,700/cu. meter at 1 k.m. height can be observed. It has economic importance as it causes plant diseases e.g. *coccidio mycosis* in the Southwest USA. Fungal diseases affect many grain crops. Diseases like corn leaf blight, corn smut, rusts etc. are some of the examples of fungal diseases of plants.

Dispersion of these microbes in air is performed by simple physical of diffusion and brownian diffusion. Accordingly, organisms can travel a significant distances. However, it is affected by motion of the air. e.g. sneezing as well as by impacts on surfaces. Even in still air, settling velocity of fungi spores was estimated as approximately 1/10 cm./sec.

Atmospheric conditions affect dispersion

— winds.

— rains-settle out particles, clear air.

organnisms may get airborne from wastewater treatment.

— air bubbles bursting from aeration.

— activated sludge and trickling filter release fewer microbial droplets to air.

— mechanical aerators stir up lots of organisms.

- wind affects on solid waste.
- wind affects on soil micro-organisms.

Klebsiella pneumonia

- biological aerosols 40% to lung, 60% to alveoli.
- at 5 ft. away, breathing in viable *Klebsiella* every 2nd breath.

Diseases incidence among waste treatment workers

- measured by determining what antibodies are among workers that the general population doesn't have by blood sample.
- a person can be infected by a disease without the disease symptoms.

Organisms are affected by the airborne state

- results disagree on survival kinetics.
- experimental growth conditions may affects the organism more than being in the aerosol state-there is no standard method for experimentation.

Survival is a terminology problem

- an organism may be "living" but not active
- differences in ability to withstand and conditions of being in the atmosphere
  - measles-infect at close distance
  - some organisms survive atmosphere better than others
    - intrinsic factors
    - different survival mechanism
- ability to withstand drying
  - enteric organisms have poor resistance to drying.
  - geometric die-off with time in the atmosphere.
  - organisms in sneeze droplets can survive.
- physiological condition of the culture
  - survive air less in log growth phase than in stationary phase

- temperature—especially relative humidity
  - relative humidity below 30% little difference in survival at 15°C change
  - above 30% survival decreases with increase in temperature e.g. Temperature constant but varying humidity.

Percentage of survivors high below relative humidity of 30% and above 80%

- rapid changes in humidity affect organisms.

Radiation—sunlight

- 580μ can inactivate bacteria, virus and spores.
- a sunny room will have fewer organisms.
- ultraviolet light may be a factor to use in controlling organisms.

Chapter -6

MICROBIAL GROWTH : MEASUREMENT AND KINETICS

Generally, growth in a single organism refers to a general increase in size; every part grows, doubles in size until replication. If every necessary cell component is being duplicated at the same time, it is called *Balanced Growth.* Growth of a bacteria *population* increases in geometric progression Growth = 2^n

parent 2^0

generation 1 2^1

Generation 2 2^2

Generation 3 2^3

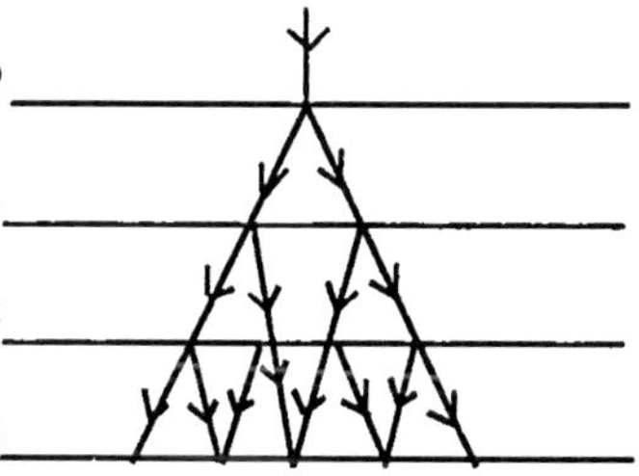

Where N_O = initial concentration of bacteria :

$$\frac{N_t}{N_n} = N_O \cdot 2^n$$

$$\frac{N_n}{N_O} = 2^n$$

$$--\log_2 \frac{N_n}{N_o} = n$$

general form of growth

$$--\log_2 = n \frac{\log_{10} n}{\log_{10} 2} = \frac{\log_{10} n}{0.301} \quad \text{(Fig 6.1 and 6.2)}$$

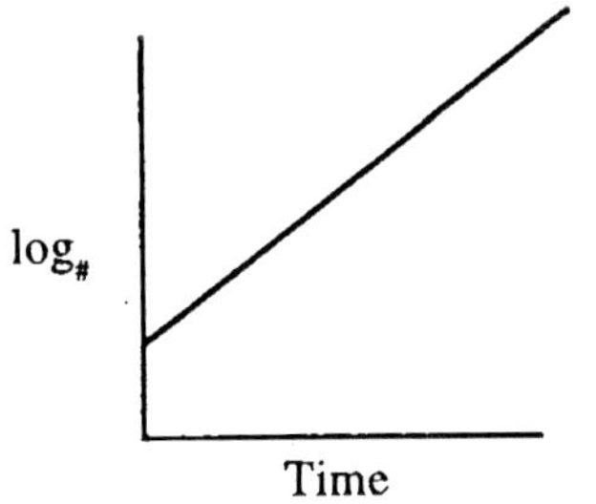

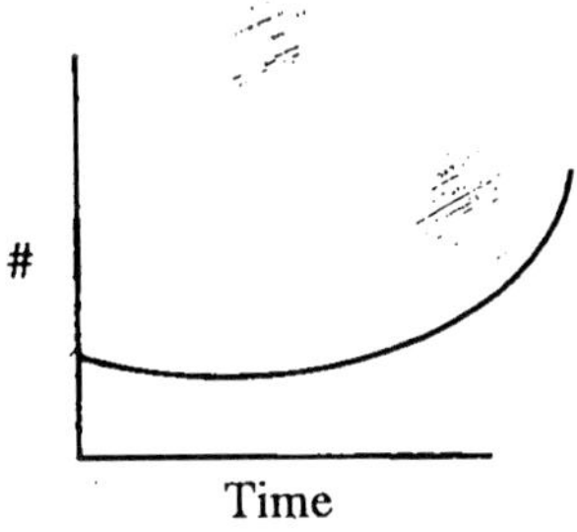

Fig. 6.1 : Growth Pattern ($\log_{\#}$)

Fig. 6.2 : Growth Pattern (#)

Generation time The time interval, required for the cell to divide or for population to double is knwon as generation time.. In geenral :

$$g = \frac{t}{n}$$

Where,

g = generation time

t = given time in experiment

n = number of generations

e.g. *E. coli* ⟶ g=20 min.

if t=60 min. then $20 = \frac{60}{n}$ or, n=3

It is important to note that :

1. generation time is not the same for all bacteria.

2. generation time for the same bacteria in different media under different environment conditions is different.

   a) based on available nutrient.

   b) based on specific physical or chemical conditions.

e.g. for *E.coli* in lab under optimum conditions, calculated generation time g = 20-35 min., whereas, it is several hrs. in intestine.

K = *growth rate constant* = inverse of generation time

$$K = \frac{1}{g} = \frac{1}{time} = \frac{1}{hr^{-1}} \longrightarrow n = Kt$$

$$Nt = N_O \cdot 2^{Kt} \longrightarrow \frac{N_t}{N_o} = 2^{Kt}$$

$$\log_2 \frac{N_t}{N_o} = Kt$$

$$\text{or, } K = \frac{\log_O Nt - 10g_{10} N_O}{t. \log_{10}{}^2}$$

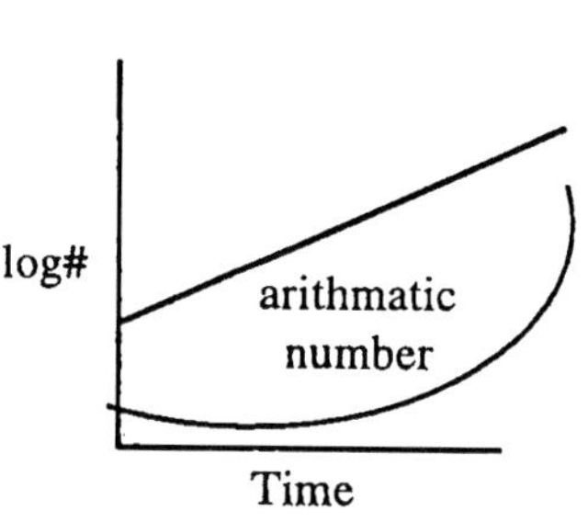

Fig. 6.3 : Growth Pattern

Examples :

1. assume $N_O = 10^2$

   $N_t = 10^9$

   t = 10 hrs.

$$K = \frac{9 - 2}{10. .301} = 2.33 \text{ hr}^{-1}$$

$$K = \frac{1}{g} = \frac{1}{2.33} = 26 \text{ min.}$$

or, g = 26 minutes.

2. assume $N_0 = 1$

$g = 20$ min.

$t = 48$ hrs.

$n_t = ?$

$$K = \frac{1}{g} = \frac{1}{20/60} = 3\ hr^{-1}$$

$$N_t = N_0 \cdot 2^{Kt}$$
$$N_t = 1 \times 2^{3.48} = 2144$$
$$N_t = 2.2 \times 10^{43}$$

Theoretical amount of bacteria if optimum conditions = 2.2×10^{31}g or 4000 × the weight of the earth.

Instantaceous growth rate constant = μ

$$\frac{dN}{dt} = \mu N \longrightarrow \frac{dN}{N} = \mu dt \longrightarrow \log\frac{N}{N_0} = \mu t \qquad \frac{N}{N_0} = e^{\mu}$$

$$\frac{Nt}{N_0} = 2^{Kt}$$

$$\frac{Nt}{N_0} = e^{\mu t} \longrightarrow 2^{Kt} = e^{\mu t} \longrightarrow Kt \log^2 - \mu t \longrightarrow \mu = 0.69\ K$$

$$g = \frac{1}{K} \longrightarrow \frac{1}{K} \quad \frac{1}{\mu} \longrightarrow \frac{1}{\mu} \longrightarrow \frac{1.45}{K}$$

GROWTH CURVE

Synchronous Growth

Assume each generation acts identical to parent; divides at exactly same time; ideal conditions; assume optimum (Fig. 6.4).

Synthetic ideal growth can be produced in the lab. e.g. synchronous growth can be simulated by producing heat shock in diplococcus; changing the temperate from the optimal 35-37° C to 25° C for 15 min. and change back again to 35-37° C (Fig. 6.5).

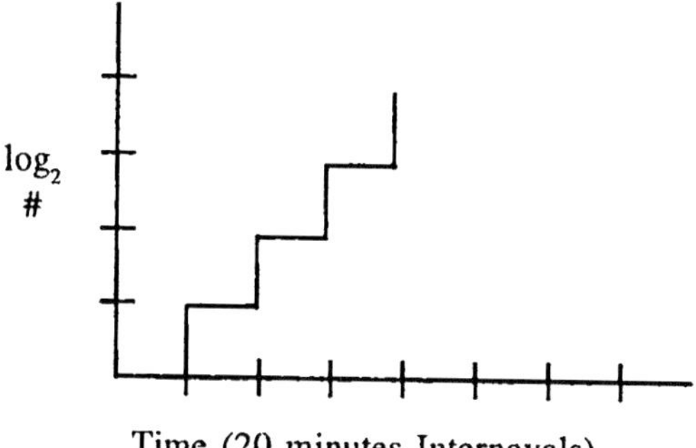

Fig. 6.4 : Synchronous Growth

IDEAL GROWTH CURVE

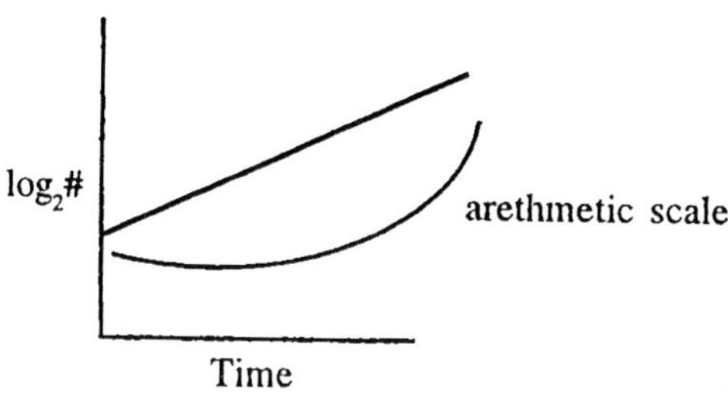

Fig. 6.5 : Ideal Growth Curve

Growth curve (Idealised) : The idealised growth curve is presented in fig. 6.6.

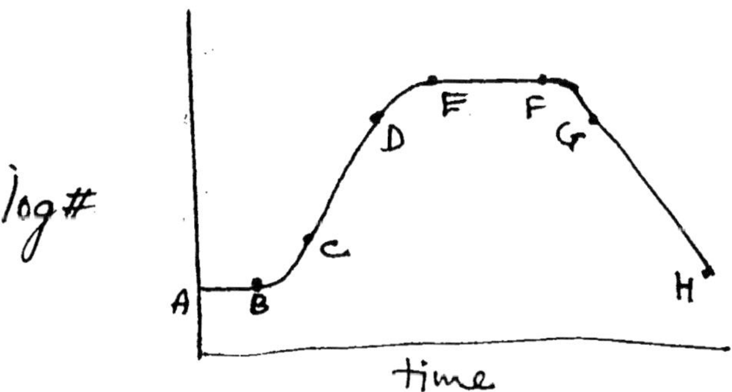

Fig. 6.6 : Idealised growth curve

A-B LAG PHASE

No change in bacterial population, adjustment phase to new media or environment.

B-C ACCELERATION PHASE

Population starts to grow slowly from lag phase minimum growth rate to maximum growth rate, *Net Constant.*

C-D EXPONENTIAL OR LOGRITHMIC GROWTH PHASE

Growth at maximum rate as long as nutrition is available; growth is constant (K) ; generation time is at a minimum.

D-E DECLINING GROWTH

Insufficient nutrients or environmental change; rate of growth is slower decreasing from (D) maximum growth to (E) minimum growth.

E-F STATIONARY PHASE

No change; Bacteria growing minus Bacteria dying = 0; insufficient food, lysed or dead bacteria provide some nutrients for living organisms.

F-G ACCELERATED DEATH PHASE

Organisms run out of food.

G-H LOG DEATH PHASE

Constant death rate, death at maximum rate; no food and accumulation of toxic material from dead cells.

Fig. 6.7 (a & b) shows bacterial growth with and without lag phase.

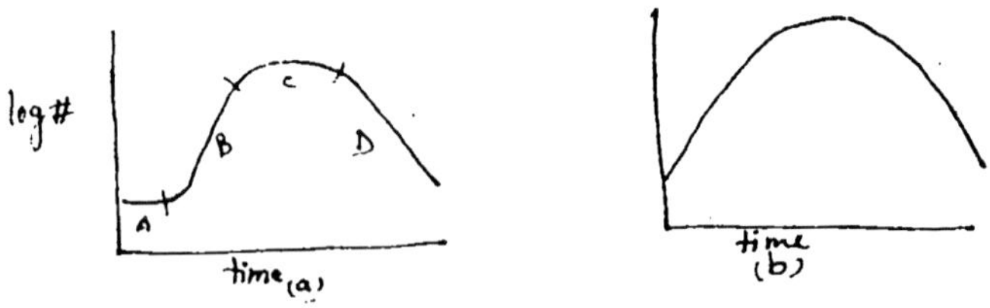

Fig. 6.7 : Growth Curve with and without lag Phase

A = Lag phase = period of adjustment

B = Log phase

C = Maximum stationary phase

D = Phase of decline or death

A = LAG PHASE

Organisms adjust to new, different nutrients and may have to produce new enzymes, *inducible enzymes*. Also, need to adjust to other environmental conditions in order to get ready for maximum. If organisms are taken from a medium while in log phase, there will be little or no lag phase.

B = LOG PHASE

Growth rates maximum ; growth rate constant assuming conditions are optimal. In Activated sludge, maintain high growth rate by providing more food than there are organisms:

$$\frac{\text{mg}}{\text{mg}} = \frac{\text{Food}}{\text{Microorganisms}} = \text{High}$$

All environmental conditions are optimal; pH, temperature, toxicity Maintain bacteria in high growth rate phase.

FACTORS AFFECTING GROWTH DURING LOG PHASE

A. ***Inherent nature of the organisms***

Some species multiply than others; the background or previous environment of the organisms; ability to produce new enzymes.

B. ***Nutrients***

Sufficient quantity or concentration; type of nutrients; organisms need optimal concentration balance, too much or too high concentration inhibits growth. (Fig. 6.8)

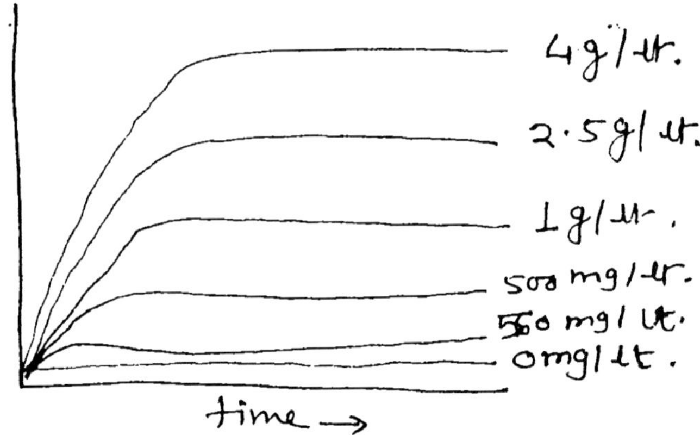

Fig. 6.8 : Effect of nutrient on log phase

GROWTH RATE

$$\frac{dN}{N} = \mu dt$$

Instantaneous growth rate

$$\frac{N_t}{N_O} = e^{\mu t} \longrightarrow \ln \frac{\frac{N_t}{N_O}}{t} = \mu$$

-----generation time ≠ doubling time

-----mean generation time = doubling time t_D

$$\frac{\ln d}{t_D} = \mu = \frac{0.693}{t_D}$$

K=1/9 g = t/n $g = t_D = 1/K$

$$\mu = \frac{0.693 \times 1}{g} = 0.693$$

CONTINUOUS CULTURES

A. Lag phase (fig. 6.9)

B. Log growth phase

(Factors affecting growth)

1. inherent nature of organisms
2. substrate, nutrient
   a. all essentials
   b. optimum concentration

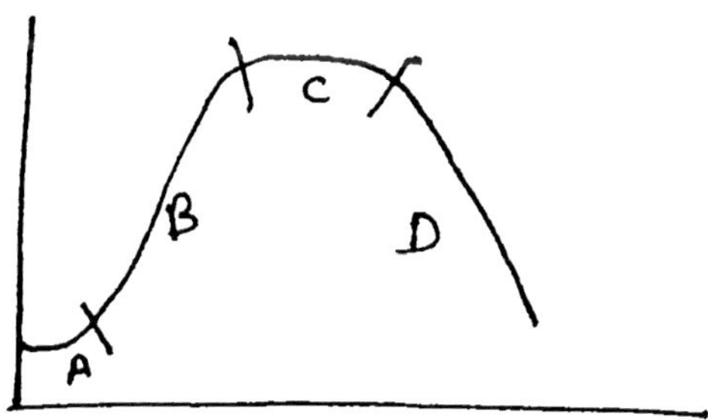

Fig. 6.9 : Continuous Culture

Minimum Concentration

Substrates concentrate on the walls of the walls of the container or on the surfaces of suspended solids. (Fig. 6.10) Example : Same concentrations of nutrients and organisms :

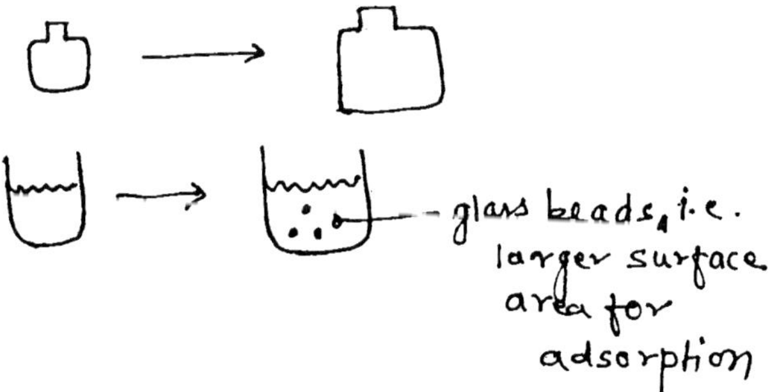

Fig. 6.10 : Microbes in Suspension

Zobell found that 1 x 10^7 bacteria / ml ≈ 1 mg/l of organic matter In marine waters, 10^2 — 10^3 bacteria/≈ 4-5 mg/1 organic matter. It is because that low concentration in sea water and little surface area for adsorption. Proved in a container that concentrated nutrients had 10^3 x more bacteria than in the sea.

Things that can be used to increase surface area: (in suspension)

— glass beads

— clay

asbestos (in digesters)

— $(Ca)_3\ (PO_4)_2$

— sand

Keeping bacteria in log growth phase (maximum growth rate)

1. Repeated transfer to fresh media (after using up nutrients *but* impractical).
2. Add fresh media to same reactor

a. chemostats

b. turbidostats

CHEMOSTAT

External control of culture to produce a steady state at a high rate of growth, there is an excess of all nutrients except one in the culture vessel. One essential nutrient, for example glucose, is controlled in the reservoir (Fig. 6.11).

Overflow contains bacteria and unused substrate in the same concentration as in the culture vessel.

Turbidostats — internal control i.e. nothing is limited, all substrates, nutrients in excess. Population is controlled by dilution culture. Internal control is achieved by detecting changes in the light passing through the culture. In a given wavelength of light, the scattering of light is proportional to the population of bacteria. (Fig. 6.12)

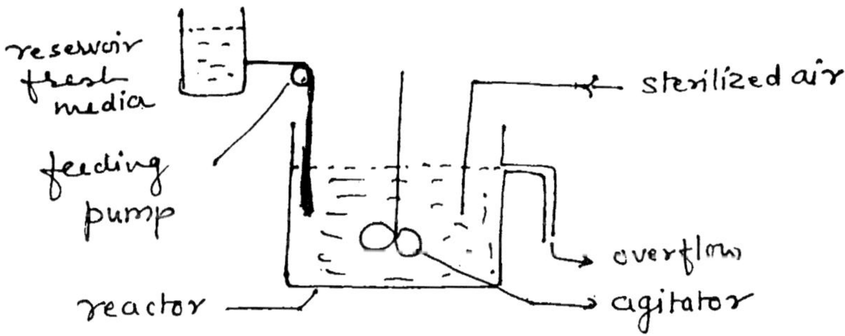

Fig. 6.11 : Chemostat

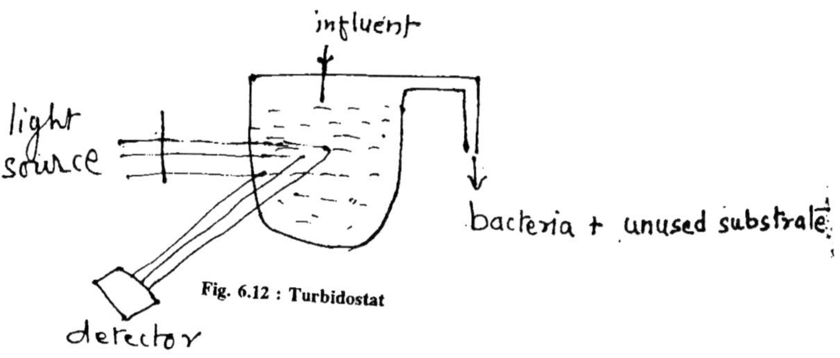

Fig. 6.12 : Turbidostat

Fig. 6.12 : Turbidostat

Difference between Chemostat and Turbidostat.

Following Table 6.1 gievs an idea about the difference between chemostat and turbidostat and Fig. 6.13 present the same diagramatically.

Table 6.1 : Difference between chemostat and turbidostat

Chemostat	*Turbidostat*
— control external	— control internal
— one substrate limiting (growth rate is below maximum because one nutrient is always limited)	— substrate is not limiting
— growth rate varies	— growth rate varies
— maximum rate not usually maintained	— growth rate maximum or near maximum constituently

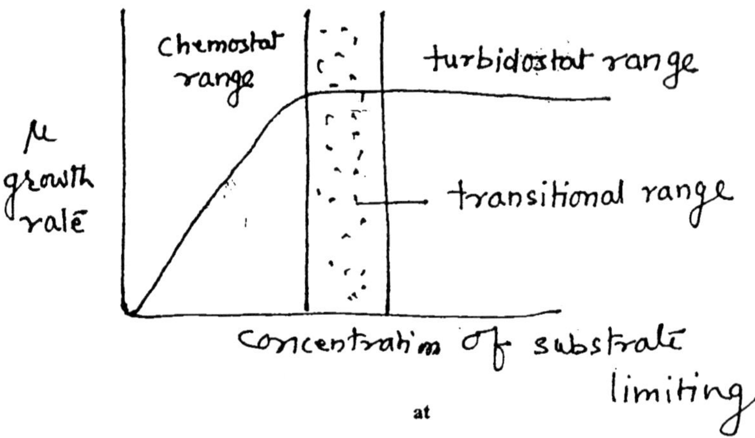

Fig. 6.13 : Growth in Chemostat and Turbidostat

FACTORS AFFECTING RATE OF GROWTH

A. Nature of bacteria

B. Nutrient

C. pH

D. Temperature

pH

It is generally neutral internal bacteria, some kinds of bacteria can be found in pH 0-13. e.g. SO_4^{--} reducing bacteria range pH 0 and E. coli can survive pH 12.5. optimal range for most bacteria at pH 6.5-7.5. As an average maximum tolerable range for most bacteria pH 4-9. Bacterial cell membrane is semi-permeable for H and OH ions.

Temperature

Extreme range for most bacteria - 5°C — 80°C. some fungi can survive. Similarly, 12°C — 104°C at 1000 atm. pressure in the temperature range between i.e. SO_4^{--} reducing bacteria can survive 104°C. Growth and chemical reactions are affected by temperature (Fig. 6.14). It can be seen that approximately, for each 10 increase the growth rate doubles (up to the maximum rate).

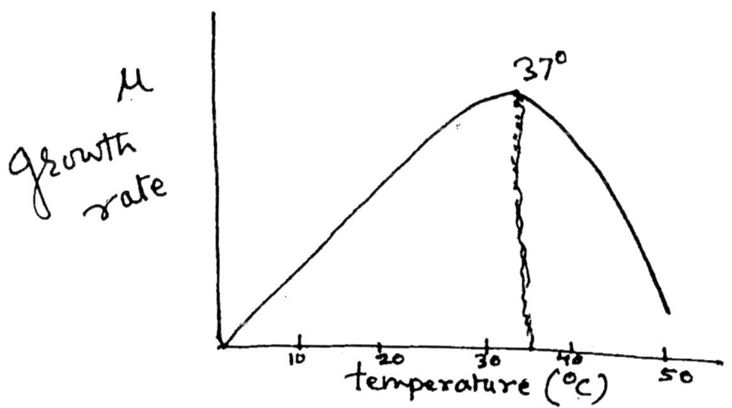

Fig. 6.14 : Minimum (10°-12°C), optimum (37-45°C) and maximum (46°C) temperature ranges for E. coli

Van-t-hoff Law

$$\log_{10} V = \frac{-\Delta H}{2.303\ RT} + C$$

Where :

V = reaction rate

ΔH = activation energy

R = constant

T = absolute temperature and,

C = constant

Division of bacteria on the basis of temperature

1. Mesophilic 20-45°C optimum 30-40°C
2. Thermophilic above 45°C ➔ above 60°C - Thermotolerance
3. Psychrophilic below 20°C

Factors negatively affection log phase growth

1. concentration of nutrient (Fig. 6.15 and Fig. 6.16)
2. accumulation of toxic material
3. Physical environment
   — pH
   — temperature
   — osmotic pressure
4. Crowding—competition
5. Presence of predation
6. Radiation — change in intensity of light

C — Growth minus death = 0

D — nutrient completely exhausted or toxic material accumulates logrithmic rate of death

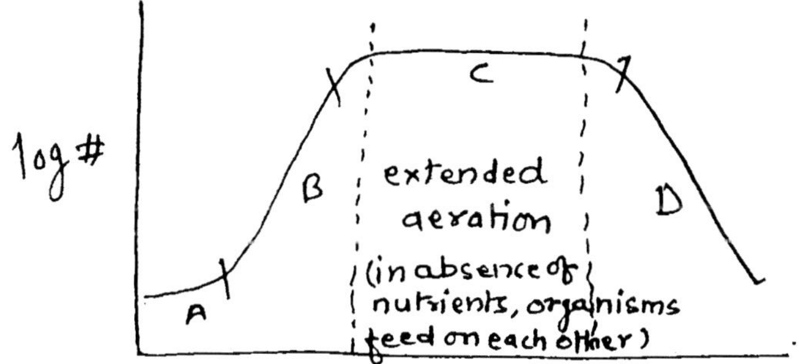

Fig. 6.14 : Effect of nutrients on growth rate of microbes

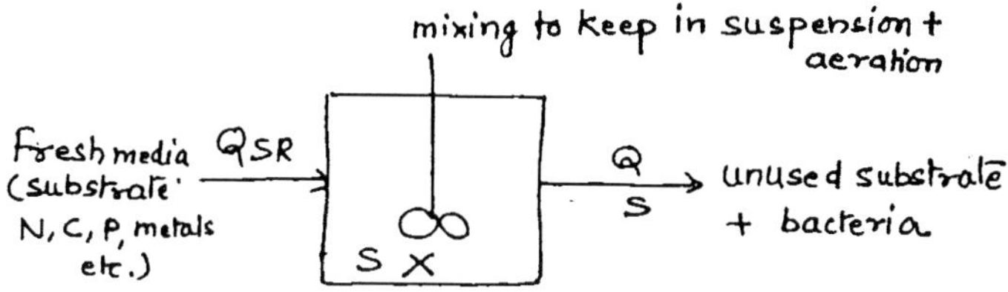

6.15 : Growth in continuous system

In this figure,

V = volume

Q = flow

X = concentration of microorganisms and

S = Concentration of substrate

1. $$\frac{dx}{dt} = \mu x$$

2. $$\mu = \mu \max \left(\frac{S}{K_S + S} \right)$$

x = weight of microganisms

t = time

μ = specific growth rate applied to bacterial growth

s = substrate, (Fig. 6.17)

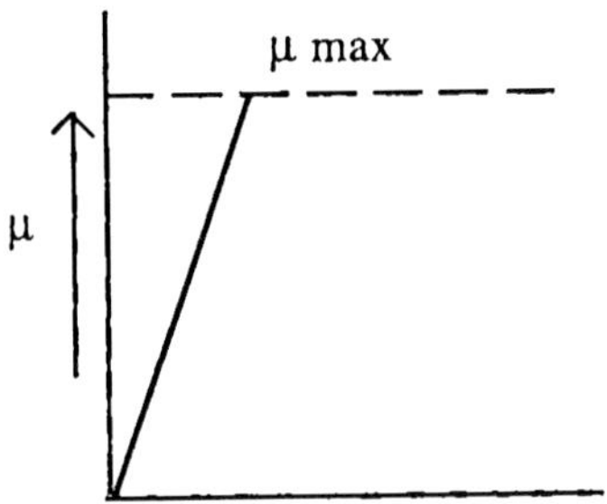

Fig. 6.17 : Microbial Growth

increase in substrate won't change growth rate

when $\mu = 1/2\ \mu max$, ($K_S = S$)

3. $$\frac{dx}{dt} = \mu \max \times \left(\frac{S}{K_s + S} \right)$$

4. amount of bacteria/amount of substrate

1 mg/lt. = 0.5 bacteria

yield :

$$-\frac{dx}{ds} = Y$$

5. divide 4. by 3.

$$-\frac{ds}{dt} = \mu\max\frac{X}{Y}\left(\frac{S}{Ks+s}\right)$$

6. D - dilution

$$D = \frac{Q}{V} \qquad \text{time}^{-1}$$

7. net change in microbial population
net increase =

$$X\,\mu\max\left(\frac{S}{K_S+S}\right) - DX \qquad \text{microbes is effluent}$$

Steady state = net increase = O $X\,\mu\max\left(\frac{S}{K_S+S}\right) - DX$

8a. $D = \mu\max\left(\frac{S}{K_S+S}\right)$ or $D = \mu$

8b. $S = Ks\left(\frac{D}{\mu\max + D}\right)$

9. Change in substrate = Influent
minus sbustrate utilized in growth
minus substrate washed out

influent = S_r effluent = S

Change in substrate = $DS_r - \mu\max\frac{X}{Y}\left(\frac{S}{K_s+S}\right) - DS$

if net change = O is steady state :

$$DS_r - D_s\,D\frac{X}{Y} = 0$$

$$S_r - S = \frac{X}{Y}$$

10. $X = Y\ (S_r - S)$

11. $X = Y \left[S_r - K_s \dfrac{D}{\mu\ max - D} \right]$

. amount of bacteria coming out weight

$$DX = Dy \left[S_r - K_s \frac{D}{\mu\ max - D} \right]$$

y = system is constant (Fig. 6.18 a & b)

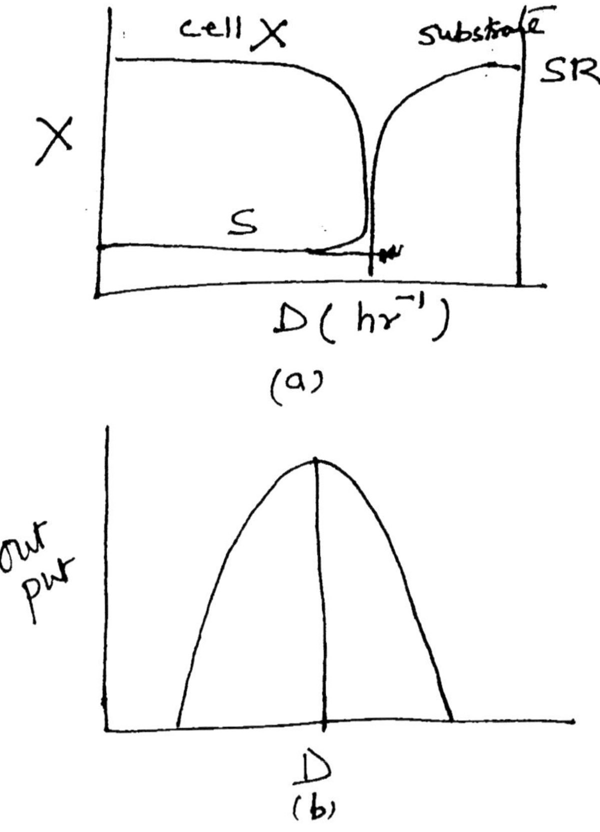

Fig. 6.18 (a & b) : Substrate bading, dilution and output (in terms of biomass production)

Chapter -7

AROMATIC HALIDES, PHENOLS AND ISOMERISM

When hydrogen atom(s) of a benzene ring is substituted by halogens the product is known as aromatic halides. They are significant because they are used as a raw material in the synthesis of organic compounds & appear in the industrial wastes. The pesticide industry used to use inorganic compounds such as sulphinates, cynanides & mercury. Now they use organic compounds, which are specific to target organisms. Some of these are resistant to biological degradation. There are following types of chlorinated hydrocarbons:-

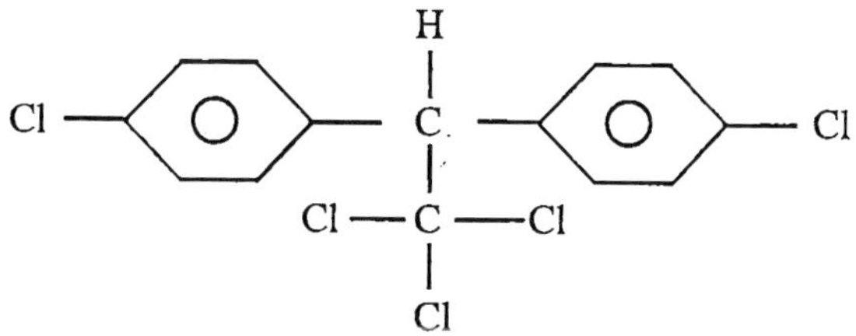

H
Cl — C — Cl
Cl — C — Cl
Cl

Dichloro diphenyl trichloro ethane or DDT.

Phenoxy acetate

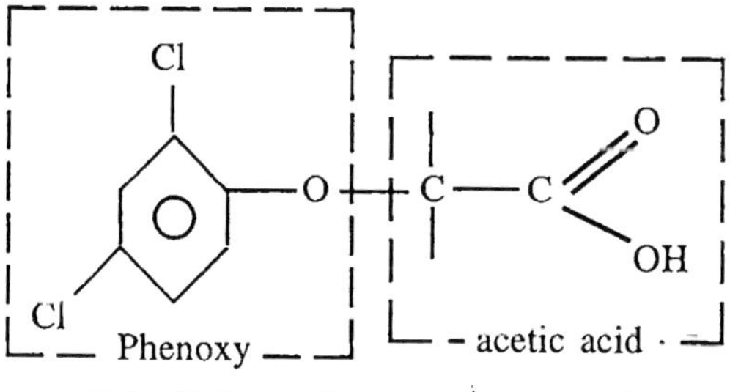

Cl
O — C — C
O
OH
Cl
Phenoxy
acetic acid

2.4 dichlorophenoxy acetic acid

Commercial names of phenoxy acetate : 2.4 D., Aldrin, Endrin.

Some chlorinated hydrocarbons are good herbicides & are effectively used to kill plants in reservoirs. The problem with this and other pesticides is not know how long they last actively in the

Organo-phosphorus pesticides

H H

O—C—C—H

H H

O_2N — ⬡ — O — P = S

H H

O—C—C—H

H H

Parathione

environment.

This is widely used at a great deal in Citrus production. It is known to be extremely toxic to mammals in direct contact, but it is not known how long it remains active or, if it accumulates in animal tissues with toxic effect.

H O CH_3

R—N—C—O—C—H

CH_3

CARBAMATES

R might be a phenol ring with chlorine substitution. The ultimate question or objective for a pesticide is whether it can be stable or stationary in the environment & degrade in a reasonable amount of time & still selectively kill target organisms. Environmental Biochemists in this area study the relation between the structure of the molecule & its activity & biodegradability e.g. with phenoxy acetates, substitutes, on the benzene ring were studied for their activity.

It was found that :

1. Para > ortho > meta (position of substitute)
2. Chlorine > Iodine
3. Chlorine > amine> nitrite (NO_2)
4. Chlorine > Bromine

Desirable attributes of organic pesticides :

1. Can control or eradicate a particular pest.
2. Non toxic to non target organisms.
3. Will degrade to ecologically acceptable end products in a reasonable amount of time.
4. Relatively stationary.
5. Low potential for accumulation in any part of the environment.

Problems of organic pesticides :

1. They are widely used & have had a tremendous impact on agricultural production.
2. Controlling use calls for legislation with its accompanying problems of opponents on both sides.
3. Acute toxicity compared to long term toxicity.
4. Transfer from one phase of the ecosystem to another soil to water etc.
5. Degree of response to biodegradation.

Biodegradation of a phenoxy acetate

Mono substituted chlorine in para position is more likely to degrade, cleaving ring. Chlorine in ortho position is inhibited in degradation.

Disubstitutes & chlorine in para position will compensate for one in the ortho position.

AROMATICS

Ring structure with 6 carbons & alternating double bonds.

The parent compound of the aromatics is benzene.

They are often substituted with halogens Iodine, Bromine & Chlorine

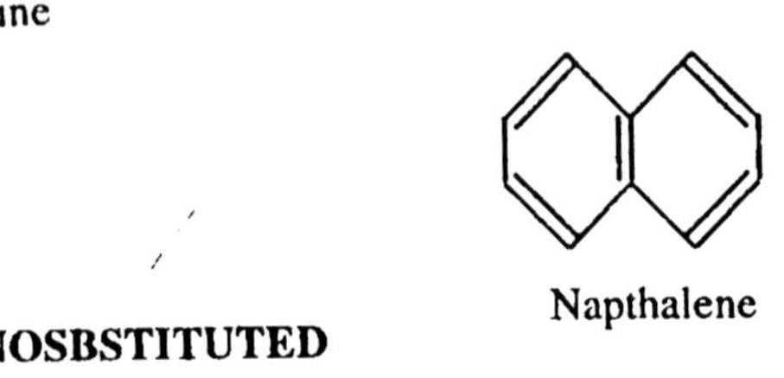

Napthalene

MONOSBSTITUTED

Cl

OH

H
C—H
H

Monochlorobenzene Phenol Toluene

DISUBSTITUTED

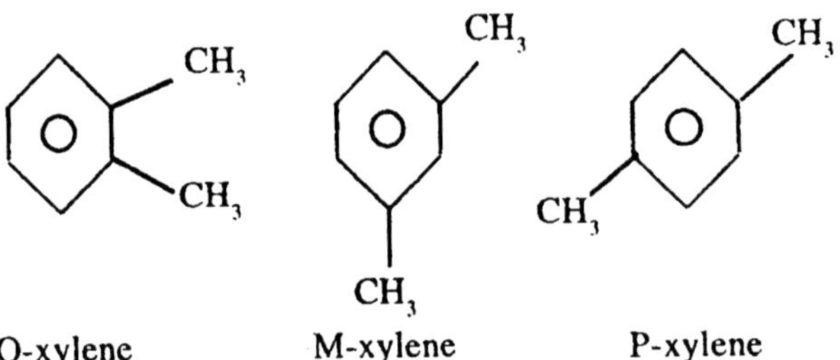

O-xylene M-xylene P-xylene

Sources of Benzene

— From coal :

1 ton coal $\xrightarrow{\Delta}$ { 1500 lb coke, 100 lb tar, 12,000 ft. gas, H_2O } $\xrightarrow{\text{distillation}}$ Toluene, cresol, Benzene, Phenol, xylene

— From Petroleum :

Synthesis—less important e.g.

$3\ C_2H_2 \longrightarrow C_6H_6$

CHEMICAL REACTIONS OF BENZENE

1. Reduction

$$C_6H_6 + 3H_2 \xrightarrow[Ni]{200° C} C_6H_{12}$$

(hydrogen breaks double bond) Cyclohexane

2. Oxidation - generally resistant. COD test doesn't show benzene. It occurs only partial, not complete oxidation.

$$C_6H_6 + O_2 \longrightarrow \text{maleic anhydride}$$

maleic anhydride

Halogenations :

$$C_6H_6 + Br_2 \xrightarrow{FeBr} C_6H_5Br + H^+$$

Substitutional bromobenzene

In aliphatic compounds, halogenation of unsaturated compounds breaks the double bond to make them saturated. The double bonds does not break in aromatics.

$$CH_3-C=C-H + HOCl \longrightarrow CH_3-C(Cl)-C-$$

1 propane

Nitration (substitution reaction) :

$$C_6H_6 + HONO_2 \longrightarrow H_2O + C_6H_5NO_2$$

Nitrobenzene

Polysubstitutions depend on concentration of nitrite ions.

$$HONO_2 \xrightarrow{H_2SO_4} H_2O + NO_2^+$$

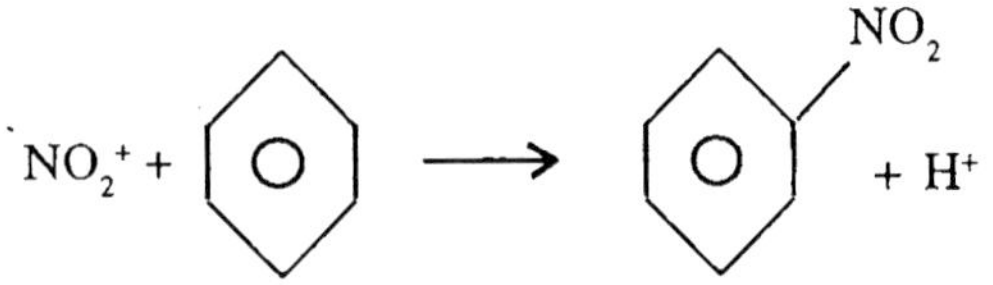

$$NO_2^+ + C_6H_6 \longrightarrow C_6H_5NO_2 + H^+$$

Sulfuric acid (same mechanism as nitration) :

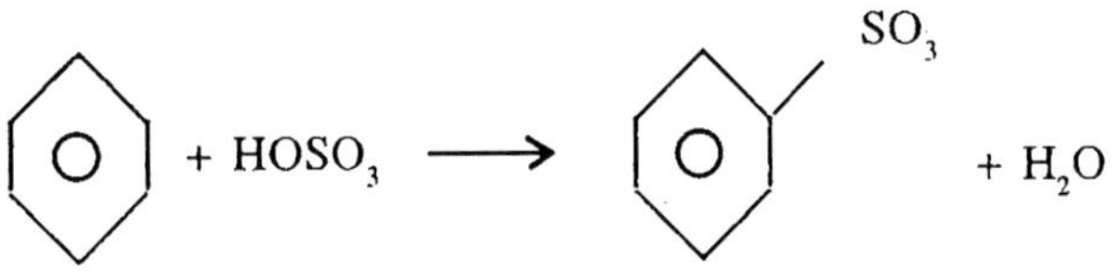

$$C_6H_6 + HOSO_3 \longrightarrow C_6H_5SO_3 + H_2O$$

There can be more than one substitution in halogenation, nitration & the reaction with sulphuric acid. e.g.

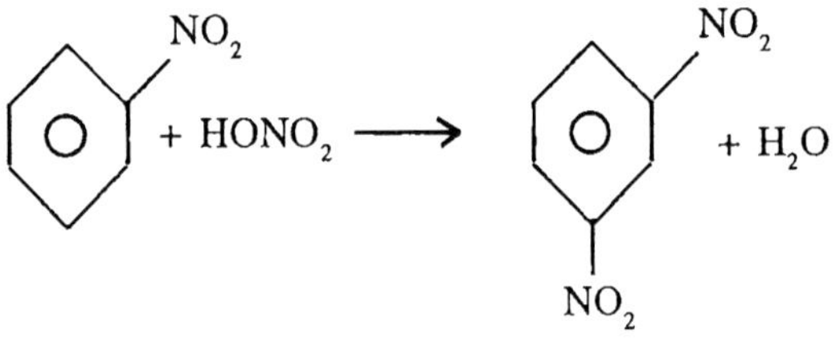

$$C_6H_5NO_2 + HONO_2 \longrightarrow C_6H_4(NO_2)_2 + H_2O$$

m-dinitrobenzene

For the most part, there are only monosubstitutions. More depends on the concentration (poly-substitute - e.g. - TNT).

Some substituting compounds prefer only cerain positions. e.g. The nitro group prefers meta positions.

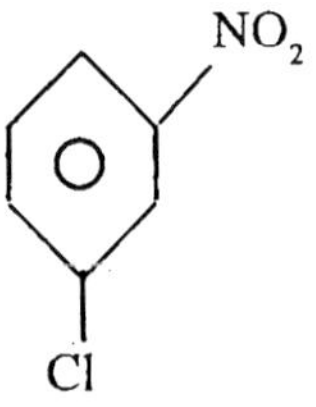

NO_2

Cl

Most other compounds prefer either para or ortho positions. e.g. NO_2 & halogens.

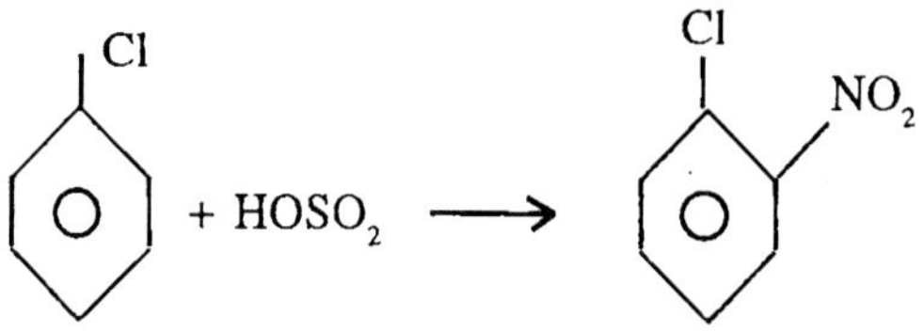

Cl + $HOSO_2$ ⟶ Cl, NO_2

PHENOL

It is known as hydroxy benzene. The general formula of phenol is :

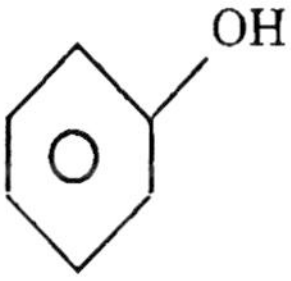

OH

In acts more like an acid than an alcohol:-

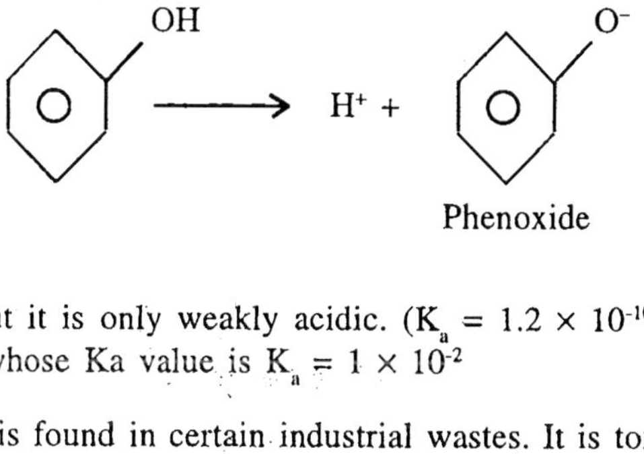

OH ⟶ H^+ + O^-

Phenoxide

But it is only weakly acidic. (K_a = 1.2 × 10^{-10}) compared to H_2SO_4 whose Ka value is K_a = 1 × 10^{-2}

It is found in certain industrial wastes. It is toxic to bacteria. Concentration of 25 mg/lt. can not be biologically degraded but recent research has shown that certain bacteria can be acclaimed to attack as much as 500 mg/lt. e.g. *Candida sp., Pseudomonas sp., Acinetobacter sp.*

Since it is a germicide, it is used as a disinfectant. Other disinfectants used to be rated in strength compared to phenol to give their phenol coefficient.

e. g. 5% phenol = 1% disinfectants A in the same concentration & volume of the same organisms in a given time.

Therefore, disinfectant A is 5x stronger than phenol. In other words, disinfectant A has a phenol coefficient 5.

Cresol

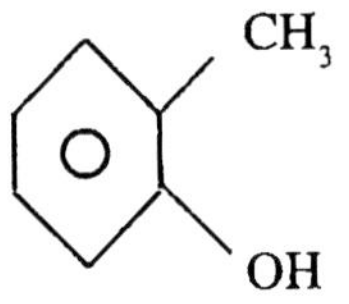

CH_3

OH

It is a stronger disinfectant than phenol. It is the active ingredient in Lysol which is not bio-degradable. It comes in three forms :

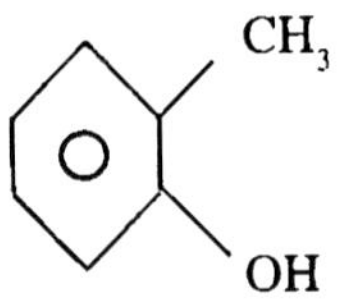

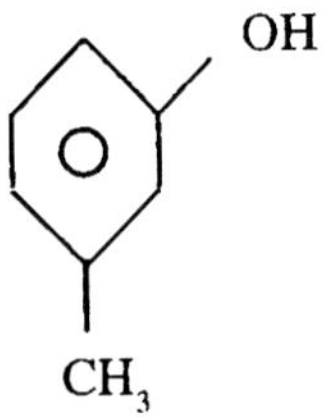

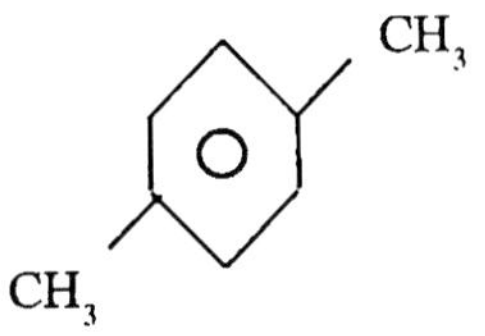

CH_3 OH CH_3

OH CH_3 CH_3

O-cresol M-cresol P-cresol

Polyhydric phenol

More than one functional -OH group

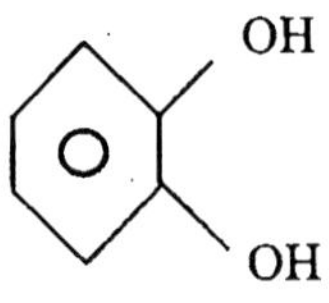

OH OH

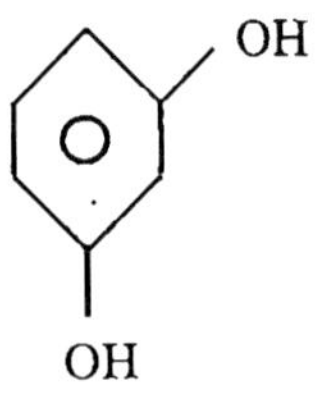

OH OH

Benzene - 1.2. diol Benzene - 1.3. diol

Polyhydric phenols can be treated with activated sludge. They will undergo biological degradation with acclimatised bacteria.

Certain substitutions change the characteristics of the parent benzene e.g. pyrogallic acid, which is used in photography & is found in that industry' waste, is easily oxidized & absorbs O_2 gas.

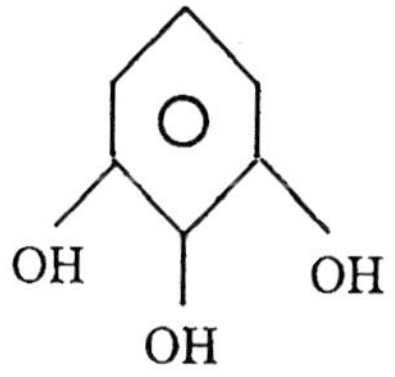

OH OH
OH

1.2.3 trihydroxy benzene or pyrogallic acid

Aromatic hydrocarbons, alcohols, acids, aldehydes & ketones are derived from alkaline groups.

AROMATIC ALCOHOLS

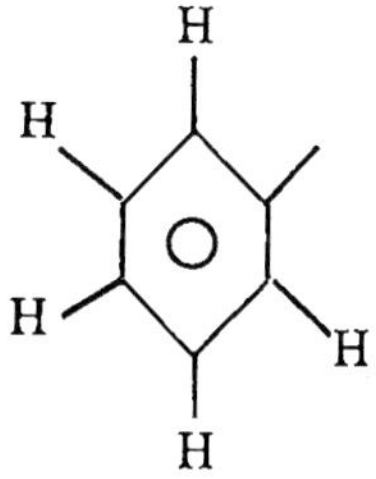

H
H
H
H
H

Phenyl group in all aromatic compounds

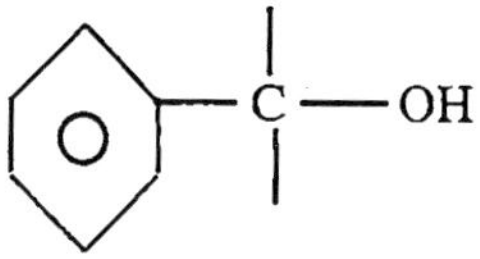

C —OH

benzyl alcohol or phenyl methyl alcohol

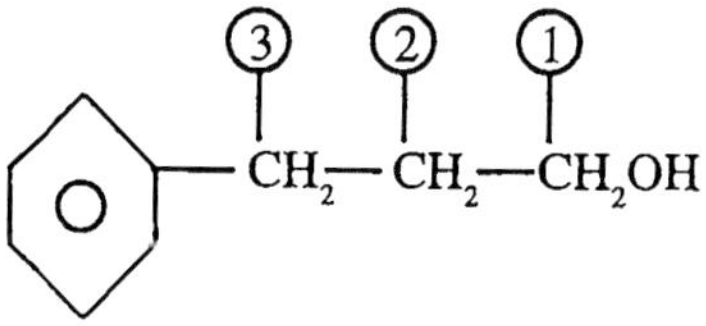

3 2 1

CH_2— CH_2—CH_2OH

3. Phenyl propinol or phenyl n-propyl alcohol

AROMATIC ALDEHYDES

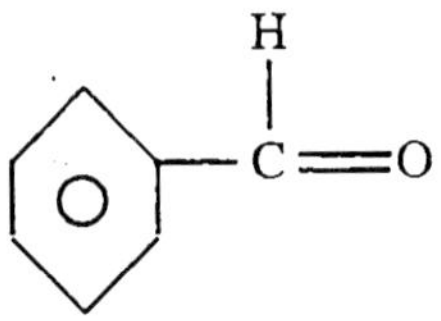

Phenyl methyl aldehyde or benzeldehyde

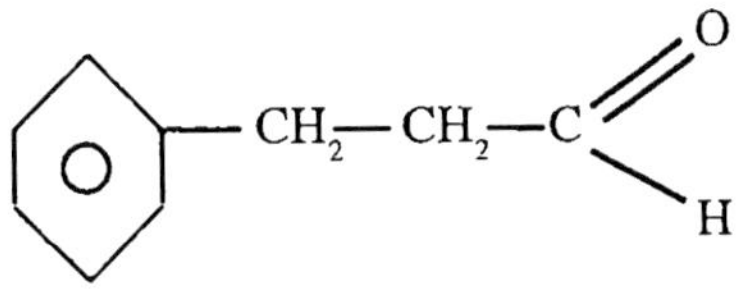

Phenyl propyl aldehyde or phenyl ethane

AROMATIC KETONE

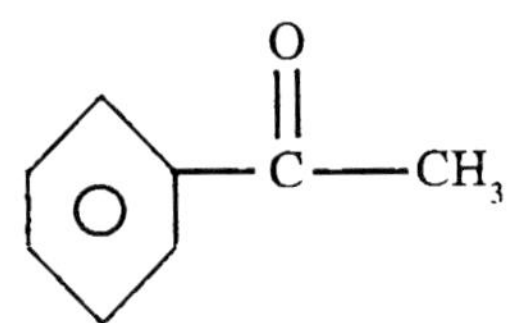

Acetophenone

AROMATIC ACIDS

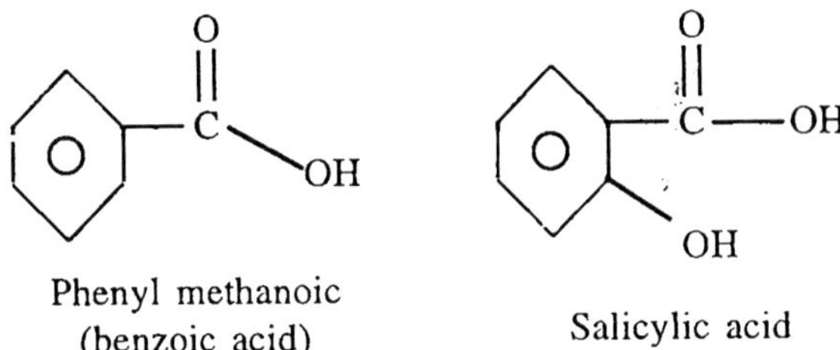

Phenyl methanoic (benzoic acid)

Salicylic acid

HETEROCYCLIC COMPOUNDS

Ring structure with atleast one component is not a carbon.

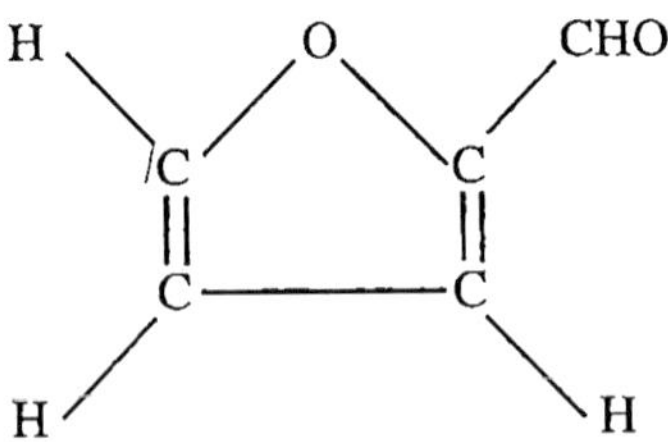

furaldehyde or furfural (from cereal industries)

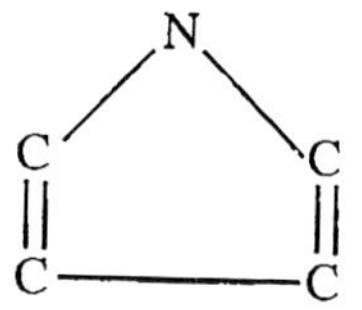

Pyrole

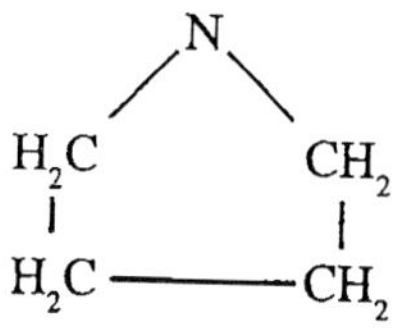

Pyrolidine

Pyrole & pyrolidine are found naturally in nicotine & cocaine.

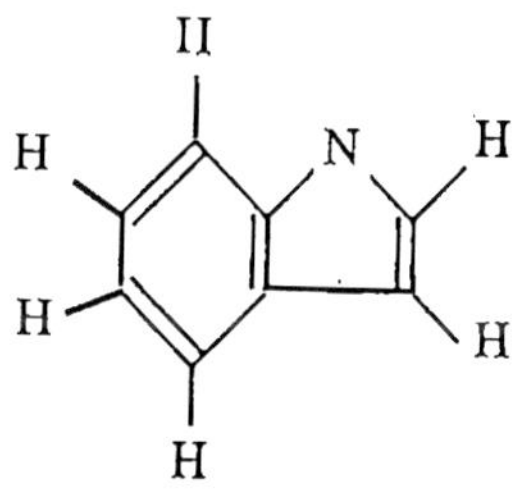

Indole

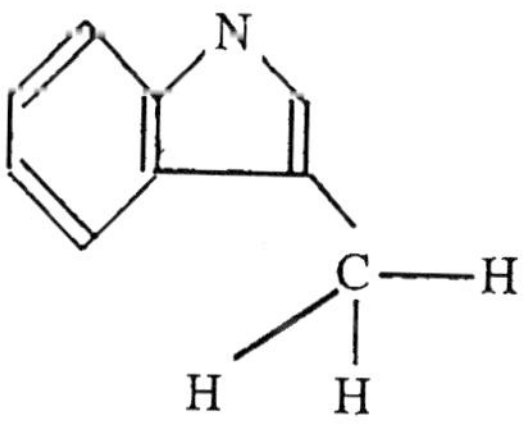

Skatole

Skatole & indole are important in sludge digestion. They produce a bad odour if the digester is improperty working.

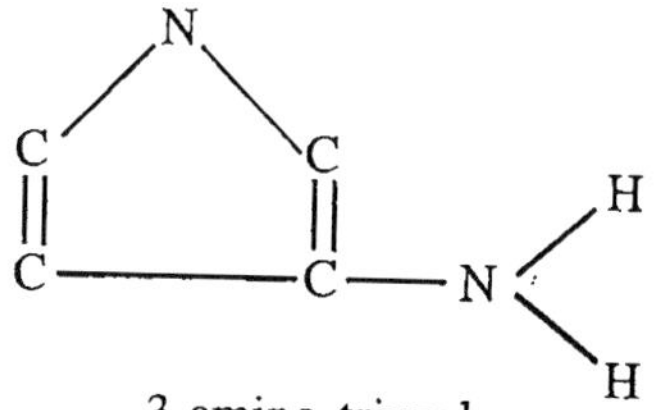

3-amino triazole

This compound resists biological oxidation even at 0.5 mg/lt.

ISOMERISM

Compounds having the same molecular formula but have different structural formulae:

1. *Chain isomerism*

Substances differ in the arrangement of carbon atoms in the chain e.g. butane & isobutan are both C_4H_{10} but -

$$CH_3 - CH_2 - CH_2 - CH_3$$

n-butane

$$CH_3 - \underset{\underset{CH_3}{|}}{CH} - CH_3$$

Iso butane

2. ***Position isomerism***

Compounds which differ in the position which a substituent occupies on the chain of Catoms e.g. alcohols

$$-\overset{|}{\underset{|}{C}}-\overset{|}{\underset{|}{C}}-\overset{|}{\underset{|}{C}}-\overset{|}{\underset{|}{C}}-OH$$

Primary alcohol -butanol

$$-\overset{|}{\underset{|}{C}}-\overset{|}{\underset{|}{C}}-\overset{\overset{OH}{|}}{\underset{|}{C}}-\overset{|}{\underset{|}{C}}-$$

Secondary alcohol

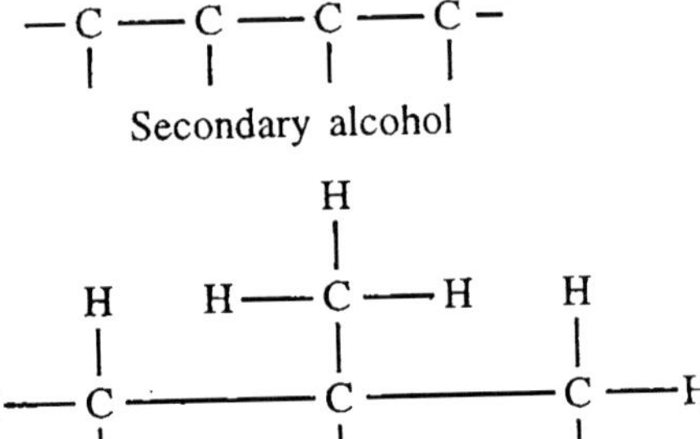

tertiary alcohol

3. ***Functional group isomerism***

The atoms are arranged into different functional groups e.g. methyl ether & ethyl alcohol have same molecular formula C_2H_6O but-

$$H-\overset{\overset{H}{|}}{\underset{\underset{H}{|}}{C}}-O-\overset{\overset{H}{|}}{\underset{\underset{H}{|}}{C}}-H$$

methyl ether

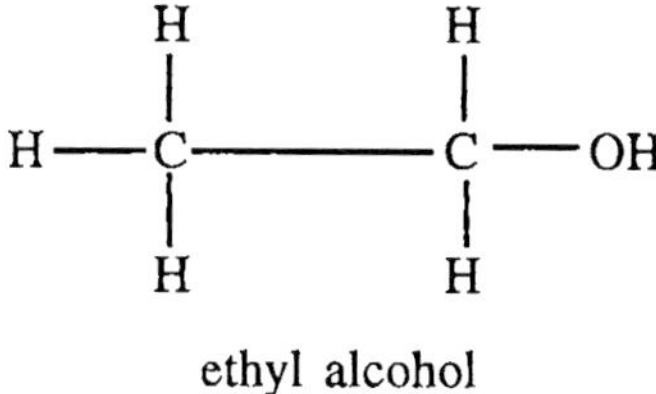

ethyl alcohol

4. ***Tautomers***

Functional groups which are readily interconvertible. e.g

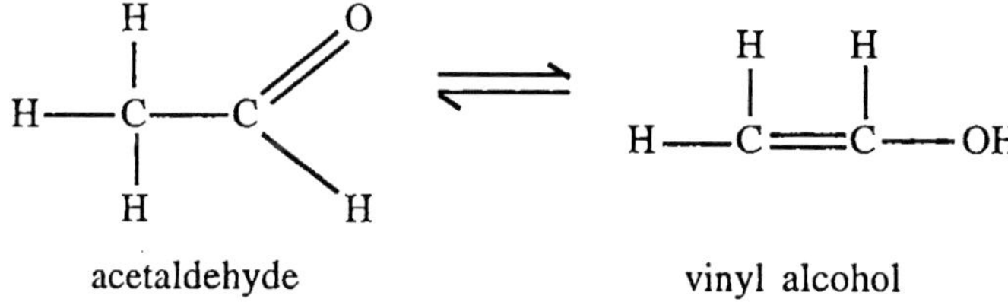

acetaldehyde vinyl alcohol

5. ***Stereo Isomerism***

Deals with differences in spacial arrangement of atoms or groups of atoms

A. ***Geometrical or Cis-trans Isomerism:-***

Compounds have a plane of demarcation

Cis = same elements in same side of plane

trans = same elements in different side of plane

e.g.

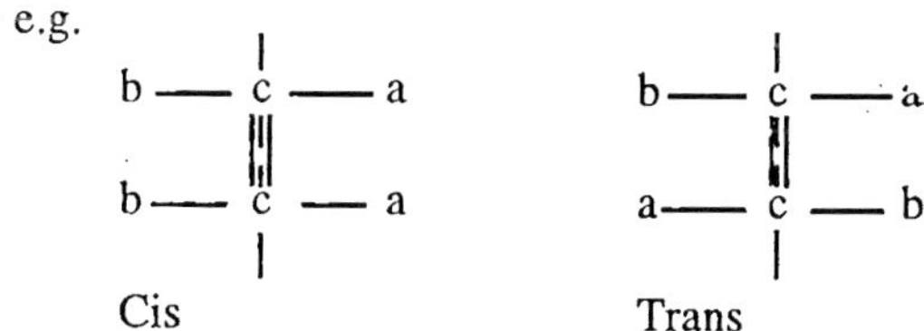

Cis Trans

B. ***Optical Isomerism***

Compounds which are the mirror image of another. They are named for the left or right sdide hat is mirrored.

e.g. Dextro or D-isomerism (dex-tro=right)

Levo or L-isomerism (Levo=left)

Optical isomers have the characteristic of being able to rotate polarized light.

To test for optical isomerism

Source of Light	Prism 1	Solution	Prism 2	Analyser

Some compounds bend light right or clockwise(+)

Some compounds bend light left or counter clockwise(–)

OPTICAL ISOMERISM

Compounds which are assymetrical & which rotate the plane of polarized light in one or other direction are known as optical isomers and phenomenon is called optical isomerism. The degree & direction that polarized light in one or another direction measured by a polarimeter or polariscope.

White light is composed of many different wavelengths which vibrates on many different planes. The light is directed thru a nickel prism so that it will all vibrates in one planes. It is important for categorizing compounds for optical activity. A solution of the compound to be measured in beam of light. Measure the degree & direction compound bends or rotates light.

Right or dextrorotatory (+)

Left or Levorotatory (-)

In order to be optically active, the compound must have at least one assymertic carbon atom, a carbon atom with 4 dissimilar groups attached e.g.

$$\begin{array}{c} C_1 \\ | \\ C_1 - C - a \\ | \\ b \end{array}$$

Hydroxy acid or poly hydroxy carboxylic acids are likely to have assymetric carbon atoms.

The two forms of lactic acid have the same chemical & physical properties but rotate polarized light differently and some organisms will only attack D-lactic acid.

HO—C(=O)—C*(HO)(H)—C(H)(H)—H

L-lactic acid

HO—C(=O)—C*(H)(OH)—C(H)(H)—H

D-lactic acid

* Assymetric carbon atom

Some compounds have a mirror image these are meso or M-compounds. e.g. tartaric acid occurs in the form or D, L as well as meso. Meso form sometimes called recemic mixture

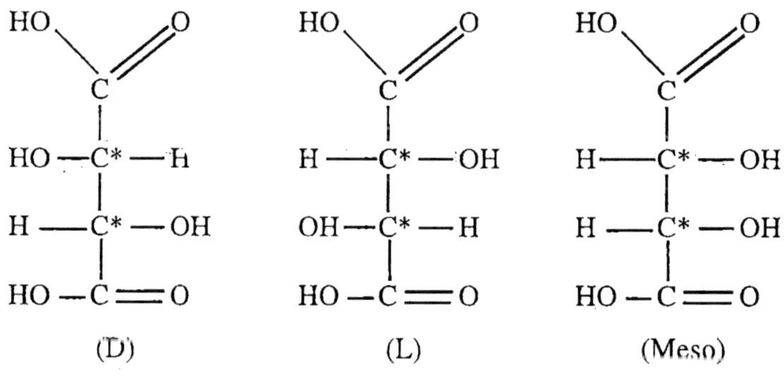

Tartaric acid

Chapter -8

CARBOHYDRATES : GENERAL INTRODUCTION

Sugars, Starch, Cellulose are generally knwon as carbohydrates. This is one of three classes of foods with fats & protein. It is also found in industrial wastes, both in textile & paper industries using cellulose fibres. It is important in municipal waste (BOD). They are found in packing house wastes along with fat & proteins. They are also present in dairy wastes in the form of lactose. There are many carbohydrates in effluent from pulp and paper manufacturing industry. Carbohydrates are related to polyhydroxy alcohols e.g.

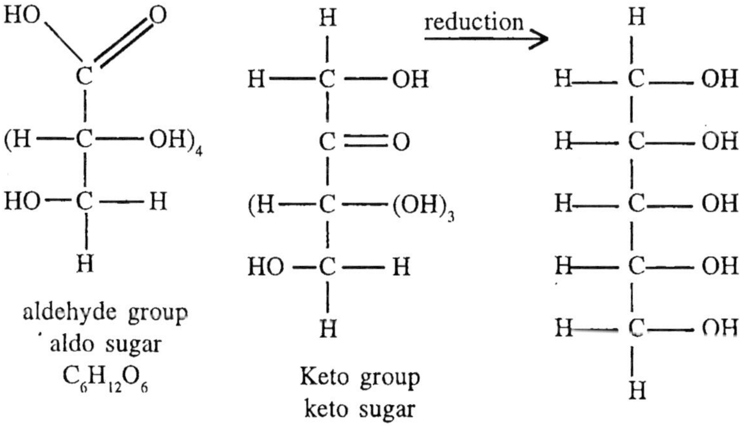

```
HO     O            H       reduction        H
  \   //            |      ---------->       |
    C          H —— C —— OH             H —— C —— OH
    |               |                        |
(H — C —— OH)4      C == O              H —— C —— OH
    |               |                        |
HO — C —— H    (H —— C —— (OH)3         H —— C —— OH
    |               |                        |
    H          HO —— C —— H             H —— C —— OH
                    |                        |
                    H                   H —— C —— OH
                                             |
                                             H
```

aldehyde group
aldo sugar
$C_6H_{12}O_6$

Keto group
keto sugar
$C_6H_{12}O_6$

Sorbitol
polyhydroxy
alcohol

Carbohydrates are all C, H & O. They typically have the formula:

$$\mathbf{(CH_2O)_n \text{ or } Cn(H_2O)_n}$$

The latter demonstrates why these are named for hydrates why these are named for hydrates of carbon. Other compounds may have empirical formulas with the same ratio but are not carbohydrates. e.g. formaldehyde or acetic acid.

Formaldehyde - HCHO CH_2O

Acetic acid - CH_3COOH $C_2H_4O_2$

They have suffix—ose.

Sugars are also called saccharides

SACCHARIDES

Monosaccharides

Simple sugars, they can no longer undergo hydrolysis.

Disaccharides

2 monosaccharides tied together.

Trisaccharides

3 monosaccharides.

Polysaccharides

Starch

All carbohydrates are soluble in water since they are primarily made up of water but polysaccharides resist dissolving except under certain conditions. Many carbohydrates are sweet to taste and this is the reason that they are known as sugars. All monosaccharides are classified as either aldo or keto types.

Classes-

Aldohexoses - 6 carbons

Aldopentoses - 5 carbons

ketohexoses - 6 carbons

ketopentoses - 5 carbons

Most carbohydrates for commercial use come from natural sources. Carbohydrates can be synthesized but there is usually a low yield.

Monosaccharides are the simple sugars and cannot be hydrolized to fewer carbons.

Aldoglucose goes to other forms in solution tautomers.

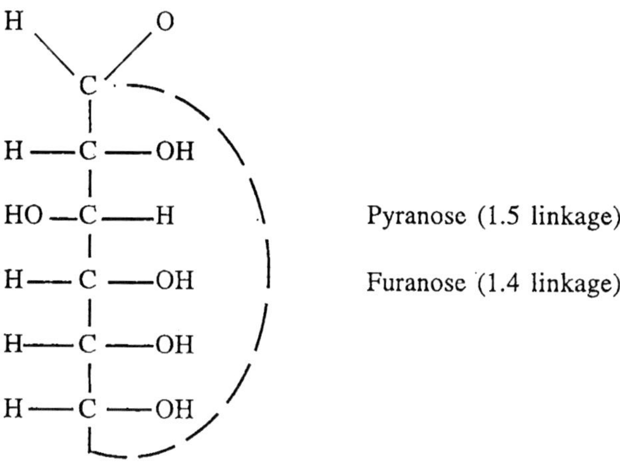

Aldo glucose

Carbohydrates usually have one or more assymetric carbon atoms, but they are difficult to identify. The simplest sugars have the empirical formula $C_3H_6O_3$. The two forms of glyceraldehyde each have one assymetric carbon & are derived from its isomeric ketone, which is the reference for fixing the direction of other carbohydrates. e.g.

D or L is determined from the configuration of the parent compound by the farthest assymetric carbon from the aldo group.

Where n = the number of asymetric carbons. 2^n = the numbers of epimers or the number of possible optical isomers.

+ or - indicates optical rotation.

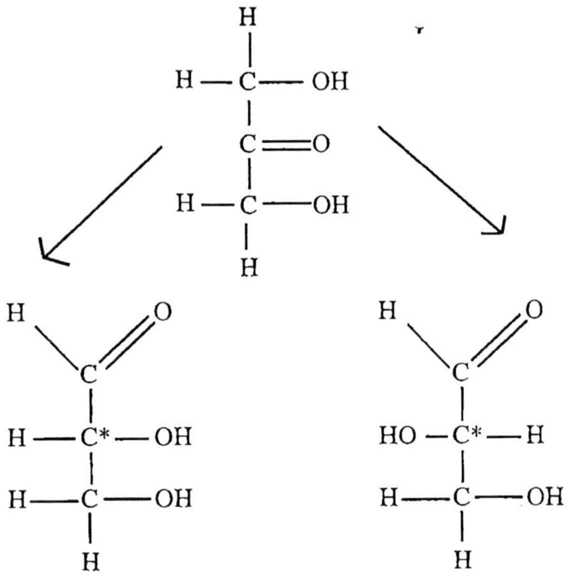

```
          H
          |
     H — C — OH
          |
          C = O
          |
     H — C — OH
          |
          H

   ↙                 ↘

  H       O            H       O
   \     //             \     //
     C                     C
     |                     |
H — C* — OH          HO — C* — H
     |                     |
H — C — OH            H — C — OH
     |                     |
     H                     H
```

D-glycereldehyde L-glycereldehyde

Pentoses ($C_5H_{10}O_5$)

There are formed by the hydrolysis of polysaccharides. These are rarely occur naturally. They are significant to environmentalists because they are not attacked by yeast under anaerobic conditions. They are generally attacked under aerobic conditions. Some bacteria will ingest it either way. Paper pulp is high is carbohydrates, has high COD & has 10,000-25000 BOD. Waste sulphite liquor from paper industry is of particularly high strength.

In looking for a way to recycle the compounds found in the waste, it was discovered that synthetic vanilla could be made. However, the market for vanilla is not large enough to take up the huge amount of vanilla produced from the sulphide liquor. Another scheme was to propogate a variety of yeast *Torula* in the organic waste" harvest the yeast which is high in protein and amino acids & sell it is animal feed producers to supplement grain feed. This is a practical enough idea except that the yeast will not attack the pentosans so there is still an organic waste which is more difficult to remove.

Yeast = single cell protein

glucose = + dextrose = blood sugar=grape sugar.

Keto sugars

Only few are known. Most important is fructose, found in the D(-) form in nature.

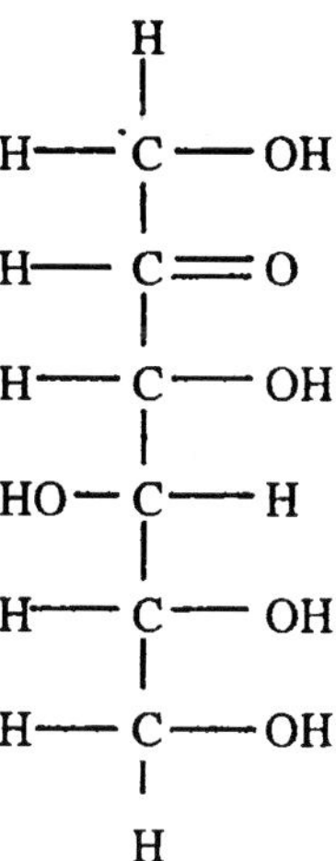

fructose

Disaccharides

Most complex sugars, have the general formula $C_{12}H_{22}O_{11}$. Their natural forms are lactose, sucrose & maltose.

Dihexose - H_2O.

On hydrolysis of 2 monosaccharides, it is formed. The two monosaccharides can be the same or different e.g. maltose contains 2 glucose, sucrose contains 1 (aldo) glucose + 1 (keto) fructose, Lactose contains 1 glucose + 1 glactose.

Reducing sugars

Those which have a free aldo or keto group are known as reducing sugars e.g. all monosaccharides.

Non-reducing sugars

Have a linkage at the aldo or keto group are called non reducing sugars e.g. sucrose.

Sucrose:

- comes naturally in sugar cane or sugar beets.

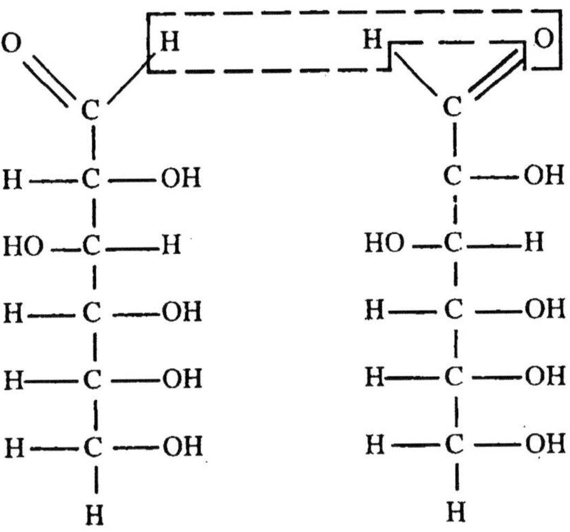

Maltose = 2 glucose minus water

- 15% of the weight of sugar cane is sucrose.
- Table sugar is 99% sucrose.

Lactose

It is present in milk waste especially cottage cheese. It is sometimes called milk sugar. It is found in all mammals. Whey is high in lactose & is used for pig feed.

Carbohydrates

- hydroxy aldehydes or ketones.
- aldose or ketose have the general formula $(CH_2O)_n$

Tautomers of D-glucose

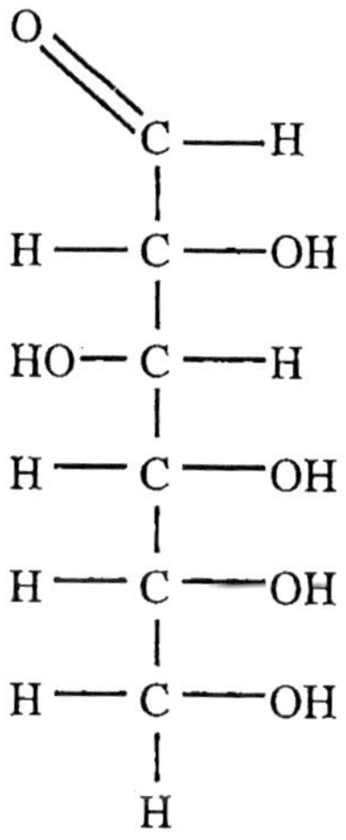

```
O
 \\
  C — H
  |
H — C — OH
  |
HO — C — H
  |
H — C — OH
  |
H — C — OH
  |
H — C — OH
  |
  H
```

Aldose form

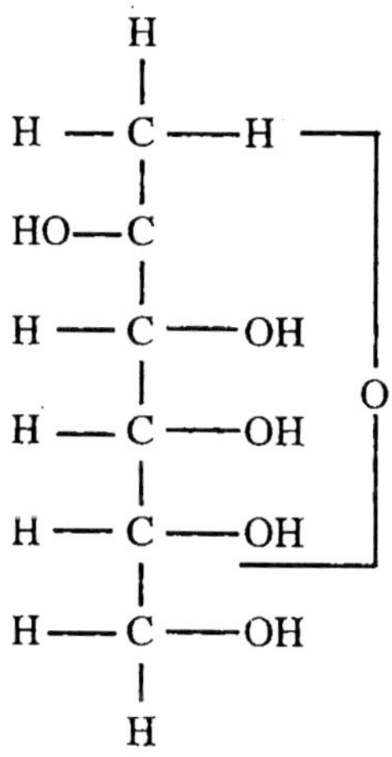

```
      H
      |
H — C — H ────┐
      |       │
HO — C        │
      |       │
H — C — OH    O
      |       │
H — C — OH    │
      |       │
H — C — OH ───┘
      |
H — C — OH
      |
      H
```

Fischer Projection
(cyclic)

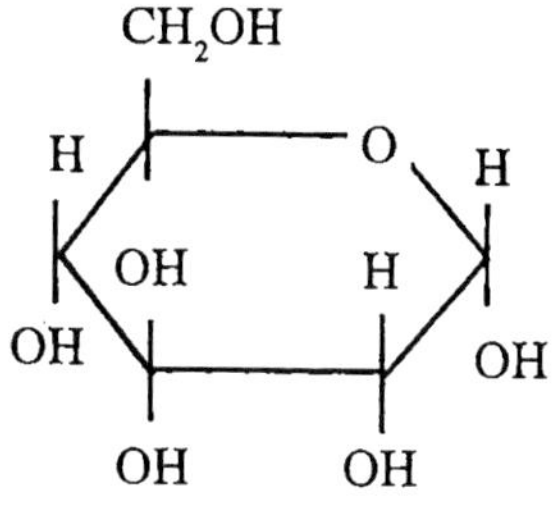

```
        CH2OH
        |
  H    /‾‾‾‾‾‾‾‾‾O\     H
  |   /            \    |
  |  /  OH      H   \   |
 OH  \              /   OH
      \____________/
       |          |
       OH         OH
```

Howarth projection

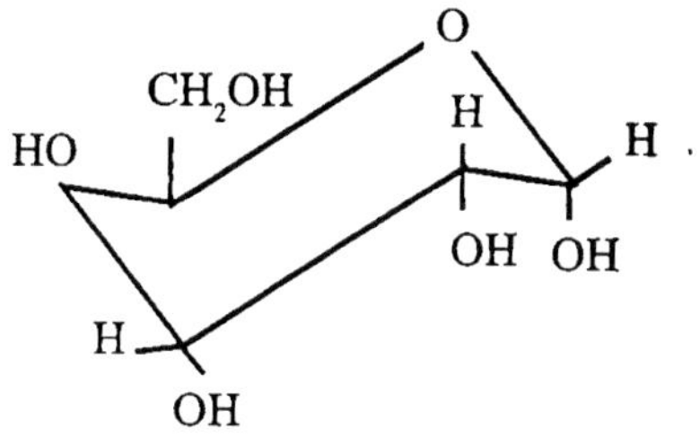

```
                                O
          CH2OH       ___/       \      H
  HO       |     ___/         H   \  __/
   \_______|____/             |____\/
    \               ___/      |     |
     \        ___/            OH    OH
   H__\___/
       \
        OH
```

Conformational form

The Howrath projection is merely a simpler form of the conformational form. The crystallization of glucose is closest to the conformational form. All substituents to the right (on the aldose form) are below the ring (on the Howarth projection). All the substituents to the left are above the ring.

Hexoses have two forms :

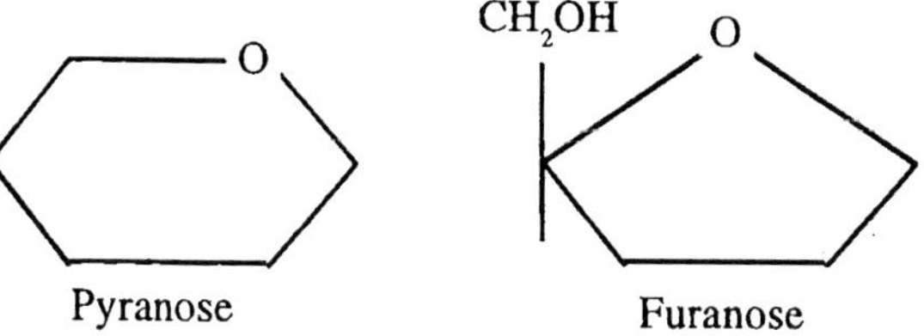

D-fructose is also found in 5-or-6 member rings, although predominantly, in 5 member rings.

keto group

D-fructose

Furanose form of fructose

STEREOSOMERISM OF CARBOHYDRATES

Carbohydrates they are optically active if they have one or more assymetric carbon atoms which turn or rotate the plane of polarized light. In this biochemicals, an assymetric carbon is attached to 4 different functional groups.

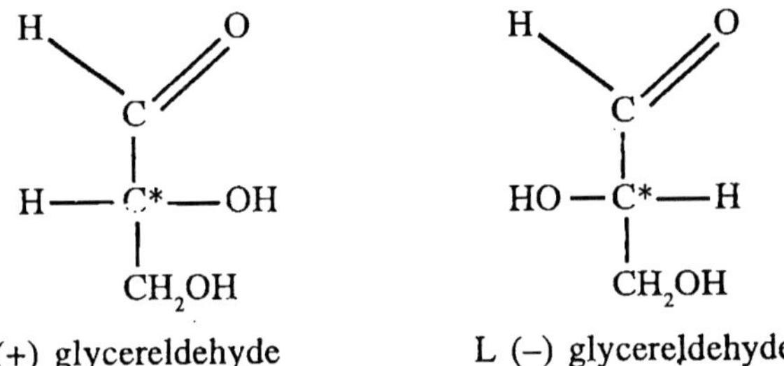

D (+) glycereldehyde L (–) glycereldehyde

ENANTIOMERS

Compounds which are the mirror image of each other, called enantiomers. Here '+' or D = indicate the rotation of polarized light to right and '–' or L = indicate the rotation of polarized light to left.

On the carbon atom before the last carbon atom in the chain, if the -OH group is attached to the right it is knwon as D and if the -OH group is attached to the left it is knwon as L. However, There is no relationship between the symbolism of '+' or '-' & D or L.

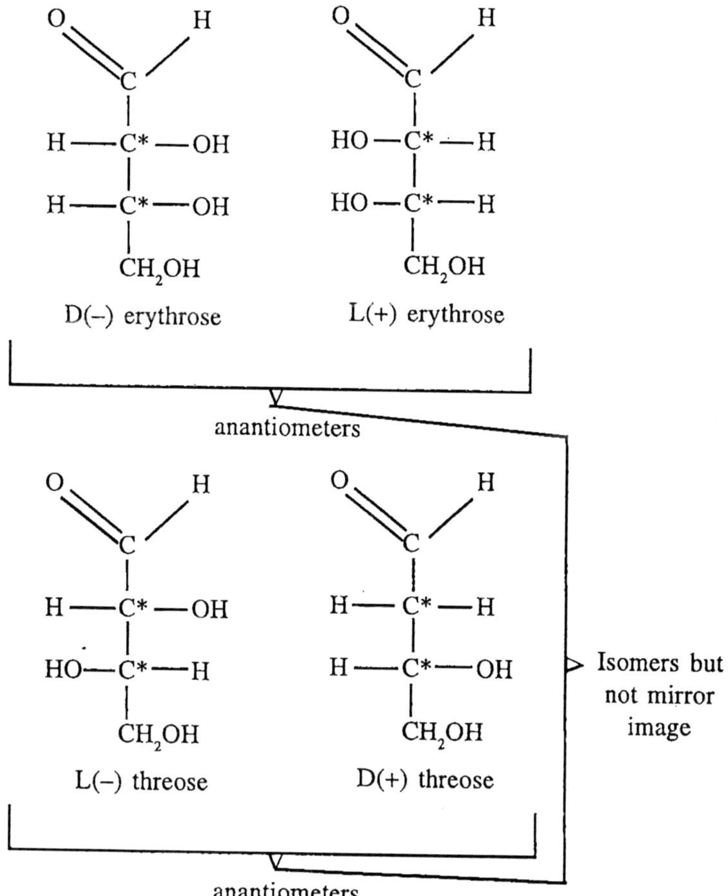

Diastereo isomers or diasteremers have two assymetric carbons & therefore has 4 rather than 2 isomers.

When 'n' = the number of assymetric carbons.

2^n = the maximum number of isomers.

ALPHA ISOMERS

-OH group to the right of the first carbon after the aldo or keto group.

BETA ISOMERS

-OH group to the left of the first after the aldo or keto group.

– In the ring form-

right = down = alpha

left = up = beta

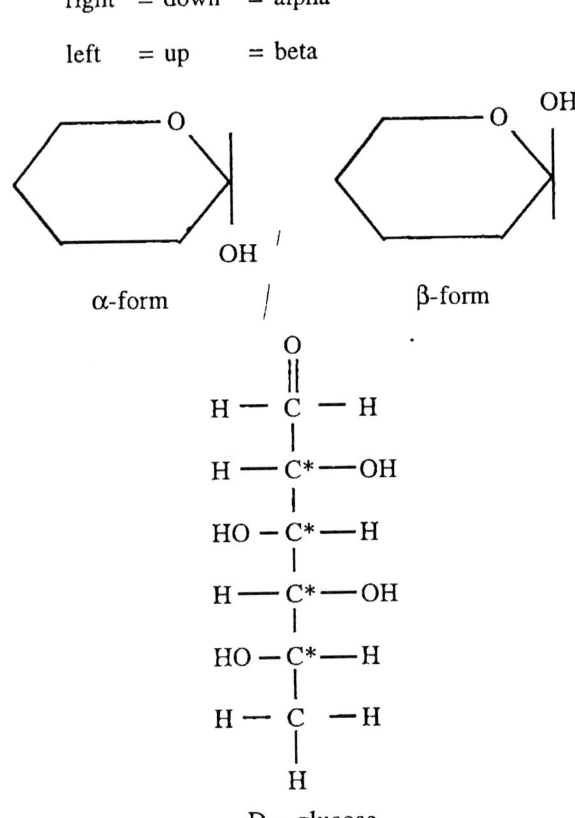

D - glucose

D-glucose has 4-assymetric carbons. In ring form the number of asymmetric carbons increases & therefore so do the number of isomers.

Isomers have difficult physical characteristis. They have different solubility & different melting or boiling points.

DISACCHARIDES

2 monosaccharides joined together. The joining is different for α or β group.

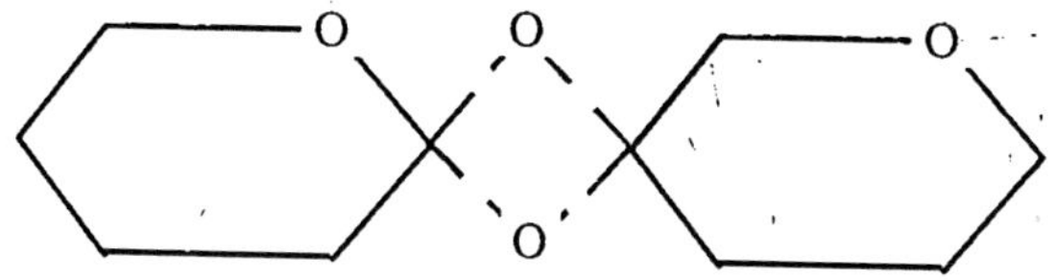

General structural formula for disaccharides

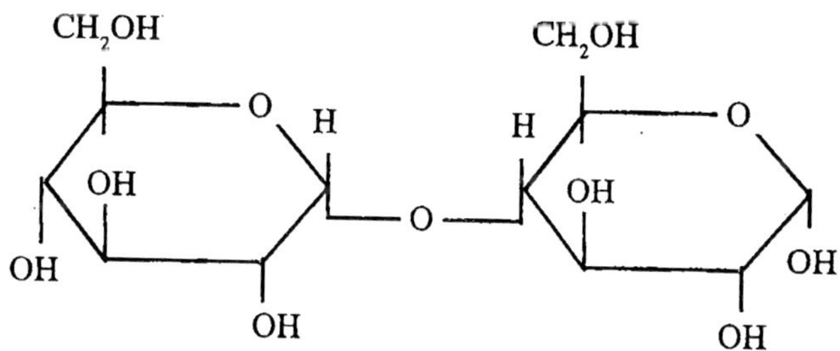

Maltose = 2–D–glucose–α–linkage

Maltose comes form the hydrolysis of starch. Enzyme diastase found in barley malt.

$$\text{Starch} \xrightarrow[\text{hydrolysis}]{\text{diastase}} \text{Maltose} \xrightarrow[\text{hydrolysis}]{\text{Maltase}} 2\ (\text{D-glucose})$$

CELLOBIOSE

β-D-glucose + α D-glucose

- produced from the hydrolysis of cellulose.

$$\text{Cellulose} \xrightarrow{\text{hydrolysis}} \text{cellobiose} \xrightarrow[\text{emulsion}]{\text{hydrolsis}} 2\ \text{D-glucose}$$

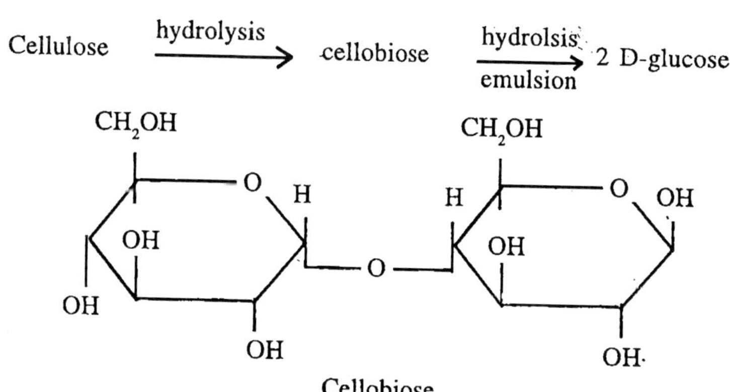

Cellobiose

LACTOSE

-D-galactose + D-glucose.

-lactose is 5% of cow's milk.

$$\text{lactose} \xrightarrow[\text{lactase}]{\text{hydrolysis}} \text{D- galactose + D-glucose}$$

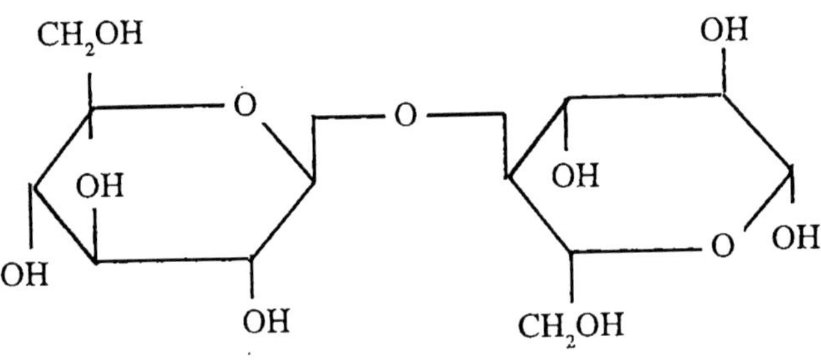

Lactose - β - linkage

SUCROSE

- L glucose + 1 fructose.
- sucrose comes from beets & sugar cane.
- table sugar is 98.9-99.9% pure sucrose.

$$\text{Sucrose} \xrightarrow[\beta\text{-maltase or } \beta\text{-fructosidase}]{\text{hydrolysis}} \text{glucose + fructose}$$

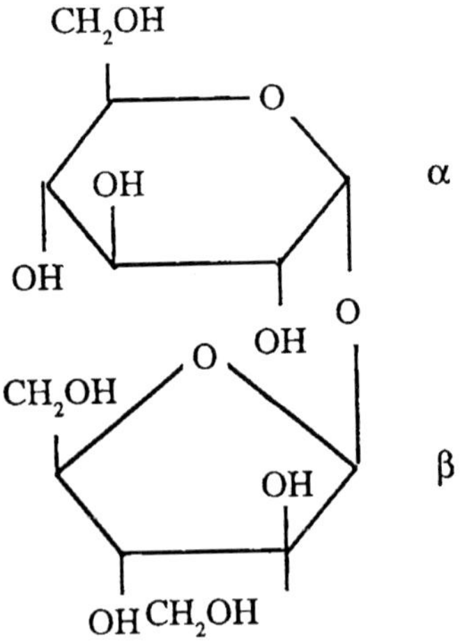

Sucrose

Classifying Disaccharides and Naming

- designate the component monosaccharides.
- distinguish the kind of ring (pyranose-6 or furanose-5).
- type of linkage i.e. α or β

e.g. Sucrose

α - D - glucopyranosyl (1⟶2) β -D-fructofuranoside

Reducing sugars have a free aldo or keto group, while non reducing sugars have no free aldo or keto group.

TRISACCHARIDES

Raffinose $C_{18}H_{32}O_{16}$

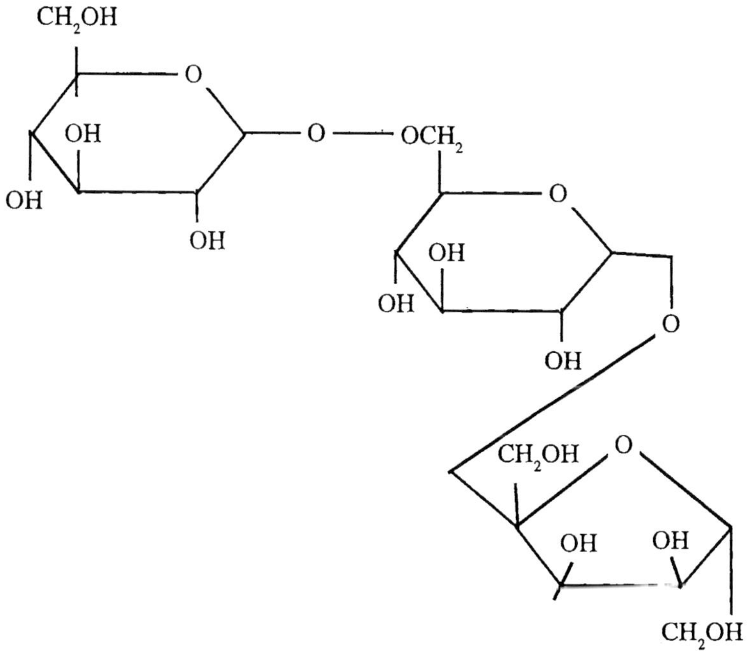

Raffinose = D (+) glucose + D (–) fructose + D (+) galactose

POLYSACCHARIDES

Cellulose

It has many hexose groups. It has molecular weight of 30,000-500,000. It has 2,000-3000 glucose units. Cellulose is an example of polysaccharide which, is made up of many of follwing units & links:

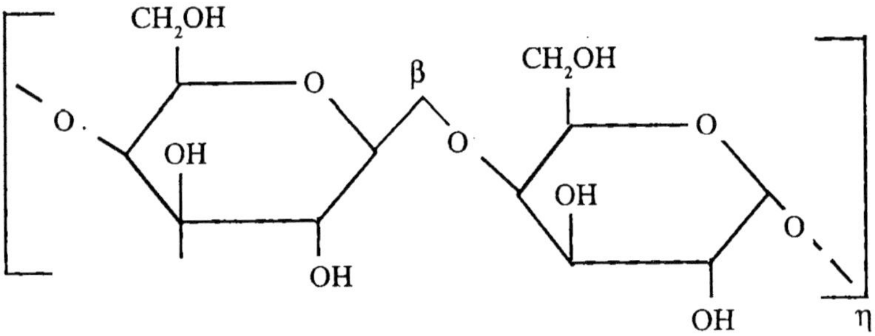

CH_2OH O β CH_2OH O OH O OH O OH OH η

Cellulose is found in plants primarily linen & cotton are 90+% pure cellulose. Similarly, wood is 60% cellulose humans & animals do not have enzymes that will break the β-linkage so that they can use cellulose as food termites themselves do not produce enzymes that will break β linkage but do have bacteria in their alimentary tract that do produce the necessary enzyes. Such bacteria are also present in the rumen of ruminants, such as cows, sheep, camels & oxen. Cellulose can be treated biologically under aerobic or anaerobic conditions (usually anaerobically). It is not soluble but stays in suspension most of the time going into dissolution only very slowly.

HEMICELLULOSE

It is made up of units of a hexose (glucose) + a pentose (usually xylose). It is found in the young shoots of growing plants. It is also found in the outer coat of peas or beans. It can be hydrolysed to 2 units hexose + 2 pentose. Both the glucose & the xylose can be treated aerobically to eventually CO_2 + H_2O. Yeast can not treat Xylose anaerobically.

STARCH

It is made up of a-glucose which is readily useable as food by humans & other animals. It is made up of many of following units & links.

Starch - α - linkage

Starch = amylose + amylopectin

AMYLOSE

It a straight chain polysaccharide. Ithas 50-300 glucose units. It forms a blue coloured complex with iodine.

AMYLOPECTIN

It is a branched form saccharide. It has 500-5000 glucose units. It does not form a blue complex with Iodine. The source of starch is corn, plants, potatoes, rice etc.

GLYCOGEN

It is a polyglucose with a linkages. It is in animal cells as reserve food. It is used by the body for manufacturing blood. It is a source of food for microorganisms.

Two forms of lactose

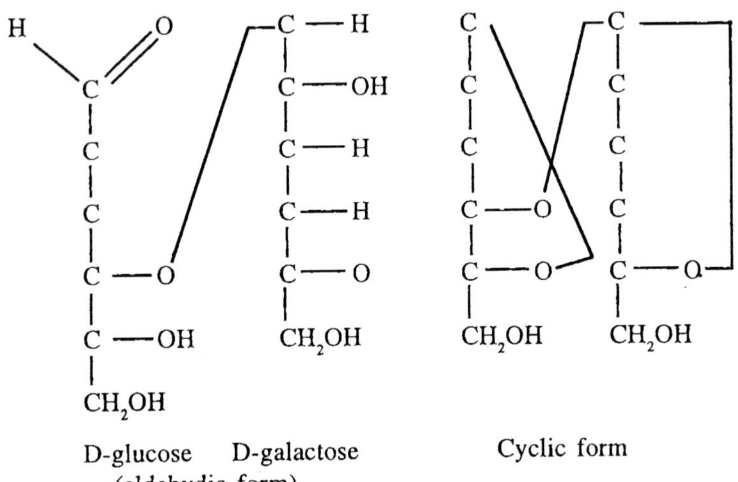

D-glucose D-galactose
(aldehydic form)

Cyclic form

Chapter -9

CARBOHYDRATES : BASIC BIOCHEMICAL ASPECTS

Carbohydrates are synthesized from CO_2 and H_2O in chlorophyll containing plants during the process of photosynthesis. They are made up of carbon, hydrogen, oxygen and were originally represented as hydrates of carbon with empirical formula $C_n(H_2O)_n$. Not all the carbohydrates, however, have this empirical formula e.g. deoxyribose ($C_5H_{10}O_4$) and glucosamine ($C_6H_6O_3$) with this general formula is not a carbyhydrate. Carbohydrates are defined as the ***compound having*** either an aldehyde or a ketone group or a modificied aldehyde or ketone group and other carbonatoms with alcohalic (-OH) groups.

The simpler carbohydrates are known as sugars. These are mostly soluble in water and sweet in taste. The sugars are named according to the carbonyl (aldehyde or ketone) group present in them. Those containing aldehyde (-CHO) group are called aldoses, while those containing ketone (=CO) group are called ketoses.

Glycereldehyde is an aldose and dihydroxy acetone a ketose. Ketose sugars are not as common in nature as aldoses barring an important ketohexose, the fructose. The sugars are also named according to number of carbon atoms present. Glycereldehyde and dihodroxyacetone are 3-C sugars and hence called trioses. Tetroses, pentoses, hexoses and heptoses contain 4,5,6 and 7 carbon atoms respectively. Such sugars which are made up of only one unit are called monosaccharides.

The word saccharide has its origin from the Greek Word "Sakcharon" meaning sugar. The monosaccharides can not be hydrolysed into simpler units. When two monosaccharides combine together with the elimination of water, a disccharide unit is formed. Few units make oligo-saccharides. Similarly large number of monosaccharides units condense together to form polysaccharides.

MONOSACCHARIDES

Monosaccharides generally have the empirical formula $C_n(H_2O)_n$. The value of n ranges normally from 3 to 8. Glycereldehyde and dihydroxyacetone are the smallest monosaccharides abundant in the living cells. Although about 70 monosaccharides are known, only about 20 of them are found in natural systems. Some important monosaccharides are listed in Table 9.1.

Table 9.1 : Some important natural monosaccharides

Monosaccharide	*Type*	*Name*
Triose	Aldose,	glycerekldehyde
	Ketose	dihydroxy acetone
Tetrose	Aldose	erythrose, threose
Pentose	Aldose	arabinose, ribose
		xylose
	ketose	ribulose, xylulose
Hexose	Aldose	glucose, galactose,
		mannose, fructose, sarbose
Heptose	Aldose	mannoheptulose perpesenulose
	ketose	sedoheptulose

PROPERTIES OF MONOSACCHARIDES

PHYSICAL

Monosaccharides are sweet tasting colourless solids. They are soluble in water, sparingly soluble in alcohol and insoluble in ether.

They contain asymmertic carbon atom and hence exist in different isomeric forms. e.g. glycereldehyde may exist in two forms which are mirror image of each other.

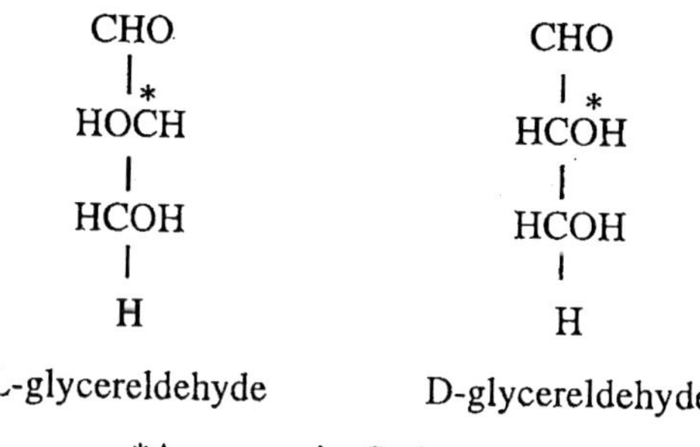

L-glycereldehyde D-glycereldehyde

*Asymmetric Carbon Atom

When a polarized light is passed through a solution of these carbohydrates, the plane of light is rotated to either right or left. All monosaccharides which can be degraded to D-glycereldehyde are called D-sugars. L-Monosaccharides can be degraded to L-glycereldehyde, Carbohydrates isolated from living system are a mixture of several isomers. The degree of optical rotation may change due to interconversion of isomers. Fresh solution of glucose gives an optical rotation of + 112° which changes to + 527° on standing. Similarly fructose gives an optical rotation of -113° which changes to -92°. This change in optical rotation is called **mutarotation**.

CHEMICAL

The properties of monosaccharides are, generally based on -CHO or =CO and —OH groups present in them.

Oxidation

Monosaccharides are easily oxidized by the oxidising agents. Glucose yields gluconic acid after oxidation with mild reagents such as Tollen's reagent (ammonical Ag_2O) or Fehling solution (alkaline $CuSO_4$)

$$\begin{array}{l} CHO \\ | \\ (CHOH)_4 \\ | \\ CH_2OH \end{array} + Ag_2O \xrightarrow[\text{Tollen's reagent}]{} \begin{array}{l} COOH \\ | \\ (CHOH)_4 \\ | \\ CH_2OH \end{array} + 2Ag$$

Gluconic acid Silver mirror

$$\begin{array}{l} CHO \\ | \\ (CHOH)_4 \\ | \\ CH_2OH \end{array} + 2Cu(OH)_2 \xrightarrow[\text{Fehling Solution}]{} \begin{array}{l} COOH \\ | \\ (CHOH)_4 \\ | \\ CH_2OH \end{array} + Cu_2O + H_2O$$

Gluconic acid

The reduction of Tollen's reagent yields silver as polishing on the surface of the tube whereas with Fehling Solution, red precipitate is obtained. Similarly potassium ferricyanide can be reduced to ferrocyanide. These reactions are used in the estimation of glucose and other sugars containing free carbonyl groups. Such sugars are called ***reducing sugars***.

When strong oxidizing agents like conc. HNO_3 are used, gluconic acid is ultimately oxidized to dicarboxylic saccharic acid.

$$\begin{array}{l} CHO \\ | \\ (CHOH)_4 \\ | \\ CH_2OH \end{array} \xrightarrow[O^-]{\text{Conc. } HNO_3} \begin{array}{l} COOH \\ | \\ (CHOH)_4 \\ | \\ CH_2OH \end{array} \xrightarrow[O]{\text{Conc. } HNO_3} \begin{array}{l} COOH \\ | \\ (CHOH)_4 \\ | \\ COOH \end{array}$$

Gluconic acid saccharic acid

In the animal body, slow oxidation of glucose may yield glucuronic acid. In glucuronic acid, aldehyde group is intact and the primary alcohol group on cabon 6 is oxidized to a carboxylic group

$$\begin{array}{l} CHO \\ | \\ (CHOH)_4 \\ | \\ CH_2OH \end{array} \xrightarrow{O^-} \begin{array}{l} CHO \\ | \\ (CHOH)_4 \\ | \\ COOH \end{array}$$

Glucuronic acid

Glucuronic acid in the animals combines with hormones and also with some toxic substances and the conjugation products are excreted out. It is also a major component of hyaluronic acid, heparin chondoitine and mucoitin sulphate, which are found in blood, skin and cartilage.

Reduction

Free aldehyde and ketone groups of Monosaccharides are reduced to alcohalic hydroxy groups by sodium-mercury amalgam and water. Thus, D-glucose after reduction yields a mixture of polyhydroxy alcohols, sorbitol and mannitol.

$$\begin{array}{c} CHO \\ | \\ (CHOH)_4 \\ | \\ CH_2OH \end{array} \xrightarrow{+2H} \begin{array}{c} CH_2OH \\ | \\ (CHOH)_4 \\ | \\ CH_2OH \\ \text{Sorbitol} \end{array} + \begin{array}{c} CH_2OH \\ | \\ (CHOH)_4 \\ | \\ CH_2OH \\ \text{Mannitol} \end{array}$$

Electrolytic reduction or catalytic reduction with hydrogen and nickel also yield sorbitol and mannitol. Sorbitol is used to prepare surface active agents, as a humectant and mannitol is used in the manufacture of explosives.

Condensation reactions

Aldehyde groups of Monosaccharides condenses with primary amines to form a ***Shiff's base***.

$$\begin{array}{c} CHO \\ | \\ (CHOH)_4 \\ | \\ CH_2OH \end{array} + \begin{array}{c} CH_3 \\ | \\ CHNH_2 \\ | \\ COOH \\ \text{Alanine} \end{array} \longrightarrow \begin{array}{l} HH \qquad\quad CH_3 \\ | \qquad\qquad\quad | \\ C = N - CH \\ | \qquad\qquad\quad | \\ (CHOH)_4 \quad COOH \\ | \\ CH_2OH \end{array}$$

Shiff's base
(Brownish colour).

When treated with hydroxylamine, glucose forms glucose oxime.

$$\begin{array}{l} CHO \\ | \\ (CHOH)_4 \\ | \\ CH_2OH \end{array} + NH_2OH \longrightarrow \begin{array}{l} H \\ | \\ C{=}NOH \\ | \\ (CHOH)_4 \\ | \\ CH_2OH \end{array} + H_2O$$

Glucose oxime

Free aldehyde and ketone group and of sugars react with phenyl hydrazine to form a special type of Schiff's base, hydrazone, which is then converted to a double **Shiff's osazone.**

$$\begin{array}{l} CHO \\ | \\ CH_2OH \end{array} + C_6H_5NHNH_2 \longrightarrow \begin{array}{l} H \quad\;\; H \\ | \quad\;\;\; | \\ C{=}N.N.C_6H_5 \\ | \\ CH_2OH \end{array} + H_2O$$

Phenyl Hydrazone

$$2C_6H_5NHNH_2 \downarrow$$

$$\begin{array}{l} H \quad\;\; H \\ | \quad\;\;\; | \\ C{=}N.N.C_6H_5 \\ | \\ C{=}N.N.C_6H_5 \\ \quad\;\;\;\;\; | \\ \quad\;\;\;\;\; H \end{array}$$

$$+ C_6H_5NH_2$$

Phenyl Osazone

The osazones of reducing sugars are yellow crystalline compounds which are insoluble in water.

REACTION WITH HCN

Addition of HCN to sugars containing free aldehyde or ketone groups yields ***cyanohydrin.***

$$\begin{array}{l} CHO \\ | \\ CH_2OH \end{array} + HCN \longrightarrow \begin{array}{l} CN \\ | \\ CHOH \\ | \\ CH_2OH \end{array}$$

Cyanohydrin

ESTERIFICATION

The hydroxyl groups of alchols in the carbohydrate may bo converted to easters by treatment with appropriate acetylating agents e.g. When D-glucose is treated with acetic anhydrine in the presence of pyridine or anhydrous $ZnCl_2$, pentacetyl glucose is formed.

$$\begin{array}{l} CHO \\ | \\ (CHOH)_4 \\ | \\ CH_2OH \end{array} + 5\ (CH_3CO)_2\ O \xrightarrow[ZnCl_2]{anhydrous} \begin{array}{l} CH_2OCOCH_3 \\ | \\ (CHO.COCH_3)_4 \\ | \\ CHO \end{array}$$

Pentacetyl glucose

$+ 5\ CH_3COOH$

Polysaccharides can also be acetylated, some acetylated products have valuable commercial use. The acetate esters produced from cellulose can be spun into rayon like fibres. These fibres are fire resistant.

Glucose can be esterified with phosphoric acid also yielding glucose 1 phosphate or glucose 6-phosphate. The phosphate esters of glucose and fructose are important compounds in carbohydrate metabolism.

METHYLATION

Methylating agents such as $CH_3I + Ag_2O$, $CH_3 + HCl$, CH_3Cl, $(CH_3)_2SO_4$ etc. react with monosaccharides to yield glycoside.

$$C_6H_{11}O_5OH + CH_3OH \xrightarrow{Dry\ HCl} C_6H_{11}OCH_3 + H_2O$$

methyl glycoside

The glycosides may further rect to form tetramethyl esters. Polysaccharides also undergo methylation.

FERMENTATION

Monosaccharides like glucose and fructose undergo alcohalic fermentation by microorganisms such as the yeasts and produce ethanol and CO_2. The yeast supplies the enzyme for fermentation.

$$C_6H_{12}O_6 \longrightarrow 2C_2H_5OH + CO_2$$

STRUCTURE OF MONOSACCHARIDES

Naturally occurring sugars exist in two structural forms: either straight chain whereas the remaining bulk in the form of closed ring. The properties which lead to the confirmation of straight chain structure is glucose were, its conversion to polyhydroxy heptylic acid via cyanohydrin. Since, during this reaction HCN was added to terminal carbon which was then hydrolysed to a -COOH group, the carboxyl group of glucose must occupy a terminal position and hence it will be an aldehyde. The acetylation properties of glucose have proved it to be a pentahydroxy molecule.

The correct position of asymmetric carbon in a glucose molecule was determined by Emil Fisher in 1896. These methods were used for the determination of structure of other monosaccharide also. The structural formula (linear chain) of some important monosaccharides are as follows :

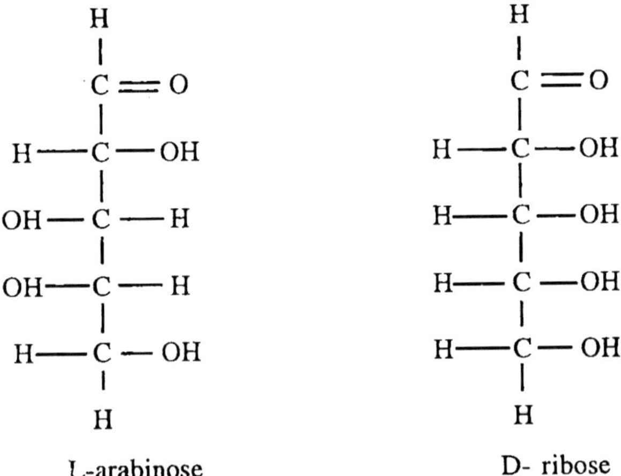

```
        H
        |
        C == O
        |
  H --- C --- OH
        |
 OH --- C --- H
        |
  H --- C --- OH
        |
  H --- C --- OH
        |
        H
```

D- xylose

```
        H
        |
        C == O
        |
  H --- C --- OH
        |
 OH --- C --- H
        |
  H --- C --- OH
        |
  H --- C --- OH
        |
  H --- C --- OH
        |
        H
```

D-glucose

```
        H
        |
  H --- C --- OH
        |
        C == O
        |
 OH --- C --- H
        |
 OH --- C --- H
        |
  H --- C --- OH
        |
  G --- C --- OH
        |
        H
```

D-fructose

```
        H
        |
        C == O
        |
  H --- C --- OH
        |
 OH --- C --- H
        |
 OH --- C --- H
        |
  H --- C --- OH
        |
  H --- C --- OH
        |
        H
```

D-galactose

After the advent of X-ray crystallagrophy technique, it was shown that all these sugars indeed do have structures assigned to them by Emil Fishers and other workers.

In many reactions, the aldehyde sugars do not behave as aldehydes but as hemiacetals. This may happen e.g. if the carbon 1 of glucose approaches near carbon 5 and the aldehyde and hydroxy groups react to form a hemiacetal. In fructose, hemiacetal can be formed by reacting carbon 2 and 5. Two types of hexose rings may be formed in this way, a 5 membered ring with one oxygen atom in the ring or a six membered ring with one oxygen atom. 5 membered cyclic compound is furan while 6 membered is pyrane and hence the correspond structure are furanose and pyranose rings. Furanose and pyranose rings of hexose sugars are given below:-

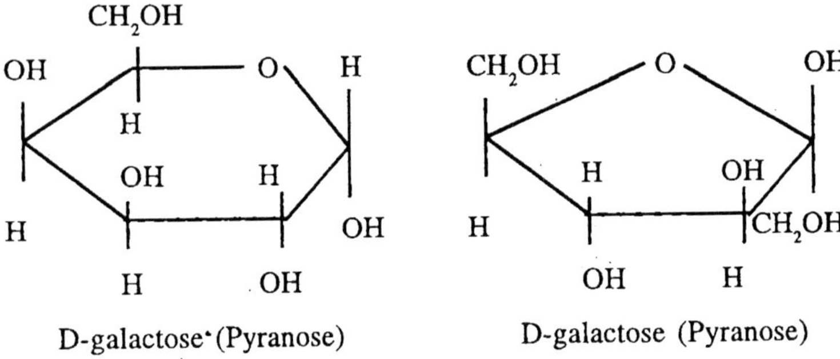

D-galactose (Pyranose) D-galactose (Pyranose)

Most of the hexose sugars occur in the pyranose form. Fructose, however, occurs in furanose form in the disaccharide, sucrose etc. Pentoses occur in furanose form.

During the formation of ring, a new asymmetric carbon is created which contains, four different substitutes. Therefore, two different anomeric forms, the α and β forms, in these sugars are possible. There anomers differ in their physical constants.

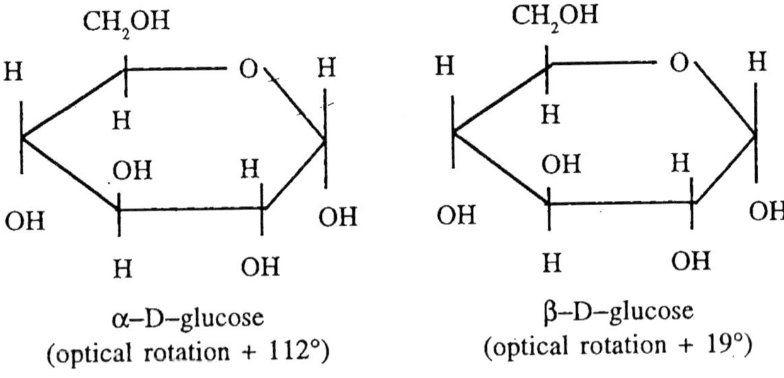

α–D–glucose
(optical rotation + 112°)

β–D–glucose
(optical rotation + 19°)

OLIGOSACCHARIDES

Oligosaccharides are compounds, composed of few monosaccharide units. During union of monosaccharide units, water molecule is eliminated and the units are linked through an oxygen bridge. It is a glycosidic linkage. Depending upon the steric configuration at carbon 1, (which is involved in the formation of glycosidic linkage) of monosaccharide unit, the bond is called α and β-bond

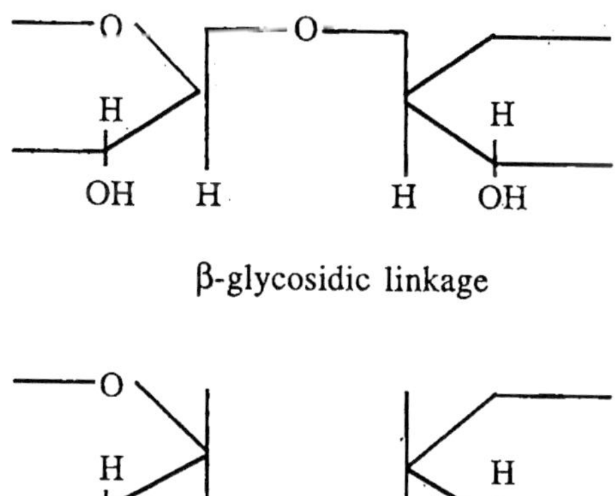

β-glycosidic linkage

α-glycosidic linkage

In the modern system of nomencalature, the oligosaccharides are named according to the constituent monosaccharide units and the positin of glycosidic bond (Table 9.2).

Table 9.2 : Systematic name of some important oligosaccharides

Common name	*Systematic name*
1	*2*
Sucrose	4-α-glucopyranoside- β- fructofuranoside
Lactose	4-β-galactosyl glucose

1	2
Maltose	4-α-D-glueopyranosyl- D-glucopyranose
Cellobiose	4-β-glucopyranosyl glucose
Gentiobiose	6-β-glucosyl-glucose α-D-galactosyl (1–>6) α-D-glucosyl (1–>2) β-D-fructofuranoside
Stachhaylose	α-D gnactopyranosyl (1–>6) α D-Galactopyranosyl (1 –>6) α-D-glucopyranosyl (1–>2) β-D-fructofuranose)

The natural source of oligosaccharides are green plants. Most of them are obtained as colourless solids. They are crystalline substances and readily soluble in water. Structure and properties of some important oligosaccharides are described as follow:

SUCROSE

It is formed by the union of α F-glucose and one β-D-fructose unit with the elimination of one water molecule. The glycosidic bond involves carbon 1 of pyranose ring of glucose and carbon 2 of fructose furanose ring.

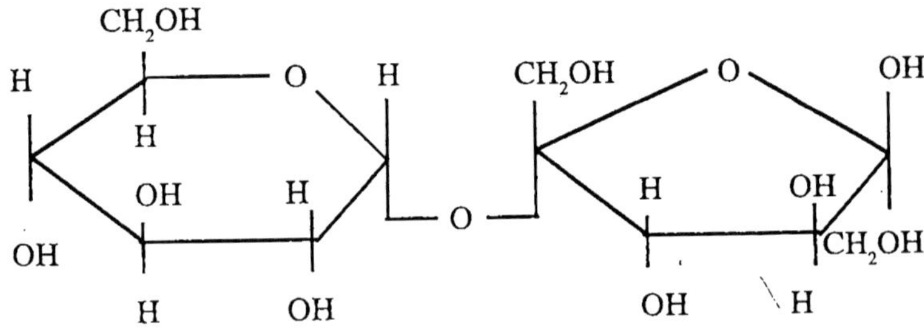

Sucrose (α-glucopyranoside and β-fructofuranoside)

Sucrose is derived from sugar beet and sugarcane and occur in the plants. It is easily hydrolysed to glucose and fructose units when warmed with dilute mineral acids. It is dextrorotory but the mixture of glucose and fructose produced after hydrolysis is laevorotatory because angle of laevorotation by fructose is glucose. This change is optical rotation of surcose due to hydrolysis is sometimes called ***Inversion***.

MALTOSE

Maltose is composed of two units of D-glucose joined together through their 1 and 4 carbon atoms. One of the units normally exists as open chain with free aldehyde group. Hence this disaccharide is a reducing sugar.

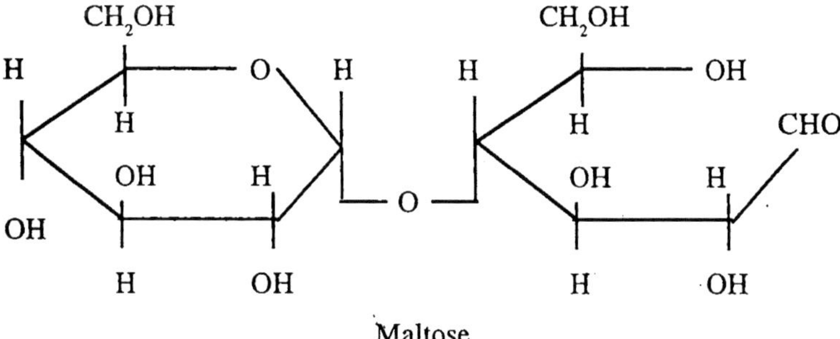

Maltose

The second unity may close also to form a ring through its carbon 1 and 5 giving rise to two forms of maltose, α maltose or β maltose.

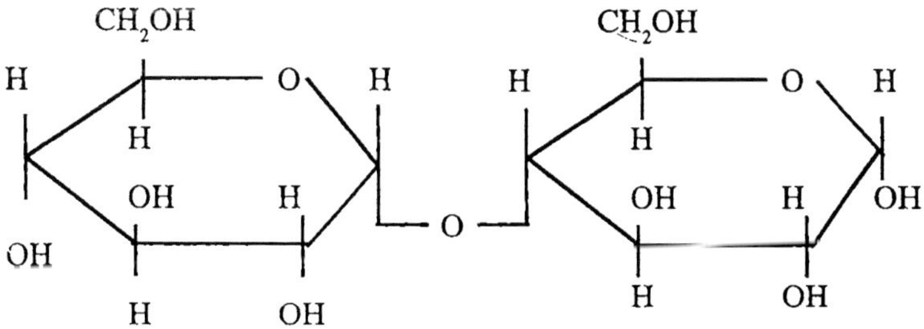

α-Maltose

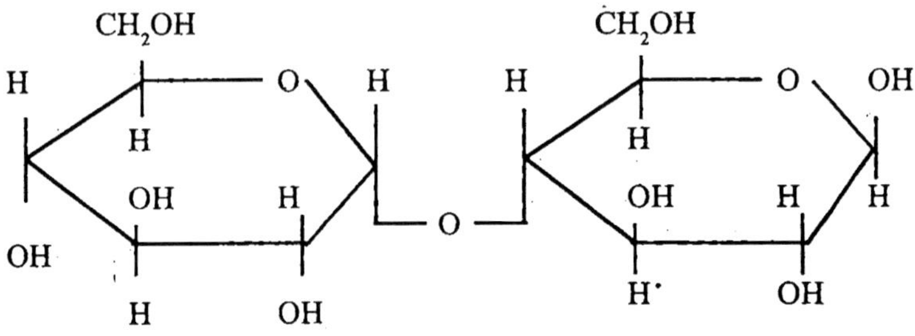

β-Maltose

Maltose occurs in plants as a degradation product of the storage polysaccharide starch. The enzyme amylase, which is responsible for this hydrolytic degradation, is abundant in germinating seeds. Commercial corn syrup, which is prepared by hydrolying corn starch, is a rich source of maltose. Free maltose has been detected in leaves.

LACTOSE

Lactose is made up of glucose and galactose units joined together through carbon 1 of galactose and carbon 4 of glucose. It is also a reducing sugar because open chain configurations also exist. Lactose occurs naturally in the milk of mammals upto 5%. It has been detected in the flowers of some plants (Porsy thias opotacea) also. Hydrolysis of this sugars yields galactose and glucose.

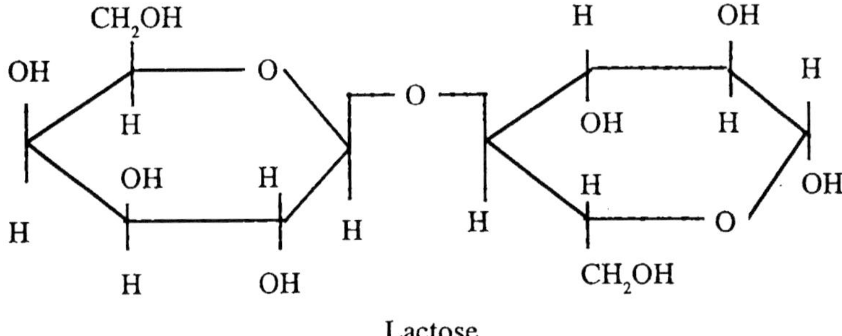

Lactose

CELLOBIOSE

Cellobiose is composed of glucose units (1-4 linkage). It may also exist in reducing form with a free aldehydic group. Alternatively, the ring may close to form α- or -β-cellobiose.

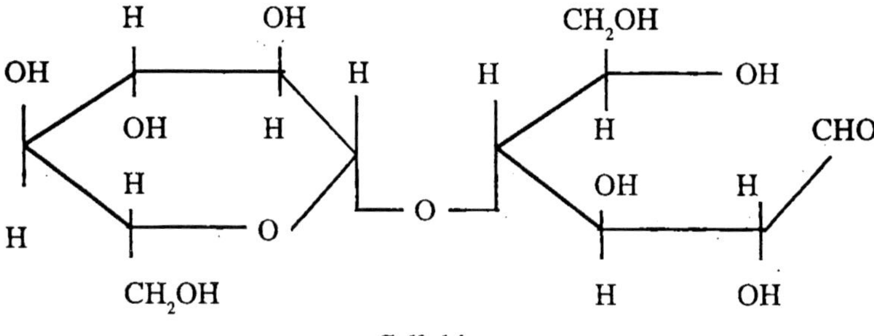

Cellobiose

Cellobiose is present only in trace amount in nature. The ellulases present in microorganisms and growing seedlings are able to hydrolyse plant cellulose to cellobiose. Its further hydrolysis yields D-glucose units.

TRISACCHARIDES AND TETRASACCHARIDES

Tri and tetrasaccharides are usually present in plants in small amounts. Raffanose (a trisaccharide) is found in sugarbeet and cotton seed meal and in some fungi. It is composed of galactose (pyranose), glucose (pyranose) and fructore (furanose) units. Another trisaccharide gentionose, occurs in rhizomes of several gentian species. It is made up by clucose (pyranose) glucose (pyranose) and fructose (furanose) units.

Stachyose is a tetrasaccharide occurring in several species including (stachystubifera). The composing units are two galactose (pyranose) one glucose (pyranose) and one fructose (pyranose) in a stachyose molecule.

POLYSACCHARIDES

Polysaccharides are composed of several monosaccharide units joined together by glycosidic bonds. The most frequent monosaccharide units in polysaccharide is D- glucose, although D-fructose, D-galactose and other hexoses also occur. The constituent monosaccharides may be either similar (homo polysaccharides) or different (hetero polysaccharides). Starch, cellulose, glucose are examples of homo polysaccharides while pectins and chitins are heteropolysaccharides.

Polysaccharides are tasteless and colour less amorphous powders which are little soluble in water, although some may form colloidal solutions. They are easily hydrolysed to their monosaccharide units.

STARCH

Starch is a high m.w. polysaccharide made up of two components, amylose and amylopectin. The amylose is an unbranched chain of 300-1000 α - D glucose units with 1 - 4 glycosidic bonds. The amylopectin is also made up of α - D-glucose units but is branched like a tree. The main chain has 1-4 linkage with 1-6 linkage at branching points.

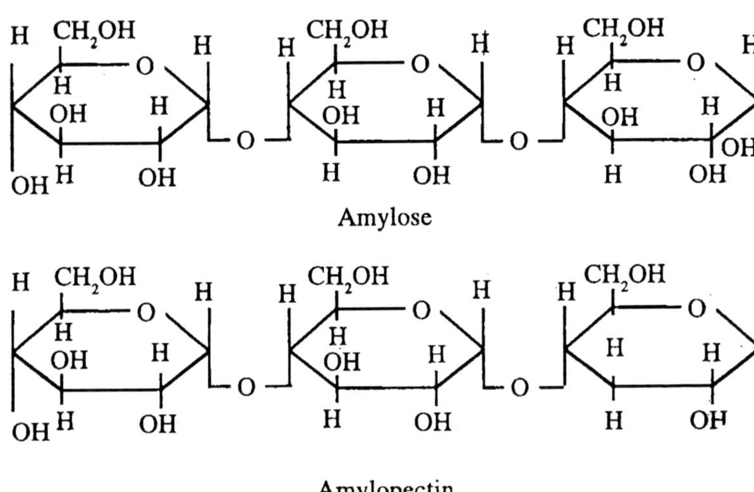

Amylose

Amylopectin

The ratio of amylose to amylopectin in starch is different for different tissues. The starch from normal maize seed contains about 20% amylose and 80% amylopectin. A waxy maize, recently developed variety of maize, contains almost 100% amylopectin.

Starch is the storage carbohydrate of plants. It occurs as grain of 1 μ (e.g. rice) to 100 μ (e.g. potato) size in the chloroplast or amyloplasts.

The grains contains about 20% of water, about 10% of which is chemically bound to starch. The shape of the grains may be oval, spherical, lens shaped or irregular. The starch in the grains is contained in the form of layers. Potato starch grains are covered with thin layers of cellulose.

GLYCOGEN

The storage carbohydrate of animals is glycogen. It is Similar to amylopectin in its structure except that m.w. is lower and the chains are shorter (fig. 9.1) The branching is also extensive, each branch containing about 12 glucose units.

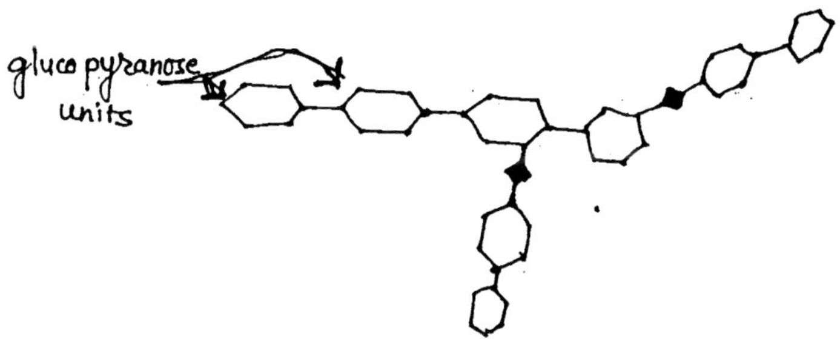

Fig. 9.1 : Glycogen molecule

CELLULOSE

Celulose molecule is composed of 1600 to 2700-β-D-glucose units with 1-4 glycosidic linkage. It is long thread like molecule.

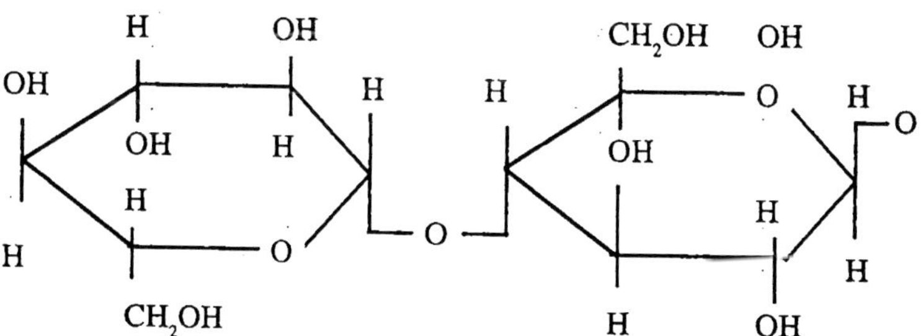

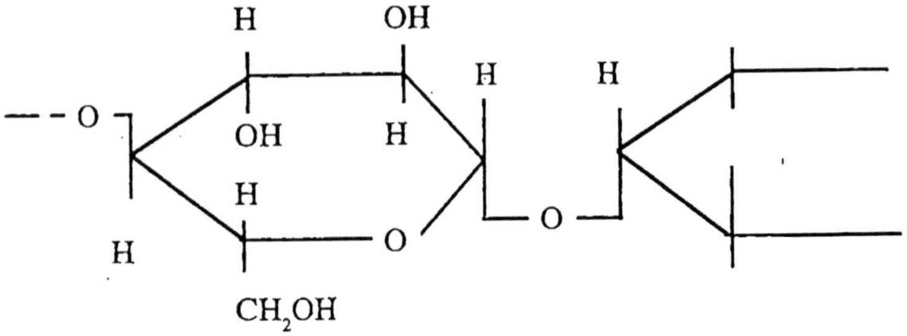

Cellulose

Cellulose is the major structural polysaccharide of plants. Careful acid hydrolysis of cellulose yields a series of oligosaccharides such as cellobiose, cellotriose, cellotetrose etc. Complete hydrolysis permits recovery of about 95% of the weight of cellulose as glucose. Enzymatic hydrolysis of cellulose is brought about by cellulose which is present in some protozoans and mites. Some of the bacteria in cattle and other herbivores also secrete cellulase and enables the cattles to digest the herbal cellulose to glucose.

INULIN

Inulin is an unbranched molecule of about 30-35 D-fructose units. m.w. is about 5000 and the glycosidic linkage is between carbon 1 and 2 of the fructose units. Inulin is obtained commercially from Jeruselum artichoke rhizome. It is found naturally with other short chain fructans. The enzyme inulase hydrolyses inulin to fructose.

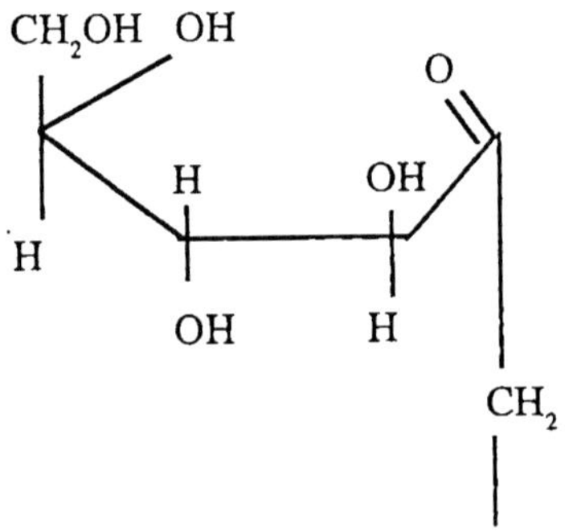

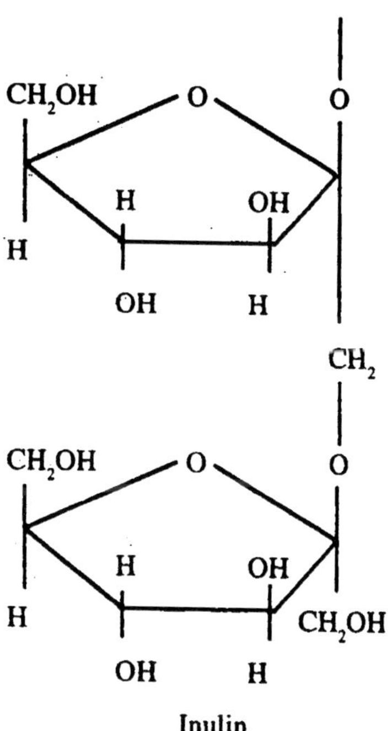

Inulin

BIOLOGICAL SIGNIFICANCE

Carbohydrates are involved in structural organisation of many tissues, both the plants and animals. Cellulose, hemicellulose (xylans and mannans) and pectic substances (galactouranans, arabans and cellose) are common structural carbohydrates in plants. The primary well wall in plants contains about 43% of cellose. Legnin is also present in the cell wall of older tissues. Chitin, hyaluronic acid and Chondroition sulphates are important structural polysaccharides of animals. They are present in the shell of crabs and insect and in the cartilage, adult bones, heart valves and cornea.

Carbohydrates are stored as reserve products in many tissues. Normally the starch is stored in seeds, tubers, and rhizomes. In some plant families, such as Compositae and Companulaceae, the stored carbohydrate in rhizomes and tubers is inulin. Sucrose (sugarcane, beet), glucose (grapes) and fructose (fruits) are also stored. Glucose is the most common carbohydrate of human blood present in a concentration of 1g/lt. In animals cells, glycogen is the storage polysaccharide.

The most important role of carbohydrates is the production of energy in the form of ATP both in plants and animals. The polysaccharides such as starch and glycogen are converted to glucose by enzymes. The glucose is finally broken down to Co_2 and H_2O and yield NADH which is eventually oxidized to produce ATP.

$$C_6H_{12}O_{66} + 6CO_2 \longrightarrow 6CO_2 + 6H_2O + ATP$$

In overall oxidation of 1 molecule of glucose, 38 ATP molecules are generated.

During aerobic breakdown of carbohydrates through EMP or Kreb's cycle several organic compounds are produced which act as substrates for the synthesis of lipids and amino acids. The phosphogluconate oxidative pathways or pentose phosphate pathways provide pentoses, essential components of nucleotides and nucleic acids. NADH produced during the process is used as coenzyme in several biosynthetic processes such as fatty acid and steroid synthesis.

BIOSYNTHESIS OF SOME IMPORTANT CARBOHYDRATES

Monosaccharides

Monosaccharides are formed in green plants by photosynthesis or in other organisms by the reversal of glucolytic pathways. In the process of photosynthesis, atmospheric CO_2 is fixed-up by ribulose diphosphate carboxylase or phosphoenol pyruvate carboxylase to produce phosphoglyceric acid as first sugar. Conversion of phosphoglyceric acid to phosphoglycereldehyde is achieved through NADPH and ATP produced during the light reactions of photosynthesis in the chloroplasts. The phosphoglycereldehyde may be converted to a various monosaccharides, which maybe ultimately utilised for the synthesis of oligo and manosaccharides (fig. 9.2). Synthesis of glucose in other organisms employs reversal of glycolytic pathway. However, all the glycolytic steps are not reversible. At least 3 reactions substitute for the irreversible steps.

1. Synthesis of phosphoenol pyruvate from pyruvate occurs through following reactions catalysed by pyruvate kinase proceeds mainly for the synthesis of pyruvate.

Pyruvate + ATP + Co_2 + H_2O $\xrightarrow[\text{Carboxylase}]{\text{Pyruvate}}$

$\longrightarrow$ Oxaloacetate + ADP + ip

Oxaloacetate + GTP $\xrightarrow[\text{carboxykinase}]{\text{PEP}}$ Phosphoenol pyruvate + CO_2 + GDP

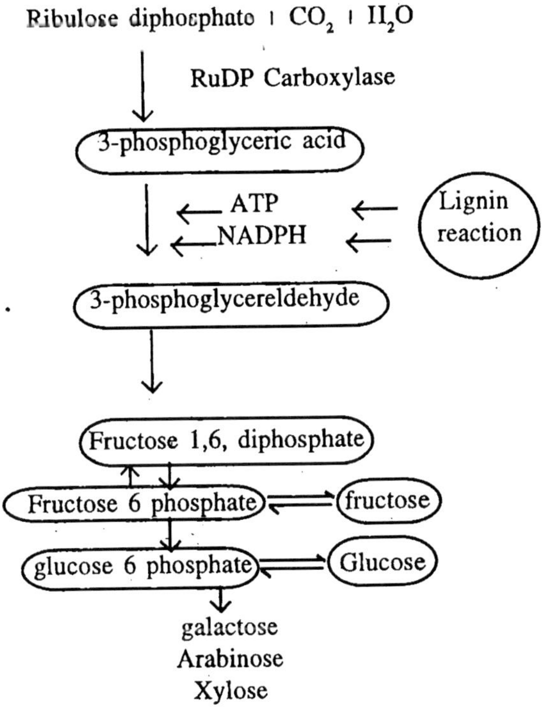

Ribulose diphosphate + CO_2 + H_2O

↓ RuDP Carboxylase

3-phosphoglyceric acid

↓ ← ATP ← NADPH ← Lignin reaction

3-phosphoglycereldehyde

↓

Fructose 1,6, diphosphate

↑↓

Fructose 6 phosphate ⇌ fructose

↓

glucose 6 phosphate ⇌ Glucose

↓

galactose
Arabinose
Xylose

Fig. 9.2 Biosynthesis of monosaccharides

The two enzymes pyruvate carboxylase and PEP carboxykinase have been detected in high concentrations in liver and kidney cells.

2. Hydrolysis of frutose 1-6-diphosphate to fructose-6-phosphate and inorganic phosphate is catalysed by fructose diphosphatasc.

Fructose 1,6, diphosphate $\xrightarrow{\text{fructose di phosphatase}}$ Fructose-6-phosphate + iP

The enzyme fructose diphosphatase is found in appreciable amounts in only those tissues which synthesize glucose or glycogen.

3. Hydrolysdis of glucose 6 phosphate is the final step in glucose synthesis. It is catalysez by glucose -6- phosphatase.

$$\text{Glucose-6-phosphate} + H_2O \xrightarrow[\text{-6- phosphatase}]{\text{Glucose}} \text{glucose} + iP$$

The glycogenic reactions are summarised in fig. 9.3.

SUCROSE

Sucrose is synthesized in plants by two reactions

1. Glucose 1 phosphate + UTP UDP-glucose + iP

$$\text{UDP-glucose} + \text{fructose} \xrightleftharpoons[]{\text{Sucrose synthatase}} \text{Sucrose} + \text{UDP}$$

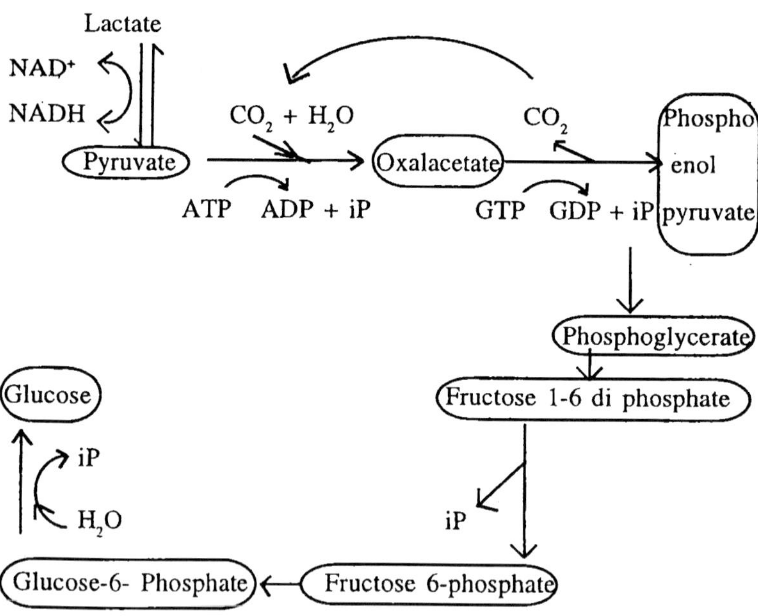

Fig. 9.3: Synthesis of glucose via reveral of glycolysis

2. UDP-glucose + Fructose-6- phosphate $\xrightleftharpoons[\text{synthatase}]{\text{Sucrose Phaphate}}$ Sucrose-P + UDP

$$\text{Sucrose - P} + H_2O \xrightleftharpoons{\text{Phosphatase}} \text{Sucrose} + iP$$

Reaction A occurs mainly in the non-photosynthetic tissues. Actual mechanism of sucrose synthesis during the process of photosynthesis appears to be by reaction II. In photosynthetic tissues, when a short exposure to $C^{14}O_2$ is given both glucose and fructose residues are strongly and equally labelled.

In the bacteria, ***Pseudomonas saccharophila***, the synthesis of sucrose takes place by another mechanism involving sucrose phosphorylase.

Glucose-1-phosphate + D - fructose

⇅ sucrose phosphorylase

Sucrose + iP

This directo mechanism of sucrose synthesis does not appear to function in higher plants.

STARCH

Starch is synthesized manner (Fig. 9.4).

It appears that the initial product of these reactions is amylose, which is rearranged to more complex amylopectin. The arrangement is catalysed 'enzyme' which has been obtained in crystalline form from potato tuber extracts.

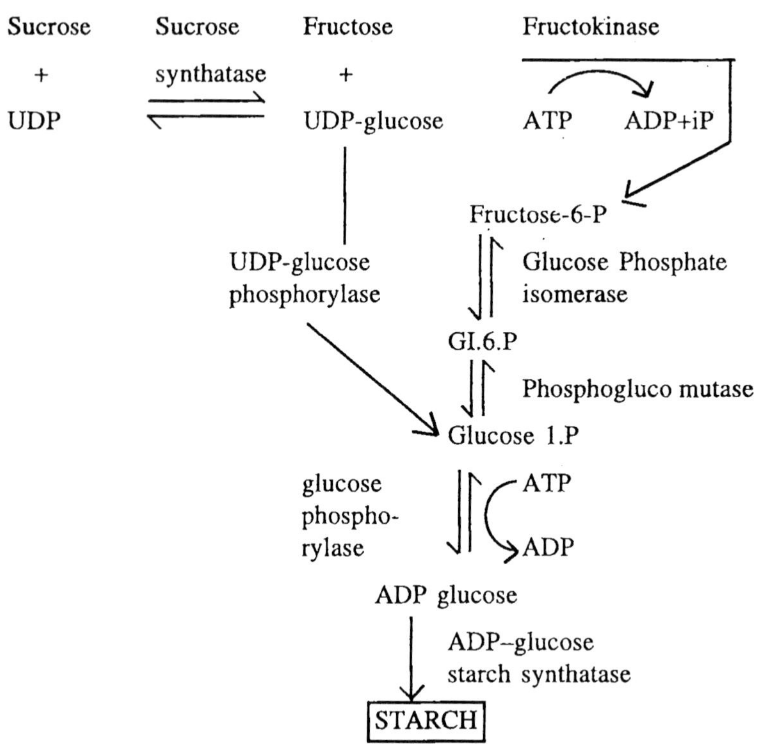

Fig. 9.4 : Synthesis of Starch

GLYCOGEN

glycogen is synthesized from UTP activated glucose units. The UTP-glucose is added to a growing chain of glucose units (glycogen primer). As a result the chain is lengthened and UDP is released. This mechanism of lengthening the chain is similar to that of starch and cellulose.

UDP-glucose is synthesized from glucose-1-phosphate and UTP by a specific enzyme UDP-glucose phosphorylase.

$$\text{UTP + glucose-1-P} \xrightarrow[\text{Phosphorylase}]{\text{UDPG}} \text{UDP-glucose+iP}$$

The activated glucose (UDPG) in transferred to a glycogen primer by the enzyme glycogen synthatase or UDP glucose : glycogen glucosyl transferase

$$(glucose)_n + UDP\text{–}glucose \xrightarrow[\text{synthatase}]{\text{glycogen}} (glucose)^n + 1 + UDPG$$

Finished glycogen is a branched chain structure. The branches of small polyglucose chains are added to the main chain by a different enzyme, the liver branching enzyme.

Separation of soluble carbohydrates from a organic matter

Homogenise a tissue piece in water and allow it to decant. Separate the supernatant solution. This contain sugars and protein. Remove the proteins from this solution by profitation method, through the following process: Add lead acetate drop by drop till a white precipitate appears. Add a few drops more for complete precipitate.

Apply test in the filtrate form the pressure of carbohydrates.

BENEDICT'S TEST

Take 1 ml. of this filtrate in a test tube and add 5 ml. of Benedit's reagent. Mix them well and boil for 2-3 min. Cool it under the tap-water. A coloured precipitate is formed due to the formation of cuprus oxide..

FEHLING'S SOLUTION TEST

Take Fehling's solution in a test tube and add 0.5 ml. of filtrate. Boil till a coloured precipitate appear.

The precipitate is due to the formation of cuprus oxide.

PICRIC ACID TEST

Take 5 ml. of filtrate in a test tube and add 2 ml. of saturated solution of picric acid and 2 ml. 4% NaOH. Heat the solution till red colour is developed.

The red colour is due to the formation of piecylic acid as a result of reduction of picric acid.

QUANTITATIVE ESTIMATION OF REDUCING SUGAR BY DNSA METHOD

Reducing sugars can be estimated by dinitro salicylic Acid method (Miller, 1959).

REAGENTS

DNSA reagent

Dissolve 30 gms. Rochalle salt (Na, K. tartrate) and 1 gm DNSA is 20 ml. 2 N NaOH and dilute to make up the volume to 100 ml. with dist. H_2O.

PROCEDURE

To 1 ml. of test solution, add 1 ml. DNSA reagent and heat for 5 min. in boiling water bath and add and ml. dist. H_2O and read at 450 nm in a colorimeter.

Draw a standard curve with the O.D. value of the glucose or maltose solution of concentration ranging from 0.1 to 10 mg/ml.

The concentration of the test solution is to be calculated by referring into standard curve.

Biodegradation of carbohydrates

Biological degradation of Glucose completes in 2 stages :

1. breakdown to pyruvic acid
2. formation of various end products

— In anaerobic environment, pyruvic acid get converted into acids, alcohols while aerobic environment, pyruvic acid get converted into CO_2 + water.

There are different 3 pathways for glucose as mentioned below:

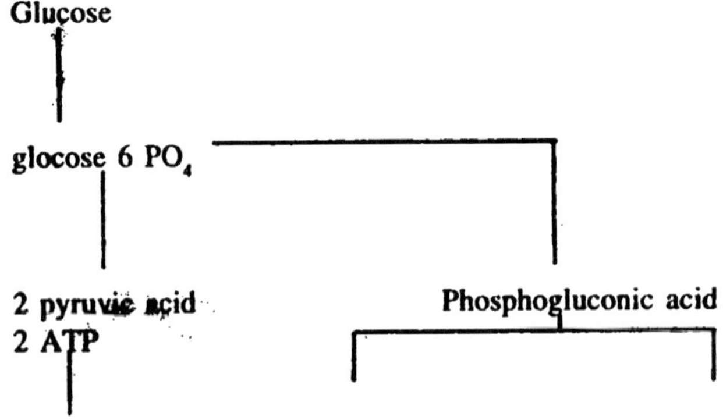

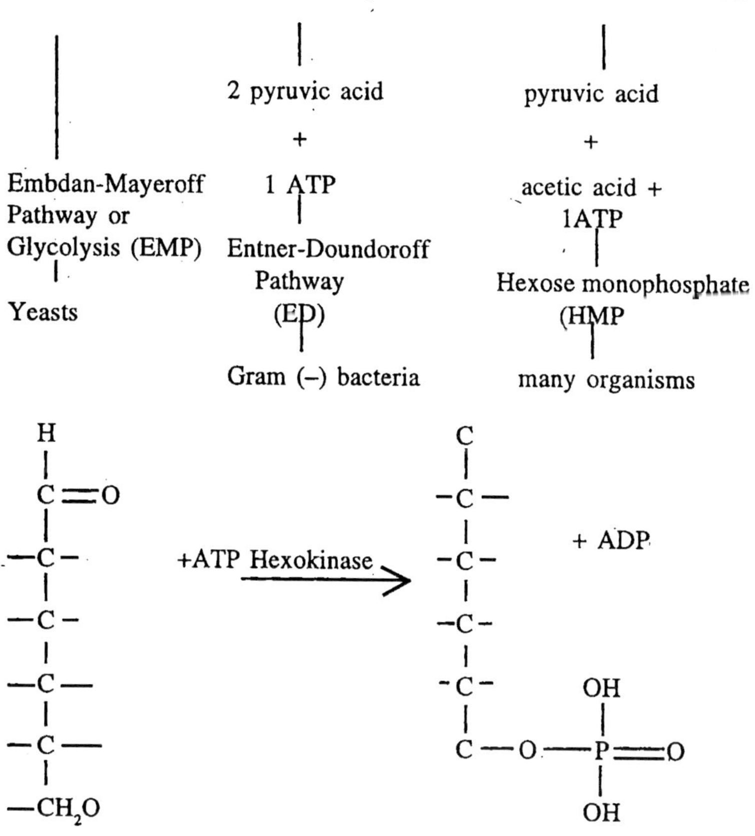

Fermentation (no external H acceptor)

1. $CH_3-C(=O)-COOH$ (pyruvic acid) $\longrightarrow$ $CH_2-C(H)=O$ (acetaldehyde) $+ CO_2$

acetaldehyde + NADH $\rightarrow$ etOH + NAD^+

2. Pyruvic acid + NADH ⟶ lactic acid + NAD^+

Aerobic oxidation

It gives H, gives electron to O_2 which is available from NADH + O_2 by converting them into NAD^+ + H_2O. The reaction mediated by enzymes

ELECTRON TRANSPORT SYSTEM

1. flavo-proteins = protein + derivation of riboflavin accepts or donates electrons.
2. cytochromes = protein + *iron-containing ring compound.*

— iron-compound oxidized or reduced.

— several types : a, b, c.

— different organisms have different cytochromes

$$\underset{\text{(reduced)}}{\text{Cyt} - \text{Fe}^{+2}} \rightleftharpoons \underset{\text{(oridized)}}{\text{Cyt–Fe}^{+3}}$$

ELECTRON TRANSPORT SYSTEM (AEROBIC)

It is presented diagramatically in Fig. 9.5

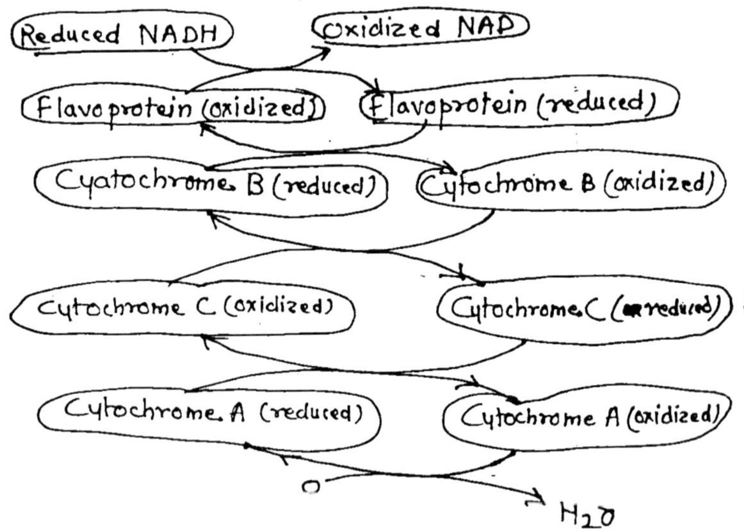

Fig. 9.5 : Electron Transport System

— ADP + energy $\xrightarrow{iP}$ ATP

— P/O (taken in) = 3 (different for different organisms)

— for each Oxygen $\longrightarrow$ 3 ATP produced

— anaerobic : recycle NAD $^{+}$ by passing electron through other pathway

KREB'S CYCLE OR TRICARBOXYLIC ACID (TCA) CYCLE OR CITRIC ACID CYCLE

It is summarised in Fig. 9.6

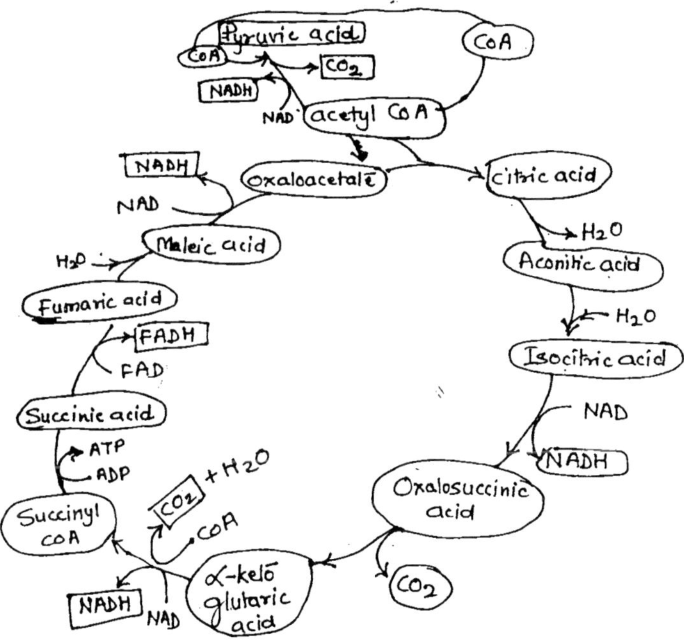

Fig. 9.6 : Kreb's Cycle

Energy in Kreb cycle

— pyruvate + 4NAD + FAD ⟶ $3CO_2$ + 4NADH + FADH

— NADH ⟶ NAD = 3ATP 4 (3ATP) = 12ATP

— FADH ⟶ 2 ATP 2 ATP

1 ATP

15 ATP

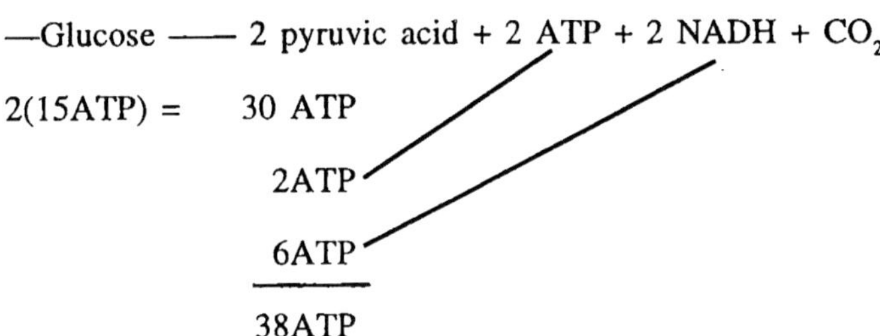

—Glucose —— 2 pyruvic acid + 2 ATP + 2 NADH + CO_2

2(15ATP) = 30 ATP

2ATP

6ATP

38ATP

— cell under aerobic conditions converts glucose all the way to CO_2 and water.

— Anaerobic conditions ⟶ 2ATP

aerobic ⟶ 38ATP

— ATP is energy used for synthesis.

— cells synthesize : protein, nucleic acids (hereditary control), lipids, carbohydrates etc.

— reserve energy from carbohydrates stored as glycogen or starch.

— Glucose + ATP ⟶ starch/ glycogen.

Rate of Enzyme reaction

— rate of chemical reaction = substrate used/unit time concentration of substrate ⟶ product

zero order reaction $-\frac{dc}{dt} = K^0$

first order reaction $-\frac{dc}{dt} = K_1C^1$

second order reaction $\frac{dc}{dt} = K_2C^2$

rate of enzyme = first order reaction

1. $E + S \underset{K_2}{\overset{K_1}{\rightleftarrows}} ES \xrightarrow{K_3} P + E$

ES = enzyme-substrate complex

Michaelis-Menton Equation

assumptions : single substrate

single product

process proceeds to completion

concentration of S >>>> than E

2. $\frac{d\,[ES]}{dt} = K_1\,([E] + [ES])\,[S]$ rate of apperance of ES

3. $\frac{d\,[ES]}{dt} = K_2\,([ES] - K_3\,[ES]$

Steady State rate of production ES (equation 2) = use of ES (equation 3)

4. $K_1\,([E]\text{-}[ES])\,[S] = K_2\,[ES] + K_3\,[ES]$

Rearrange

5. $\frac{[S]\,([E] - [ES])}{[ES]} = \frac{K_2 + K_3}{K_1} = Km$

6. $[ES] = \frac{[E]\,[S]}{Km = [S]}$ S >>> E

7. $\frac{-d[S]}{dt} = K_3\,[ES]$ /mole = /mole

[E] = [ES] max. rate

Michaelis constant

$$V = \frac{K_3 [E] [S]}{Km + [S]}$$

Autotrophs (lithotrophs) ⟷	Heterotrophs
CO_2 or bicarbonate	organic matter
inorganic compounds	
chemolithotrophs	

Autotrophs oxidize inorganic substrates for energy. H_2 from sulphides, ammonia. Their biochemical reactions are substrate specific. In case of heterotrophs or organotrophs, metabolic pathways not as well known. It is established that use ATP for energy production.

Hydrogenmonus

These types of organisms oxidizes molecular H_2 and convert them into water. They are short rods, gram (–), aerobic and do not strict lithotroph as they can use simple organic matter.

$H_2 \rightarrow 2\ [H] + 2e$

$[H] + CO_2 \rightarrow (CH_2O) + H_2O$

$H_2 + O_2 \rightarrow H_2O$

Sulphur bacteria

They are chemolithotroph, colourless variety (not purple), able to oxidize H_2S, S_2O_3 or S^0 for energy. Bacteria like ***Beggiotoa*** and ***Thiotrix*** found in ground water. Under the microscope, one can see sulphur granules right in their protoplasm.

$H_2S \rightarrow S$

$S_2O_3 \rightarrow S$

Thiobacillus thiooxidans

It can grow at pH 1. The micro organisms ar eable to turn to SO_4 as external H acceptor in absence of O_2 . It can also use organic matter + SO_4 to produce S or H_2S

In Septic or Stale sewer conditions, which occur in long sewer runs and/or warm conditions, organisms produce gaseous H_2S after the oxygen supply is exhausted. When the solution is saturated with the H_2S, the gas rises to the top of the sewer organisms living in the top of the sewer take S and convert it into sulphate.

$$H_2S + SO_4 \rightarrow H_2SO_4$$

The extremely acidic conditions can deteriorate the sewer causing ***crown corrosion***.

Desulfovibrio desulfurecan - prefers to oxidize SO^{--}_4

Iron bacteria

The energy is derived from ferrous (fe^{++}) → ferric (fe^{+++}) iron in water as low as 0.5 to 1 ppm will support these organisms ferrous form of iron is relatively soluble, whereas, ferric form of iron is relatively insoluble. The organisms will deposit precipitates of insoluble Fe in filaments e.g. Galionella. Acid mine drainage will support large build-ups of ***Ferro bacillus***

$$Fe\ (HCO_3) + H_2O \rightarrow Fe\ (OH_3) + CO_2 + \text{Energy}$$

Ferric oxide turns water red - aesthetic problem.

Chapter -10

NITROGEN AND NITROGENOUS COMPOUNDS IN THE ENVIRONMENT

Nitrogen is an important constituent of air. It is directly utilized by some specific microflora.Oxidation of ammonia to nitrate is known as nitrification

$NH_3 \rightarrow (NO_2) \rightarrow NO_3$

It is important in the soil and equally significant in stream standards. Based on its toxicity to fish and aquatic life. When NO_3 N is about 10 mg/l present in water, it is not as toxic. However, domestic and industrial waste waters contain much nitrogen. Nitrification rather than nitrogen removal will fit the easier standard for NO_3. nitrogen removal in an effort to stop *eutrophication* or over-fertilization.

NITRIFICATION

1. NH_3, $\rightarrow NO_2$ (nitrites)

   autotrophic metabolism $\rightarrow NH_3 + O_2 \rightarrow NO_2 + H_2O$ + 79 Kal

   Nitrosomonas, Nitrosococcus are some example, who derive energy in this reaction

2. $NO_2 \rightarrow NO_3$ (Nitrates)

   $NO_2 + O_2 \rightarrow NO_3$ + 22 Kcal aerobic

   Nitrobacter (a bacillus)

Reactions 1 and 2 go on simultaneously; little nitrite accumulate Nitrobacter uses the NO_2 as soon as it is produce so little accumulates. NO_2 is toxic in any concentration.

Nitrogen cycle

Conversion of atmospeheric nitrogen to usable forms for various plants and animals use is govern certain law. Nitrogen cycle is presented diagrammatically in Fig. 10.1.

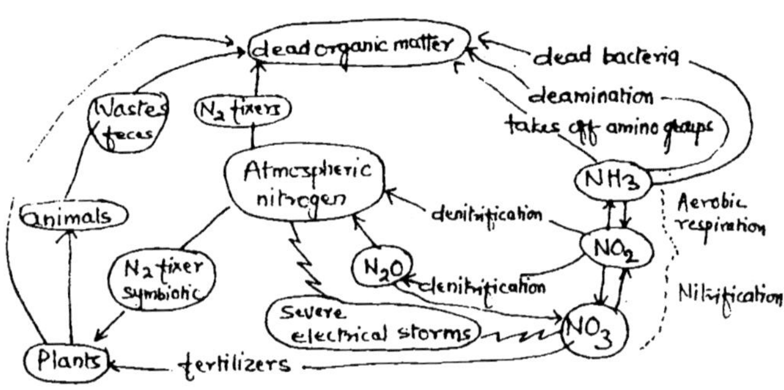

Fig. 10.1 : Nitrogen Cycle in Environment

— nitrate reduction = *denitrification*

— some organisms use NO_3 for their H acceptor

— a few organisms turn $NO_2 \rightarrow NH_3$ others produce N_2O

Operation for nitrification $NH_3 \rightarrow NO_2$

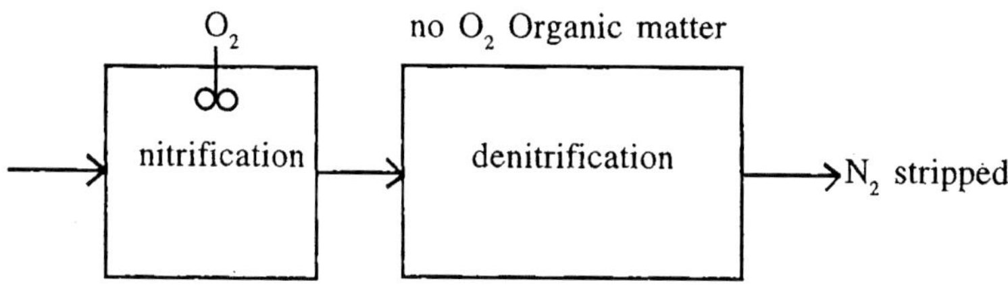

Methane formers

- methane fermentation.
- obligate anaerobic decomposition of organic matter with the production of methane.
- such organisms are widely distributed in nature.
- starting new treatment plant, organisms are naturally present.
- methane production is studied in connection with the organisms in the rume in cows, organisms which breakdown cellulose.

metabolic pathway of methane formers

$$CH_3 - CH_2OH + CO_2 \rightarrow CH_3 - COOH + CH_4$$

$$H_2 + CO_2 \rightarrow CH_4 + H_2O$$

$$CH_3\ OH \rightarrow CO_2 + CH_4$$

— in good operating digester released gas contains approaximately 70% CH_4 and around 30% CO_2.

— methane formers are heat sensitive.

—sensitive to fluctuations-need a constant temperature.

$$C^*H_3\ COOH —— C^*H_4 + CO_2$$

ORGANISMS CAUSING TASTE AND ODOUR PROBLEMS

— algae-their metabolism produces oils

— some worms, snails

— Actinomyces (research done by various investigators)

  — between true bacteria and fungi.

  — grow in long filaments.

  — called "mold-like bacteria".

— *Nocardia* - aerobic.

— actinomyces - anaerobic.

— Zooglea ramigera- one bacterial form which forms floc and settles easily.

— Nocardia - grows as big clumps of filaments even in suspension.

— settles.

— affinity for paraffinic hydrocarbons + phenols.

NITROGEN COMPOUNDS

Aliphatic groups

Amines - derived alkyl compounds of ammonia (NH_3)

Primary amines - 1 hydrogen replaced (R-NH_2) methylamine - CH_3NH_2

Secondary amines - 2 H's of ammonia replaced by alkyl

R—— NH

R /

Alkyls may be the same or different.

Tertiary amines - 3H's of ammonia replaced with alkyls.

R

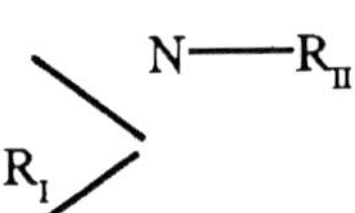

amines get more basic as they lose more hydrogen.

They are found in fish & beets. Very little is known about the biological oxidation of the amine group a amino acids are broken down by animals for food. Animals can not synthesize all the amino acids they need.

- Dispensable amino acids are those that can be synthesized.
- Indispensable amino acids can not be synthesized.
- get indispensable amino acids from plants.

- Indispensible or essential amino acids are :-

  Lycine

  Tryptophine

  Histidine

  Phenyl alanine

  Leucine

  Isoleucine etc.

Dispensable or non-essential amino acids are :-

  Glycine

  Alanine

  Serine

  Aspartic acid etc.

Structural formula of amino acid

$$R_\beta - \underset{\displaystyle H}{\overset{\displaystyle NH_2}{\overset{|}{\underset{|}{C_\alpha}}}} - COOH$$

R=alkyl, phenyl or sulphur group

Amino acids named α amino acids because the amine group is connected to the α carbon atom after the carboxyl group.

The amine group has basic properties & the carboxylic group has acidic properties so therefore the amino acid can function either way-

- acts as base :

$H_2N\text{-}CH_2\text{-}COOH + HCl \longrightarrow Cl^- \; H_2^+ \; N \; CH_2COOH$ (Chloride Salt)

- acts as acid :

$H_2N\text{-}CH_2COOH + NaOH \longrightarrow H_2O + H_2O + H_2N \; CH_2 \; COONA$

Amphoteric compounds are those having both acid & basic properties on basis of ionic charge

- cationic - having positive charge
- anionic - having negative charge
- amphoteric - having either positive or negative charge
- nonionic - having no charge

Amino acids & some polymers are amphoteric

- an amino acid may 1 amine group for every 1 carboxylic group making it neutral.
- an amino acid with more carboxylic groups than amine groups than carboxylic groups will act as a base.
- Combining a tertiary amine with a halogen under heat = a quarternary amine.

$$R_3N + RCl \xrightarrow{\Delta} \{R_4N\}^{+} + Cl^-$$

- quaternary amines have disinfecting capacity. Detailed description of various amino acids is given elsewhere in this volume.

AMIDES

- form from the action of an organic acid & ammonia under special conditions :

$$R-\overset{O}{\overset{||}{C}}-OH + HNH_2 \longrightarrow R-\overset{O}{\overset{||}{C}}-NH_2 + H_2O$$

amide bond

- amide bonds are related to peptide bonds.
- An important reaction of amides :

$$CH_3CONH_2 + Br_2 + 4\ NaOH \longrightarrow 2\ NaBr + Na_2CO_3 + 2H_2O +$$

$$\begin{array}{c} \quad H \qquad\quad H \\ \quad | \qquad \diagup \\ H - C - N \\ \quad | \qquad \diagdown \\ \quad H \qquad\quad H \end{array}$$

amine

This reaction is important in synthesizing organic compounds since it will produce an amine with one less carbon than the original amide + H_2O.

UREA

Urea is found in agricultural fertilizer & urine.

$$\begin{array}{c} O \\ \| \\ H_2O - C - NH_2 \end{array}$$

Urea

Urea is synhesized from inorganic cyanide.

Urea can not be used as food by saprophitic bacteria, but other bacteria can remove the carbon from urea.

$$\text{Urea} + H_2O \xrightarrow[\text{action}]{\text{bacterial}} CO_2 + H_2O + 2\ NH_3$$

NITRILES

- have the general empirical formula R-CN.
- R may be saturated or unsaturated.
- some are toxic to bacteria.
- are used in the synthesis of some fibres & are found in the industrial waste.

PROTEIN

Complex compound found in living material.

- Comes from the Greek Word 'protos' first important pre eminent.
- Proteins have the following concentrations of elements (Table 10.1).

Table 10.1 : Various Elements and their Percentage

Element	*Pergentage (%)*
C	50
O	25
N	16
H	7
S	1
P	0.4

Estimates of the concentration of protein in food or wastes are made with Kjeldahl technique by presuming the total nitrogen to be 16% of the protein concentration :

$$\frac{10}{16} = 6.25$$

molecular wt. of protein is $15{,}0000^{-2} \times 10^{6}$

Proteins are the union of amino acids as polysaccharides are the union of simple sugars :

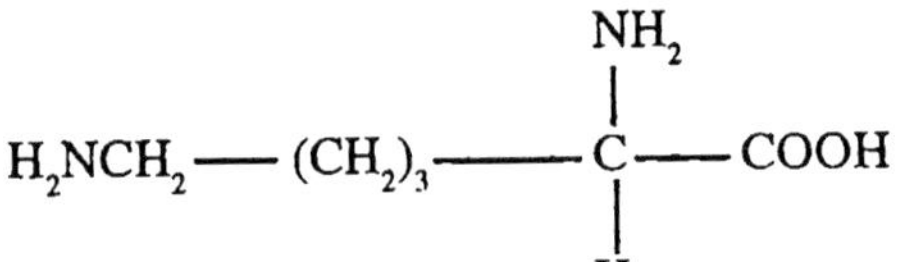

$$H_2NCH_2 — (CH_2)_3 — \underset{\displaystyle H}{\overset{\displaystyle NH_2}{\overset{|}{\underset{|}{C}}}} — COOH$$

D-Lysine

$$C_6H_5 — CH_2 — \underset{\displaystyle H}{\overset{\displaystyle NH_2}{\overset{|}{\underset{|}{C}}}} — COOH$$

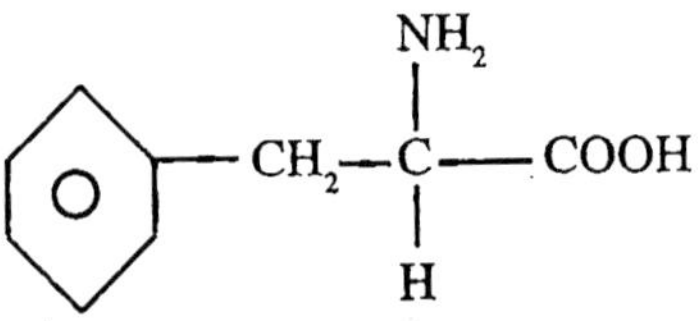

Phenyl alanine

All amino acids are optically active (except glycine), most in the L form.

Amino acids may take any of the following 3 forms in solution:

1. in high pH, carboxylic groups disassociates

$$\begin{array}{c} NH_2 \\ | \\ R \text{---} C \text{---} COO^{-}H^{+} \\ | \\ H \end{array}$$

2. in low pH :

$$\begin{array}{c} H \\ | \\ R \text{---} C \text{---} COOH \\ | \\ NH_3^{+} \end{array}$$

3. at isoelectric point, the net change is O.
   - zwitterions is a double ion.
   - self neutralizing, H from the carboxylic group goes to NH_2 :

$$\begin{array}{c} H \\ | \\ R^{-} \text{---} C \text{---} COO^{-} \\ | \\ NH_3^{+} \end{array}$$

PEPTIDE LINKAGE

- in proteins links amino acids.
- the linkage joins the carboxylic end of one amino acid to the amine end of another amino acid.

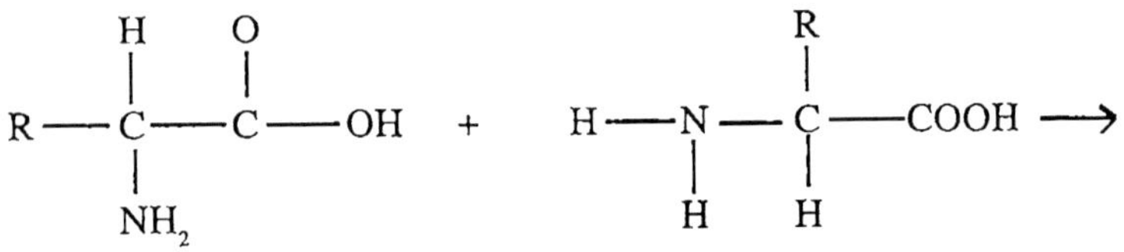

$$\underset{\text{a amino acid}}{R-\underset{NH_2}{\overset{H}{\overset{|}{\underset{|}{C}}}}-\overset{O}{\overset{|}{C}}-OH} \quad + \quad \underset{\text{b amino acid}}{H-\underset{H}{\underset{|}{N}}-\underset{H}{\underset{|}{\overset{R}{\overset{|}{C}}}}-COOH} \longrightarrow$$

$$R-\underset{NH_2}{\underset{|}{\overset{H}{\overset{|}{C}}}}-\boxed{\overset{O}{\overset{|}{C}}-\underset{H}{\underset{|}{N}}}-\underset{H}{\underset{|}{\overset{R}{\overset{|}{C}}}}\text{-}COOH + H_2O$$

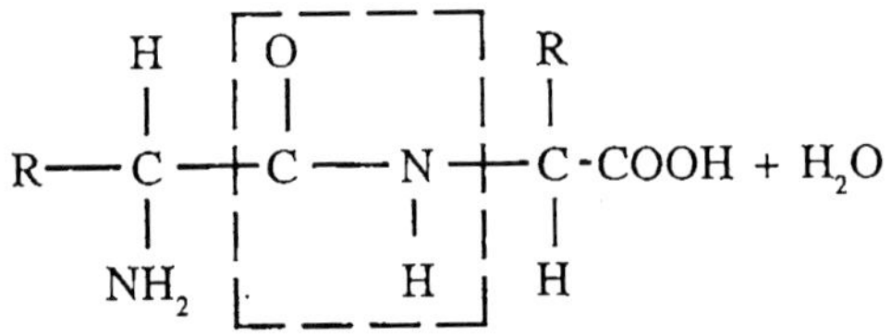

dipeptide

- the peptide linkage acts as the organic acid & the amine group of another amino acid.
- the peptied linkage is

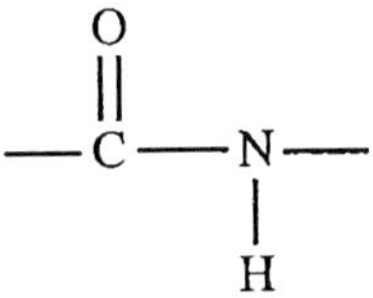

$$-\overset{O}{\overset{\|}{C}}-\underset{H}{\underset{|}{N}}-$$

the dipeptide has 1 free carboxylic group & 1 free amine group.

- the free carboxylic & amine groups keep reacting to for more new peptide linkages making polypeptides.

POLYPEPTIDE

- naming starts with amino acid with free amine group and ends with the one with the free carboxylic group. e.g. Using this system, the protein given in next page would be called :
- alanyl - leucyl - cystyl - tyrocinyl - glucine

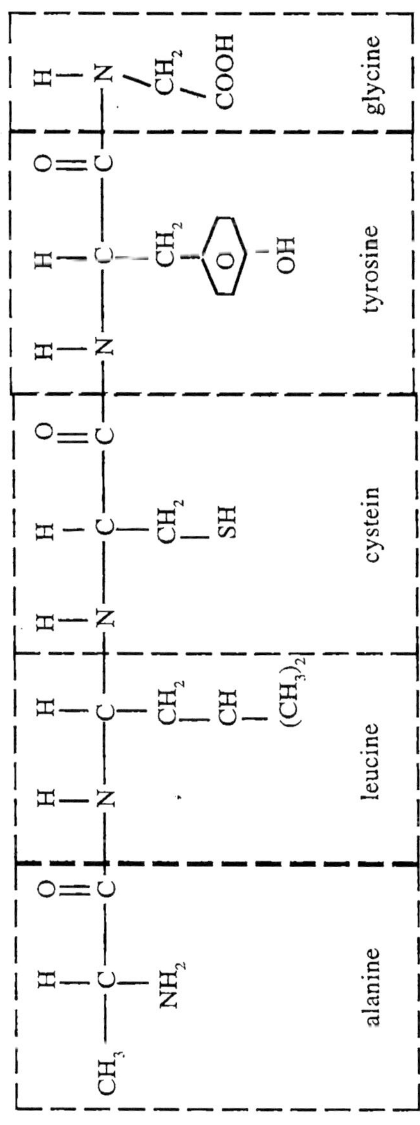
alanine
leucine
cystein
tyrosine
glycine

CHARACTERISTICS

- Odour when burned e.g. keratin of hair.
- forms precipitates with salts & heavy metals.
- precipitate with certain acids e.g. TCA or perchloric acid (HOCl).

  react with some dyes.
- give certain colour test.

COMPLEX PROTEINS

Some complex proteins and their amino acids are presented in Table 10.2.

Table 10.2 : Complex proteins & their amino acids

Complex proteins	*Amino Acids*
Insulin	- has 51 amino acid (Frederick Sanger Noble prize-1958)
Ribonuclease	- 124
TMV protein	- 158
Haemoglobin	- 574 (mol. wt. 64, 500)

- the arrangement of amino acids changes with heat.
- the protein becomes denatured & cannot be changed back to its original state e.g. albumin in egg whites.

CLASSIFICATION OF PROTEINS

- Those yielding amino acids in hydrolysis (simple proteins).
- albumin - soluble in water.
- globulin - insoluble in water but soluble in dilute acid or salt solution. albuminoids-insoluble in water, salt & solvent.

CONJUGATED PROTEINS

Joined with another compound.

- hydrolysis yields proteins plus the other compound.

1. nucleoprotein - protein + nucleic acid
2. glycoprotein - carbohydrate + proteins
3. phosphoprotein - phosphorus + proteins
4. lipoprotein - lipids + proteins.

DERIVED PROTEINS

those derived by physical forces. e.g. heat.

- meat, cheese, dairy products & some vegetables are prevalent in the human diet, therefore domestic waste will have lots of proteins.
- the proteins in domestic waste can be treated by biological process.

HYDROLYSIS OF PROTEINS

Proteins → proteoses ← peptones → polypeptides

$NH_3 + CO_2 + H_2O \leftarrow \alpha$ amino acids $\leftarrow$ di(tri) peptides

The reverse process takes place in biological synthesis.

Chapter -11

BIOCHEMICAL DESCRIPTION OF AMINO ACIDS

Amino acids are essential components of all living cells as building blocks of proteins. They are also metabolically active and supply substrates for many other biochemical reactions. In a bacterial cell, about 0.4% of the dry weight is amino acids. The amount in other living systems varies according to the age and part of the body.

Chemically, amino acids are carboxylic acids containing at least one amino groups. In protein amino acids, it is always the α-carbon atom, which is aminated and hence they are called ***α-amino acids***. The general structural formula of the amino acids is :

$$\begin{array}{c} \quad H \\ \quad | \\ R - C - COOH \\ \quad | \\ \quad NH_2 \end{array}$$

Where R may -H as in glycine or $-CH_3$ as in alanine or some other group in some other amino acid. Since α carbon atom contains four different groups on atoms (except in the case of glycine) the amino acid exist into two forms i.e. L-form or D-form,

$$\begin{array}{c} \quad NH_2 \\ \quad | \\ R - C - COOH \\ \quad | \\ \quad H \end{array} \qquad \begin{array}{c} \quad H_2 \\ \quad | \\ R - C - COOH \\ \quad | \\ \quad NH_2 \end{array}$$

(L-form) (D-form)

All the amino acids, which occur in proteins belong to L-form. D-amino acids, however, been discovered in bacteria and in antibiotics. Structure of protein amino acids are given below :

STRUCTURE OF PROTEIN AMINO ACIDS

A-Amino acids containing aliphatic hydrocarbon chains :

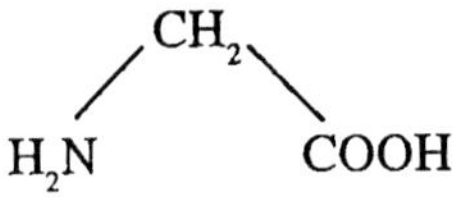

CH_2

H_2N COOH

Glycine

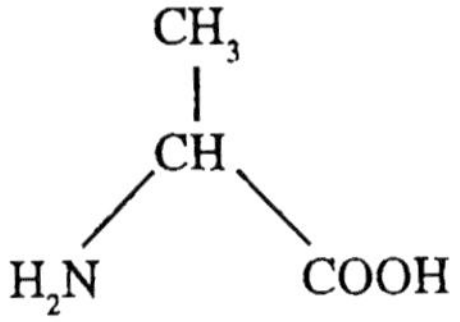

CH_3

CH

H_2N COOH

Alanine

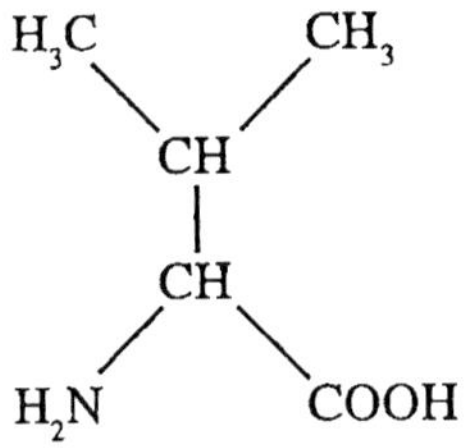

H_3C CH_3

CH

CH

H_2N COOH

Valine

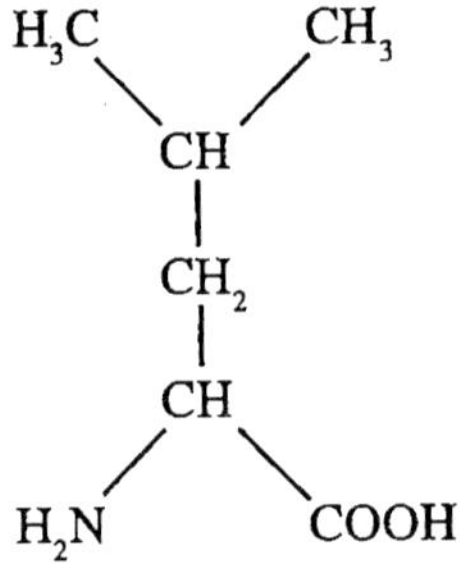

H_3C CH_3

CH

CH_2

CH

H_2N COOH

Leucine

CH_3

CH_2

H_3C-CH

CH

H_2N COOH

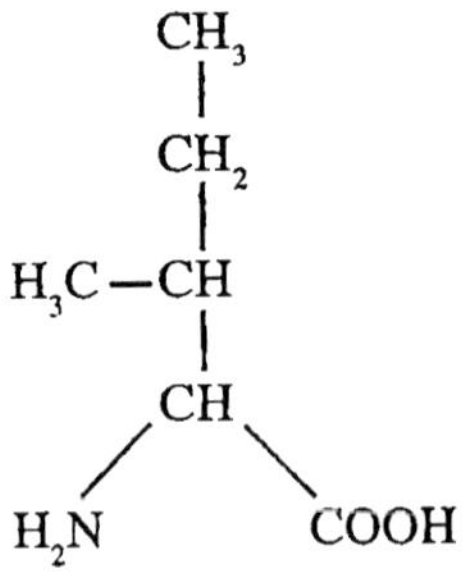

Isoleucine

B- Amino acids containing aromatic side chains :

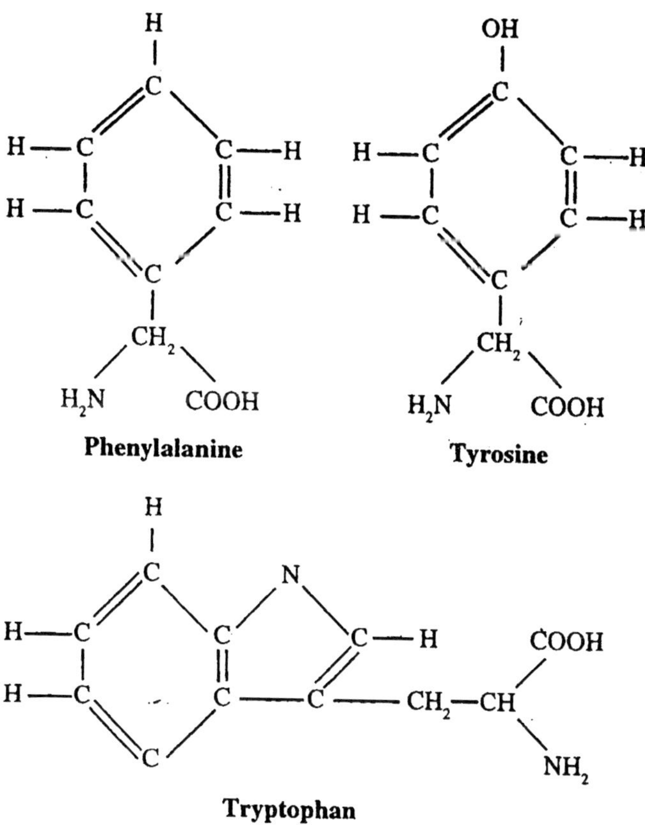

Phenylalanine

Tyrosine

Tryptophan

C. Amino acids containing acidic side chains :

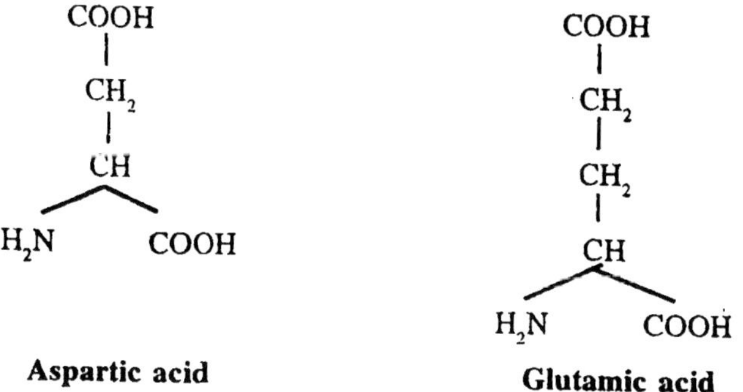

Aspartic acid

Glutamic acid

D-Amino acids containing hydroxyl groups side chains :

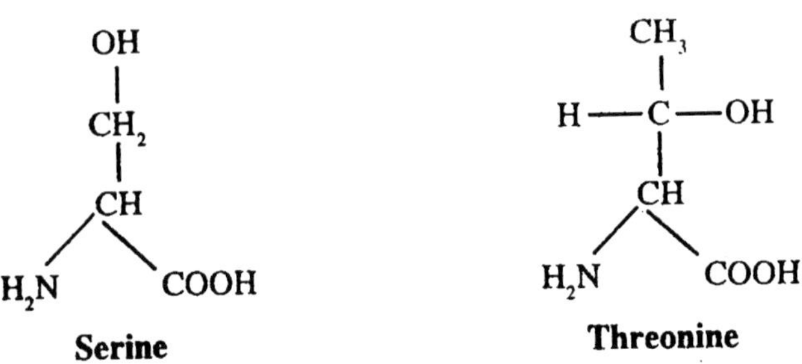

E- Amino acids containing basic side chains :

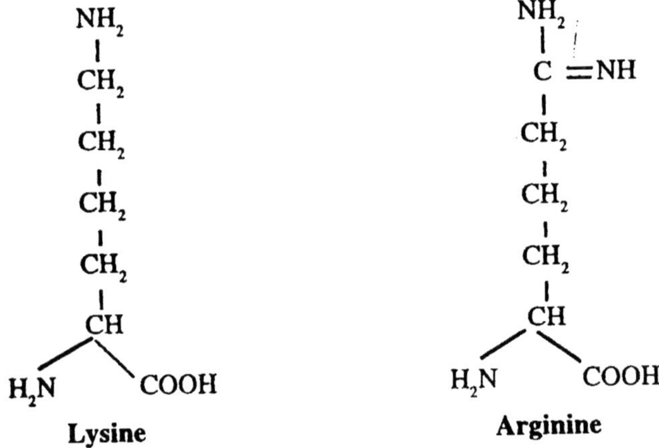

F- Amino acids containting imidazole groups in side chains:

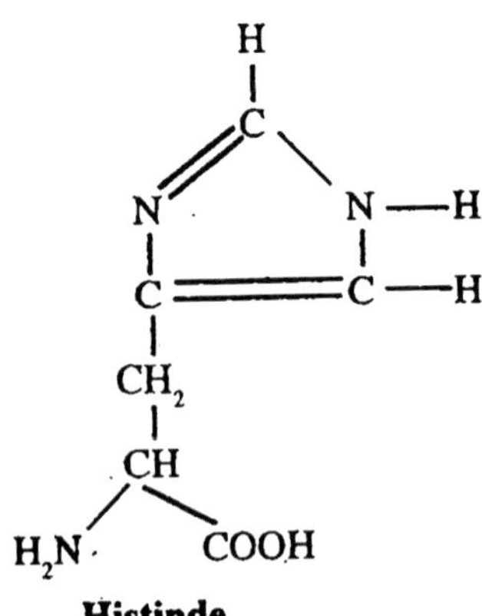

G- Sulphur containing amino acids :

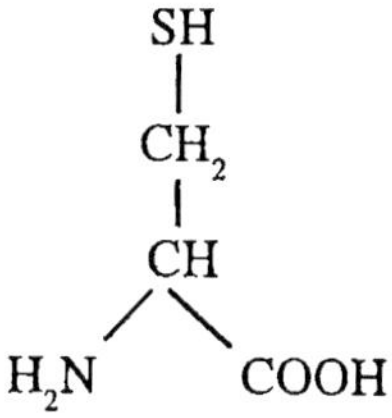

```
     SH              H2N     COOH             CH3
     |                  \   /                   \
     CH2                 C—H                     S
     |                  /                       /
     CH               CH2                     CH2
    /  \                \                      |
 H2N    COOH             S                    CH2
                         |                     |
                         S                     CH
                        /                     /  \
                      CH2                  H2N    COOH
                       |
                       CH
                      /  \
                   H2N    COOH

  Cysteine            Cystine             Methionine
```

Cysteine 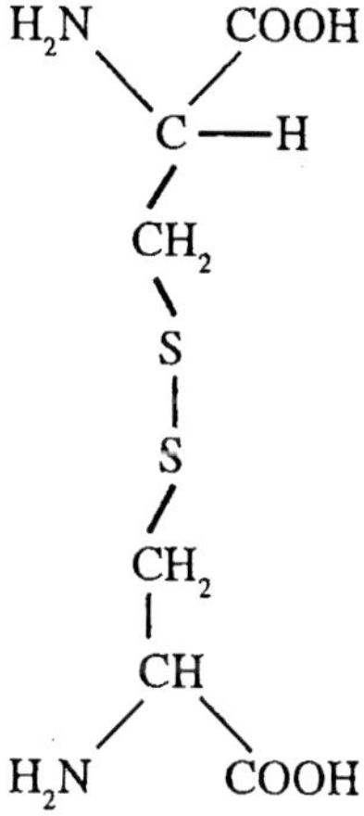 **Cystine** 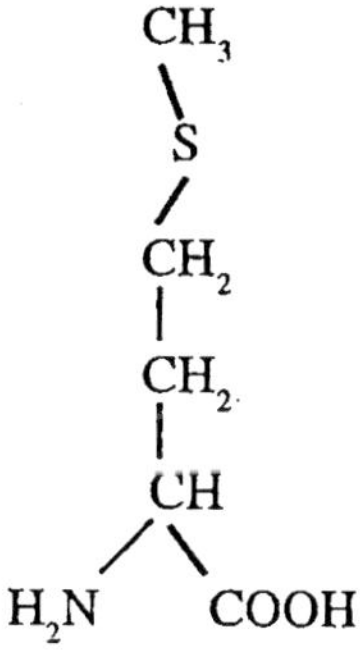 **Methionine**

H- Imino Acids containing Pyrollidine Ring :

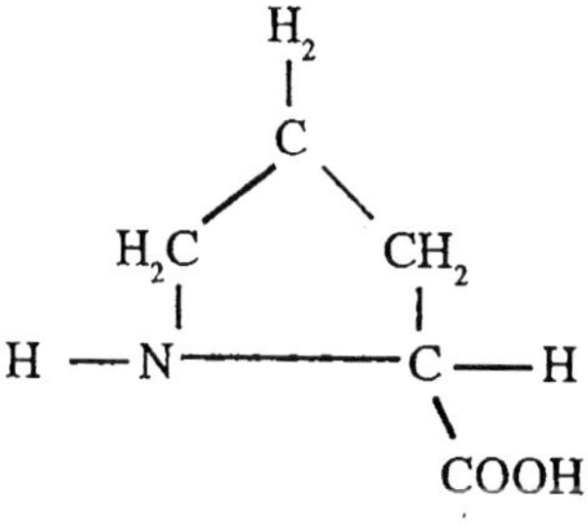

```
          H2                          OH
          |                           |
          C                           CH
        /   \                       /    \
     H2C     CH2                 H2C      CH2
      |       |                   |        |
 H —N————————C—H                 HN———————CH
              \                             \
              COOH                           COOH
```

Proline 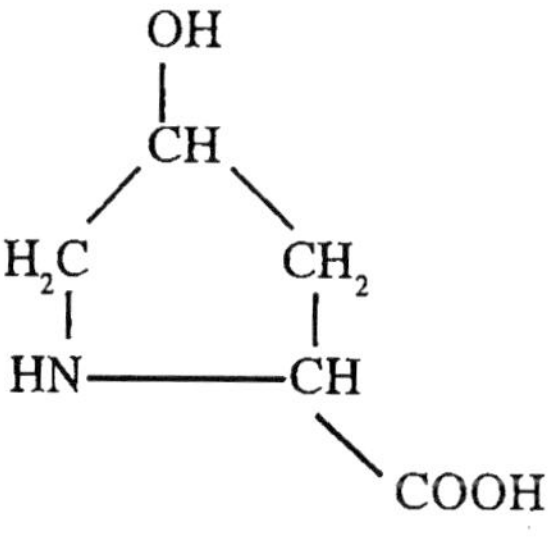 **Hydroxyproline**

GENERAL PROPERTIES OF AMINO ACIDS

Physical Properties

Amino acids are colourless, crystalline solids and soluble in water. The Solubility in water, however, varies for different amino acids. Proline for example, can dissolve up to the extent at 162 ml. per 100 gms. of water at 25°C, where as cysteinie is soluble only to the extent of 0.011 gms per 100 ml. of water. Amino acids are relatively smaller molecules, average m.w. about 120. They are highly polar and have relatively high boiling points.

CHEMICAL PROPERTIES

The chemicals properties of amino acids depend upon their reactive groups -COOH and -NH_2.

AMPHOTERIC NATURE AND ZWITTER ION FORMATION

Since amino acids contain both carboxyl and an amino group, they can react with acids and bases to form salts. Such compounds are called amphoteric compounds or ampholytes. During these reactions, the carboxylic group can either lose a proton (H^+) or amino group can accept a proton. If both of these groups are ionized, however, the solution of amino acid is in the form of zwitter ion (German, ion of both kinds) or dipolar ions.

$$R-\underset{H}{\overset{N^+H_3}{C}}-\overset{O}{\overset{\|}{C}}-OH \underset{}{\overset{-H^+}{\rightleftharpoons}} R-\underset{H}{\overset{N^+H_3}{C}}-\overset{O}{\overset{\|}{C}}-O \underset{-OH^-}{\overset{+H^+}{\rightleftharpoons}} R-\underset{H}{\overset{NH_2}{C}}-\overset{O}{\overset{\|}{C}}-O$$

Basic form Zwitter ion Acid Form

In aqueous solutions of amino acids, only one of these three forms predominates, which is determineded by the pH of the solution and the dissociation constants of COOH and -NH_2 groups. At a certain pH, amino acids is exclusively in Zwitterion and little but equal amount of basic and acid forms. This pH is called isoelectric pH (pl). Isoelectric pH of some amino acids are given in below Table 11.1.

Table 11.1 : Isoelectric pH (pl) of some amino acids

Amino acid	pl	Amino acid	pl
Aspartic acid	2.98	Glycine	6.10
Glutamic acid	3.08	Alanine	6.111
Asparagine	4.30	Proline	6.30
Cystine	5.02	Histidine	7.64
Tryptophan	5.88	Lysine	9.47
Valine	6.00	Arginine	10.76

The zwitter ion solution of amino acids do not migrate to any pole in an electric field. If acid is added to the solution, however, they become positively charged ions and move to the cathode. Addition of alkali deposits negative charges on the amino acid ions and they move to the anode.

AMIDE FORMATION

Amino acids may combine with ammonia to yield unsubstituted amides.

$$\underset{\text{Amino Acid}}{R\text{—}CH\ (NH_2)\ COOH} \xrightleftharpoons{NH_3} \underset{\text{Amide}}{R\text{-}CH\ (NH_2)\ CO\ NH_2} + H_2O$$

In the natural synthesis of amide, the ammonia may be derived from other amino acids. Two amindes, asparagine and glutamine are common in living system. In plants, the amides are stored in the seed as reserved form of nitrogen.

$$H_2N\text{—}CO\text{—}CH_2\text{—}CH_2\text{—}\underset{\underset{NH_2}{|}}{CH}\text{—}COOH$$

Glutamine

$$H_2N\text{—}CO\text{—}CH_2\text{—}\underset{\underset{NH_2}{|}}{CH}\text{—}COOH$$

Asparagine

REACTION WITH NITROUS ACID

Amino acids react with HNO_3 to liberate nitrogen gas from the amino group e.g. leucine reacts with HNO_2 to produce isocaproic acid and nitrogen gas.

$$\begin{matrix} CH_3 \\ CH_3 \end{matrix}\!\!>\!CH\text{—}CH_2\text{—}CH\ (NH_2)\ COOH + HNO_2 \longrightarrow$$

Leucine

$$\begin{matrix} CH_3 \\ CH_3 \end{matrix} \Big\rangle CHCH_2CH\,(OH)\,COOH + N_2 + H_2O$$

α-hydroxy caproic acid

The amino acids proline and hydroxyproline, which do not have a α–amino group do not react with HNO_2. Lysine also reacts very slowly. The N_2 gas formed may be measured manometrically, which is the basis for Van Slyke method for determining amino nitrogen in amino acids and proteins. The amino nitrogen may be calculated for the determination of hydroxy acid also.

NINHYDRIN REACTION

One important reaction of the amino acids is reaction with ninhydrin (triketohydrindene hydrate) which yields coloured products. This reaction is most widely used for quantitative estimation of amino acids. This is an oxidation-reduction reaction in which the amino acid is oxidized to an aldehyde with the release of CO_2 and NH_3. The reduced ninhydrin then, reacts with ammonia and some more ninhydrin to form blue violet compounds.

RCH (NH_2) COOH + [ninhydrin structure: O, OH, O, OH] $\xrightarrow{+H_2O}$ NH_3 ↗, CO_2 ↘

Ninhydrin

[structure: OH, O] —OH + RCHO

Reduced Ninhydrin

[structure: OH, O] —OH + NH_3 + [structure: O, OH, O, OH] ⟶ $3H_2O$ ↗, H^+ ↘

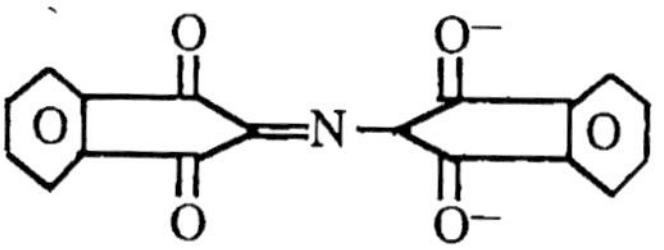

Coloured Product

The intensity of the colour is determined colorimetrically, which is proportional to the amount of amino acid present. This method is used extensively in the determination of acids by paper and column chromatography.

BIOSYNTHESIS OF AMINO ACIDS

Amino acids synthesized from pyruvate:

Serine and glycine

Phosphoglycerate is an intermediate in glycolysis. It is the parent compound for the synthesis of amino acids, serine and glycine.

$$\begin{array}{c} COOH \\ | \\ CHOH \\ | \\ CH_2O.H_2PO_3 \end{array} \xrightleftharpoons{NAD^+ \quad NADH + H^+} \begin{array}{c} COOH \\ | \\ C{=}O \\ | \\ CH_2O\ H_2PO_3 \end{array}$$

3- Phosphoglycerate   3- Phosphohydroxypuruvate

$$\xrightarrow[+NH_3]{\text{Transmination}} \begin{array}{c} COOH \\ | \\ CHNH_2 \\ | \\ CH_2O.H_2PO_3 \end{array}$$

Phosphoserine

↓ Phasphatase

$$\begin{array}{c} HCHO \\ \text{formel-} \\ \text{dihyde} \end{array} + \begin{array}{c} COOH \\ | \\ CH_2\ NH_2 \\ \text{Glycine} \end{array} \xrightleftharpoons[\text{xy methyl transferase}]{\text{Serine hydro-}} \begin{array}{c} COOH \\ | \\ CHNH_2 \\ | \\ CH_2OH \\ \text{Serine} \end{array}$$

Serine to glycine conversion is reversible and in photo synthetic tissue, glycine may be the source of serine biosynthesis.

The plants with C_3 pathway (Calvin Cycle) of photosynthesis produce large amounts of glycolic acid in high light intensities low CO_2 and high O_2 concentration. A part of glycolic acid is converted to glycine with glutamate

OPO$_3$H$_2$ O OPO$_3$H$_2$

H H

HO OH

Ribulose disphosphate

$$\xrightarrow[O_2]{CO_2} \begin{matrix} CHOH \\ | \\ COOH \end{matrix} \xrightarrow{-H} \begin{matrix} CHO \\ | \\ COOH \end{matrix}$$

Glycollic acid   glyoxylic acid

↓ glutamate → ketoglutamate

$$\begin{matrix} CH_2NH_2 \\ | \\ COOH \end{matrix}$$

glycine

Synthesis of glycine by this pathway occurs in special cell organelles called ***peroxisomes***. The glycine, thus produced diffuses out to the cytoplasm or into the mitochondria, where it is hydroxymethylated by methane tetrahydrofolate to form serine.

$$\begin{matrix} CH_2NH_2 \\ | \\ COOH \end{matrix} + 6H_4F\text{–}CH_2\text{–}\ CH_2\text{–}FH_4$$

glycine   5, 10 methylene tetrahydrofolate

$$\xrightarrow[\text{methyl transferase}]{\text{Serine hydroxy}} \begin{matrix} CH_2OH \\ | \\ CHNH_2 \\ | \\ COOH \end{matrix}$$

Serine

Alanine

Reductive amination of pyruvate to form alanine has been observed in several microorganisms.

$$CH_3.CO.COOH + NH_3 \underset{}{\overset{NADH + H^+ \quad NAD^+}{\rightleftharpoons}} CH_3CH\,(NH_2)\,COOH + H_2O$$

Pyruvate Alanine

In other living systems, however, it appears that alanine is synthesized by a transamination reaction.

$$CH_2.CO.COOH + COOHCH_2CH_2CH(NH)_2COOH$$

glutamic acid

$$\rightleftharpoons \begin{array}{c} CH_3 \\ | \\ CH.NH_2 \\ | \\ COOH \end{array} + \begin{array}{c} COOH \\ | \\ CH_2 \\ | \\ CH_2 \\ | \\ COOH \end{array}$$

Alanine

α-ketoglutaric acid

Valine and Leucine

Knowledge about the synthesis of valine and leucine from pyruvate has come out of studies with E. Coli and other microorganisms. An acetaldehyde molecule derived from pyruvate condenses with another molecule of pyruvate to yield α-acetolactate in the first step.

$$2CH_3\,CO.COOH \xrightarrow[\text{synthatase}]{\alpha\text{-aceto, }\alpha\text{-hydroxy acid}}$$

$$CH_3CO - \overset{\displaystyle OH}{\underset{\displaystyle CH_3}{\overset{|}{\underset{|}{C}}}} - COOH + CO_2$$

α-acetolactate

The enzyme reductoisomerase rearranges the product molecule involving the reduction of carboxyl group.

$$CH_3 - CO - \underset{CH_3}{\overset{OH}{C}} - COOH \xrightarrow{\text{reductoisomerase}} CH_3 - \underset{CH_3}{\overset{OH}{C}} - CHOH - COOH$$

α, β dihydroxy
β methyl Butarate

Ketovaline is a precursor for both valine and leucine. When transaminated, it produces valine.

$$\begin{matrix} CH_3 \\ CH_3 \end{matrix} \rangle CH.CO.COOH \xrightarrow[NH_3]{\text{Transaminase}} \begin{matrix} CH_3 \\ CH_3 \end{matrix} \rangle CH.\underset{NH_2}{CH}.COOH$$

α-keto β-dimethyl butaric acid or ketovaline — valine

In the synthesis of leucine, α-keto, β-methyl-butyric acid (ketovaline) is acylated first by a molecule of acetyl Co A to yield isopropyl malate.

$$\begin{matrix} CH_3 \\ CH_3 \end{matrix} \rangle CH.OH.COOH + CH_3.CO.S.COA. \xrightarrow[\text{Synthatase}]{\text{Isopropyl malate}}$$

$$\begin{matrix} CH_3 \\ CH_3 \end{matrix} \rangle CH - \underset{CH_2COOH}{\overset{OH}{CO}} - COOH + CoASH$$

α-isopropyl malate

α-isopropyl malate is then rearranged to β-isopropyl malate. This rearrangement actually proceeds into two steps. In the first step, the molecule is dehydrated and in the next, it is hydrated again.

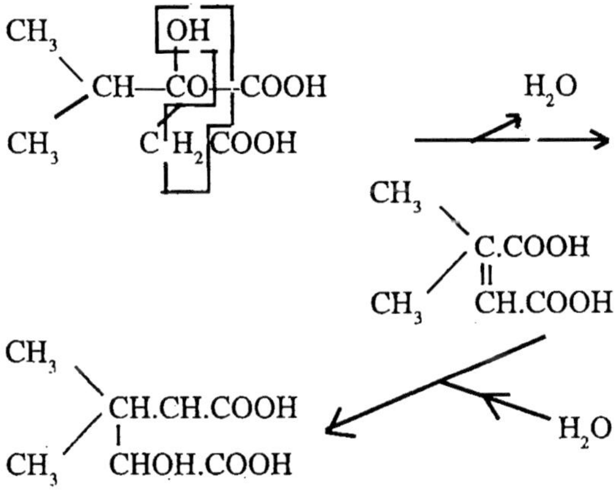

$$(CH_3)_2CH\text{—}C(OH)(COOH)\text{—}CH_2COOH \xrightarrow{-H_2O} (CH_3)_2C.COOH{=}CH.COOH \xrightarrow{+H_2O} (CH_3)_2CH.CH(COOH).CHOH.COOH$$

β-isopropyl malate

β-isopropyl malate undergoes oxidative decarboxylation to yield α-ketoisocaproate (ketoleucine) which is, then transaminated to produce leucine.

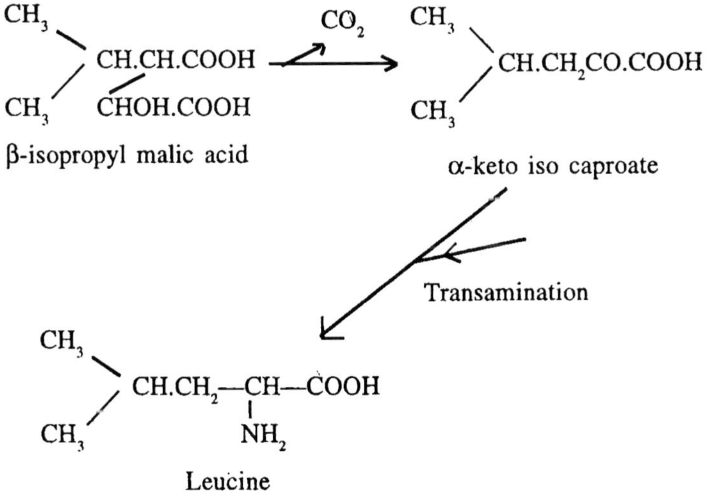

$$(CH_3)_2CH.CH(COOH).CHOH.COOH \xrightarrow{-CO_2} (CH_3)_2CH.CH_2CO.COOH$$

β-isopropyl malic acid → α-keto iso caproate

Transamination

$$(CH_3)_2CH.CH_2\text{—}CH(NH_2)\text{—}COOH$$

Leucine

Isoleucine

Synthesis of isoleucine from pyruvate involves another amino acid threonine. Threonine is dominated first by threonine deaminase to produce α-ketobutyrinc acid.

$$\underset{\text{Threonine}}{CH_3.CHOH.\underset{\displaystyle NH_2}{\underset{|}{C}}H.COOH} \xrightarrow{\text{threonine deaminase}} \underset{\text{α-ketobutyric acid}}{CH_3.CH_2.CO.COOH} + NH_3$$

α-ketobutyric acid condenses with a molecule of acetaldehyde derived from pyruvate to yield α-aceto, α-hydroxybutyrinc, acid. The enzyme, α-aceto, α-hydroxy-synthatase is the same which catalyses the synthesis of α-acetolactate from pyruvate in value biosynthetic pathway.

$$CH_3.CH_2.CO.COOH \xrightarrow[\text{α-aceto-α-hydroxy acid synthatase}]{CH_3CHO} CH_3.CO—\overset{OH}{\overset{|}{C}}(CH_2CH_3)—COOH$$

α-aceto α-hydroxy butyric aci(

α-aceto, α–hydroxy butyric acid is rearranged to dihydroxy isoleucine which is then degraded to ketoisoleucine.

$$CH_3.CO—\overset{OH}{\overset{|}{C}}(CH_2.CH_3)—COOH \longrightarrow CH_3—\overset{OH}{\overset{|}{C}}(CH_2.CH_3).C\boxed{H.OH}COOH \xrightarrow{H_2O\uparrow}$$

α-aceto, α-dhyroxy-butyric acid dihydroxy isoliecine

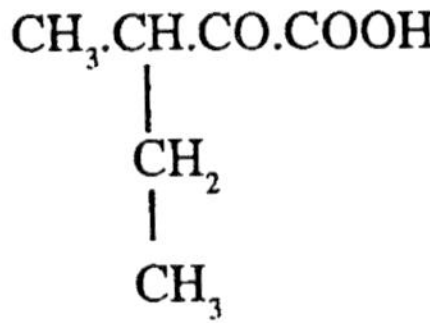

Ketoisoleucine

Ketoisoleucine is then transaminated to produce isoleucine.

$$CH_3.CH(CH_2.CH_3).CO.COOH \xrightarrow[\text{Transamination}]{NH_3} CH_3.CH(CH_2.CH_3).CH(NH_2).COOH$$

L-isoleucine

Reactions leading to the synthesis of amino acids derived from pyruvate are summarized in fig. 11.1.

GLUTAMIC ACID

Glutamic acid is the first amino acid synthesized from α-ketoglutarate. It is synthesized either by reductive amination with NH_4^+ or by transamination with aspartic or other amino donating amino group.

$$COOH.CH_2.CH_2.CO.COOH \xrightleftharpoons[\text{glutamic dehydrogenase}]{NH_3,\ NADPH + H^+}$$

α-ketoglutaric acid

$$COOH.CH_2.CH_2.CH(NH_2).COOH + NADP^+ + H_2O$$

glutamate or glutamic acid

It can also be written in the following form :

$$COOH.CH_2.CO.COOH + R\ CH(NH_2)\ COOH \xrightleftharpoons{\text{Transaminase}}$$

$$COOH\ CH_2\ CH_2.CH(NH_2).COOH + RCO.COOH$$

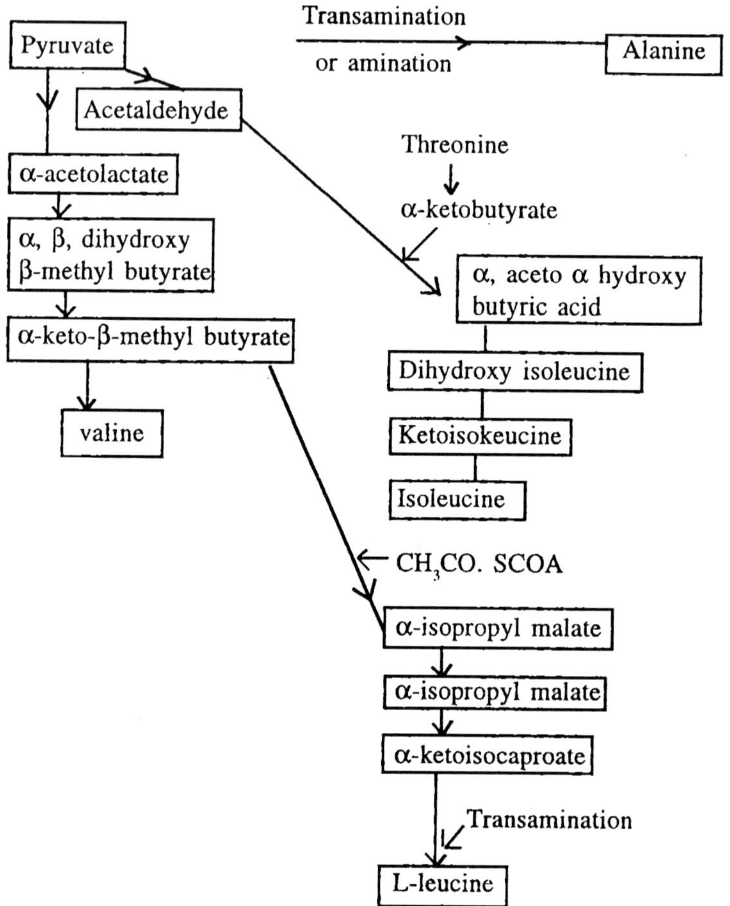

Fig. 11.1 : Summary of Pyruvate Family Amino Acid Synthesis

Synthesis of glutamic acid by reductive amination is common in plants growing in soil. They absorb inorganic nitrogen from the soil either in the form of NO_3^- or NH_4^+. Nitrate is reduced to ammonia via nitrite and possibly hydroxylamine. The ammonia then enters into amino acid and protein cycle thorough glutamic acid. The organisms which are deficient in enzyme glutamic dehydrogenase can grow only when they are supplied with an amino acid capable of donating its amino group during transamination. Recently it has been shown, however, that ammonia is incorporated primarily to glutamine instead of glutamic acid. Glutamine can donate ammonia to a-ketoglutarate and glutamic acid can be synthesized later on.

Glutamic acid synthesized may be incorporated into proteins. It is also the source of Carbon atoms, of proline and arginine and in some organisms like fungi of lysine.

PROLINE

In the proline synthesis, glutamic acid is reduced to an aldehyde first.

$$\underset{\underset{NH_2}{|}}{COOH.CH_2.CH_2.CH.COOH} \rightleftharpoons \underset{\underset{NH_2}{|}}{CHOCH_2CH_2CHCOOH}$$

Glutamyl-y-semi-aldehyde

This compound in aqueous solution is in equilibrium with a cyclic compound Δ^1 proline carboxylate, which is then reduced to proline.

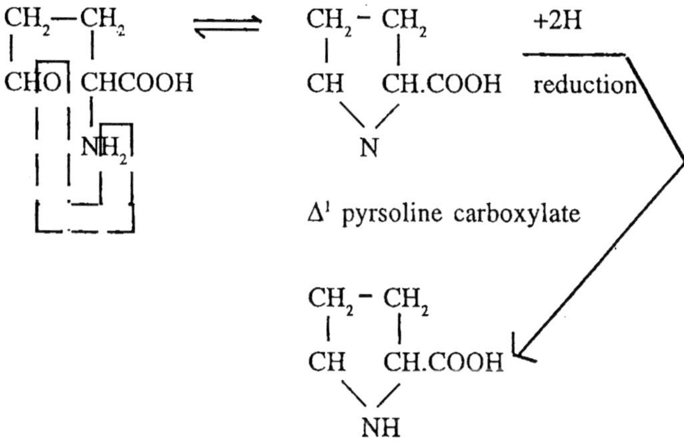

Proline

These reactions have been demonstrated in many organisms and also in pea leaves.

ARGININE

It is synthesized from glutamic acid through acylatation and transamination processes. In the first step, a specific acelylase catalyses the conversion of glutamic acid to N-acetyl glutamic acid which is, then reduced to N-acetyl glutamate semialdehyde.

$$COOH.CH_2.CH_2\underset{\displaystyle NH_2}{\underset{|}{C}}H\text{—}COOH \xrightarrow{CH_3CoA} COOH\ CH_2CH_2\underset{\displaystyle NHCO_3CH_3}{\underset{/}{C}}H\ COOH$$

N-acetyl glutamate

$$\Big\downarrow \text{reduction} \quad \rightarrow -H$$

$$CHOCH_2CH_2\underset{\displaystyle NHCOCH_3}{\underset{|}{C}}HCOOH$$

N-acetyl glutamate semialdehyde

Acetylation of this aldehyde at $-NH_2$ group prevents it from cyclization (as seen in the synthesis of proline). The N-acetyl glutamate semialdehyde is transaminated to yield N-acetyl-ornithine.

$$CHOCH_2CH_2.\underset{\displaystyle NHCOCH_3}{\underset{/}{C}}H.COOH + R\underset{\displaystyle NH_2}{\underset{|}{C}}HCOOH \quad \xrightleftharpoons{\text{Transamination}}$$

N acetyl glutamate semialdehyde

$$NH_2.CH_2.CH_2.CH_2.\underset{\displaystyle NHCOCH_3}{\underset{\backslash}{C}}H.COOH \quad + \quad RCOCOOH$$

N-acetyl ornithine

The amino acid acetyl ornithine is well distributed in plants. In some organisms, N-acetyl group is removed by specific deacylase, but in others it is transferred gain to the amino group of glutamate. Deacylation produces L-ornithine which is in turn converted to L-citrulline, utilising one molecule of carbamyl phosphate.

$$NH_2CH_2CH_2CH_2\underset{\displaystyle NHCOCH_3}{\underset{/}{C}}HCOOH \xrightarrow{CH_3COOH} NH_2CH_2CH_2CH_2\underset{\displaystyle NH_2}{\underset{\backslash}{C}}HCOOH$$

L-ornithine

$$NH_2CH_2CH_2CH_2CHCOOH \quad + \quad NH_2COO \sim PO_3H_2$$

NH_2 (on the CH carbon)

Carbamyl phosphate

$$\xrightarrow[\text{transcarbamylase}]{\text{ortithine}} NH_2CONHCH_2CH_2CH_2CHCOOH + H_2O$$

NH_2 (on the CH carbon)

L-citrulline

Synthesis of argnine from citrulline involves aspartate as amino donor and argininosuccinate as an intermediate product. Energy for the synthesis of argininosuccinate is derived from ATP. Argininosuccinate is then hydrolysed to arginine and fumaric acid by the enzyme argininosuccinase.

The presence of this enzyme and the one which synthesizes argininosuccinate has ben demonstrated in microorganisms and plants.

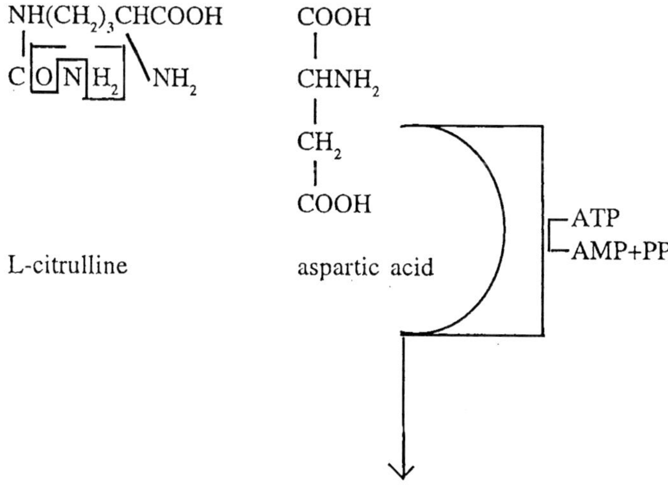

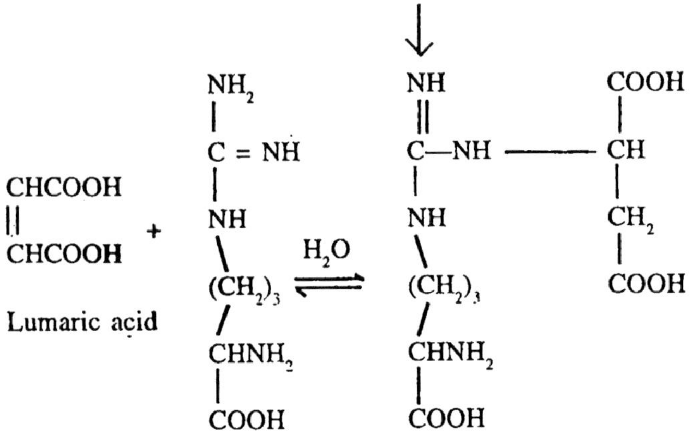

$$\begin{array}{ccccc} & & & & \downarrow \\ & & NH_2 & & NH & & COOH \\ & & | & & \| & & | \\ & & C=NH & & C—NH & —— & CH \\ CHCOOH & & | & & | & & | \\ \| & + & NH & & NH & & CH_2 \\ CHCOOH & & \backslash & H_2O & \backslash & & | \\ & & (CH_2)_3 & \rightleftharpoons & (CH_2)_3 & & COOH \\ \text{Lumaric acid} & & / & & / \\ & & CHNH_2 & & CHNH_2 \\ & & | & & | \\ & & COOH & & COOH \end{array}$$

L-arginine Arginosuccinate

LYSINE :

Synthesis of lysine in some fungi involves α-ketoglutaric acid. This acid is condensed with acetate to produce homocitrate, which is then isomerized to homoisocitrate, Homoisocitrate is then oxidized and decarboxylated to form α-ketoadipic acid.

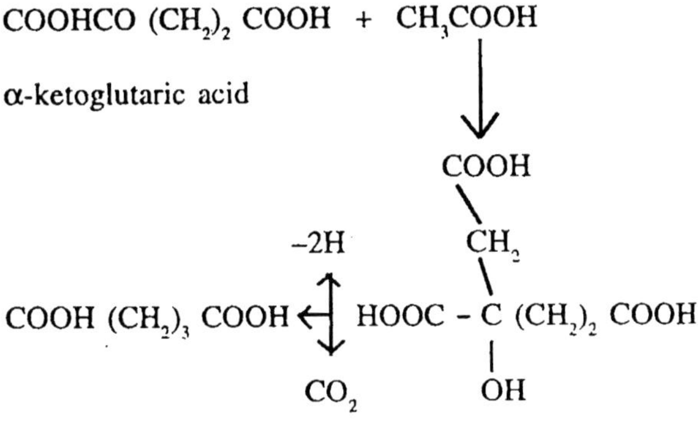

$$COOHCO\ (CH_2)_2\ COOH\ +\ CH_3COOH$$

α-ketoglutaric acid

$$\downarrow$$

$$\begin{array}{ccc} & & COOH \\ & & \backslash \\ & -2H & CH_2 \\ & \uparrow & \backslash \\ COOH\ (CH_2)_3\ COOH & \leftarrow & HOOC - C\ (CH_2)_2\ COOH \\ & \downarrow & | \\ & CO_2 & OH \end{array}$$

α-keoadipic acid homocitrate

α-ketoadipic acid is the homologue of α-ketoglutaric acid. It is transamintated to α-amino adipic acid, which is then reduced to α-amino adipate semialdehyde. This reduction is similar to the reduction of glutamic acid.

$$COOH\ (CH_2)_3\ COCOOH \xrightarrow[\text{Transamination}]{+\ RCH\ (NH_2)\ COOH}$$

$$COOH\ (CH_2)_3CH\ (NH_2)\ COOH + RCO\ COOH$$

α- amino adipic acid

↓ reduction

$$CHO\ (CH_2)_3\ CH\ NH_2.COOH$$

α-amino adipate semialdehyde

This semialdehyde condenses with a new molecule of glutamate to produce saccharopine, which is then converted to lysinc

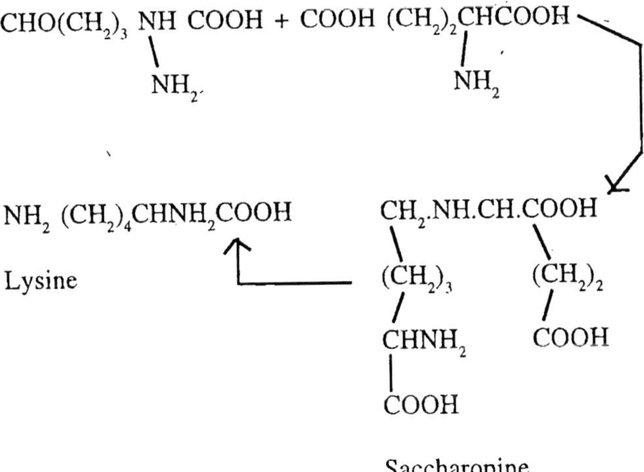

Biosynthesis routes of amino acids derived from α-ketoglutaric acid are summarised in the figure 11.2.

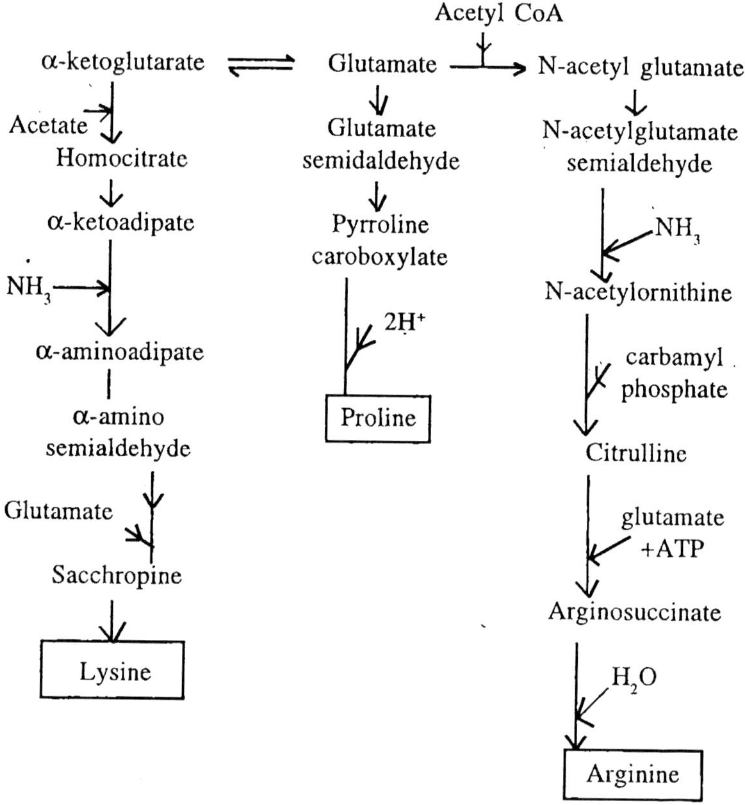

Fig. 11.2 : Synthesis of glutamate family of amino acids

AMINO ACIDS DERIVED FROM OXALACETATE

Aspartic acid or aspartate

Transamination of oxalacetic acid with glutamic acid produces aspartic acid.

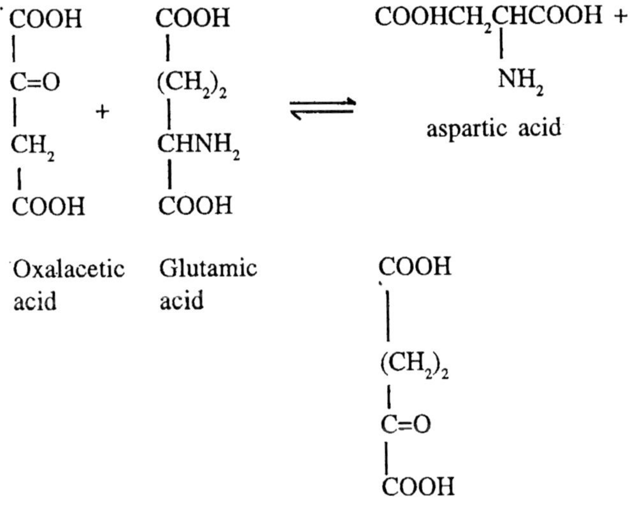

Another intermediate of Kreb's cycle, the fumarate, may also produce aspartable by direct amination.

$$\begin{array}{l} COOH \\ | \\ CH \\ || \\ CH \\ | \\ COOH \end{array} + NH_3 \rightleftharpoons \begin{array}{l} COOH \\ | \\ CH_2 \\ | \\ CHNH_2 \\ | \\ COOH \end{array}$$

Fumaric acid Aspartic acid

The enzyme aspartase catalysing this reaction is found in E.coli and other microorganisms. This reaction, is readily reversible and deamination of aspartate to furmarate occurs frequently.

METHIONINE

Aspartic acid is converted to β-aspartyl semialdehyde via its phosphate.

$$COOHCH_2CHNH_2COOH + ATP \xrightleftharpoons{\text{B-aspartokinase}}$$

$$PO_3H_2COOCH_2CHNH_2.COOH + ADP$$

β-aspartyl phosphate

$$PO_3H_2COOCH_2CHNH_2COOH \rightleftharpoons CHOCH_2CHNH_2COOH + iP$$

β-aspartyl semialdehyde

β-aspartyl semiadehyde is the common intermediate in the synthesis of methionine, threonine and isoleucine. Homoserine is the branching point for the synthesis of these amino acids. Homoserine is synthesized by the reduction of aspartyl phosphate by the enzyme-homo-serine dehydrogenase. The dehydrogenase can use either NADH or NADPH as co-enzyme.

$$CHOCH_2CHNH_2COOH + NADPH + H^+ \rightleftharpoons$$

$$CH_2OHCH_2CHNH_2COOH + NADP^+$$

Homoserine

$$CH_2\boxed{OH}CH_2CHNH_2COOH + \underset{\substack{|\\ CH_2COOH}}{CH_2COO}\boxed{H}$$

Succinic acid

$$COOH\,(CH_2)_2COO(CH_2)_2\underset{\substack{|\\ NH_2}}{CHCOOH} \quad + H_2O$$

O-succinyl homoserine

These products react with cysteine to produce cystathionine, which is then hydrolysed to homocysteine, pyruvate and ammonia.

$$COOH\,(CH_2)\,COO(CH_2)_2\underset{\substack{|\\ NH_2}}{CHCOOH} \quad \xrightarrow{COOHCH\,(NH_2)\,CH_2SH}$$

O-succinyl homoserine

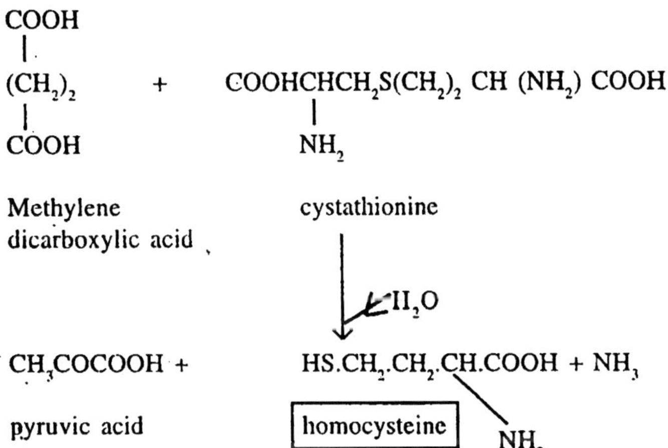

As in higher plants, homocysteine is methylated at -SH end and the product of the reaction is methionine. The methyl donor is another amino acid serine.

$$HSCH_2CH_2CH\ (NGH_2)\ COOH + \begin{array}{c} CH_2OH \\ | \\ CHNH_2 \\ | \\ COOH \end{array} \longrightarrow$$

Serine

$$\begin{array}{l} CH_3SH \\ | \\ CH_2 \\ | \\ CH_2 \\ | \\ CHNH_2 \\ | \\ COOH \end{array}$$

Methionine

THREONINE AND ISOLEUCINE

Homoserine derived from aspartate through aspartylsemialdehyde is the precursor for threonine also. In fact, threonine is an isomer of homoserine in which alcohol group is in secondary position.

$$\begin{array}{ccc} COOH & & COOH \\ | & & | \\ CHNH_2 & & CHNH_2 \\ | & \longrightarrow & | \\ CH_2 & & CHOH \\ | & & | \\ CH_2OH & & CH_3 \end{array}$$

Homoserine Threonine

The conversion of homoserine to threonine however, involves two steps. In the first step, the -OH group of homoserine is phosphorylated in the presence of enzyme homoserine kinase. Then this product is reacted upon by the enzyme homerine phosphate metaphosphatase or o-phospholomoserine lyase, which transfers alcohol group to the secondary position. This enzyme has also been called as threonine synthatase, since the product is L-threonine.

$$CH_2OHCH_2CHNH_2COOH + ATP \xrightarrow[\text{Kinase}]{\text{Homoserine}}$$

$$H_2PO_3OCH_2CH_2CHNH_2COOH + ADP$$

Homoserine phosphate

$$\downarrow \text{Homoserine phosphate metaphosphatase}$$

$$CH_3CHOH\ CHNH_2COOH + iP$$

Thereonine

In microorganisms, isoleucine synthesized from threonine, threonine is first deaminated by biosynthetic threonine deaminase to produce α - acid condenses with acetaldehyde derived from pyruvate to yield ultimately isoleucine after undergoing a series of reactions, which were mentioned earlier. Amino acids derived from oxalacetate are described in fig. 11.3.

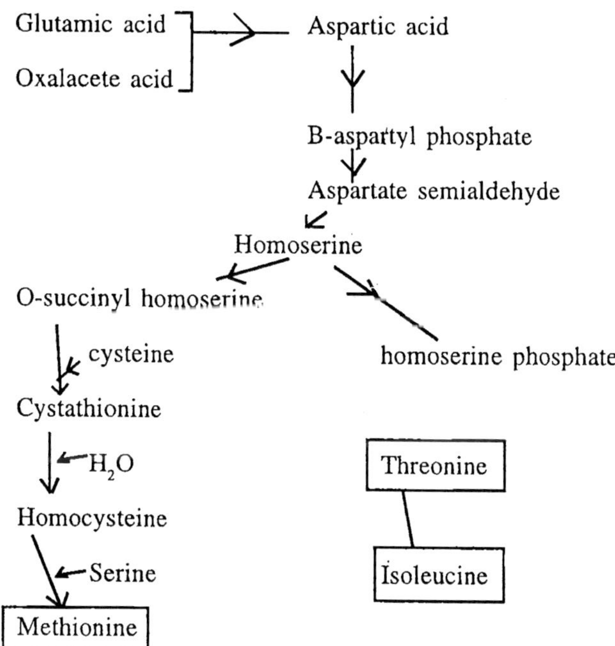

Fig. 11.3 : Summary of amino acids derived from oxalacetate

AROMATIC AMINO ACIDS

Chriosmic acid

Our knowledge of the biosynthesis of aromatic amino acids comes from investigations with microorganisms. Shikimic acid, which has been known to occur in plants as a rare chemical, is the precursor of tyrosine, phenylalanine, tryptophane and many other oromatic compounds.

Shikimic acid is synthesized from two common compounds of glucose degradation, erythrose 4-phosphate and phosphoenol pyruvate. A specific aldolase condenses them together to form 7-phospho 3-deoxy D-arabino heptulosic acid (PDAH).

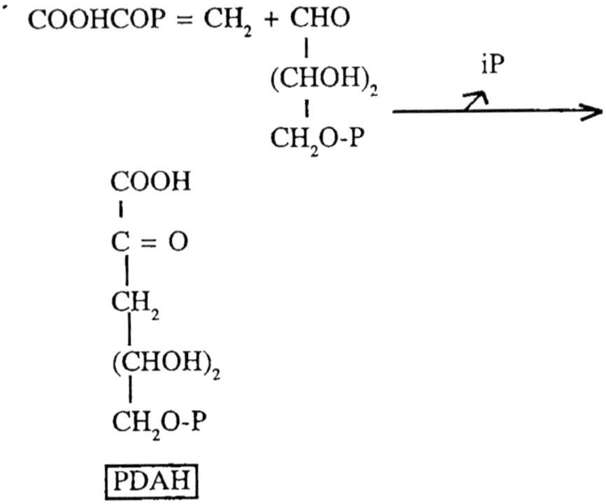

$$COOHCOP = CH_2 + \begin{array}{c} CHO \\ | \\ (CHOH)_2 \\ | \\ CH_2O\text{-}P \end{array} \xrightarrow{\nearrow iP}$$

$$\begin{array}{l} COOH \\ | \\ C = O \\ | \\ CH_2 \\ | \\ (CHOH)_2 \\ | \\ CH_2O\text{-}P \end{array}$$

PDAH

In the presence of NADH and Co^{++}, PDAH is reduced to 5-dehydroquinic acid by one or more enzymes. Dehydroquinic acid is a cyclic compound but has not double bonds and hence no aromatic properties.

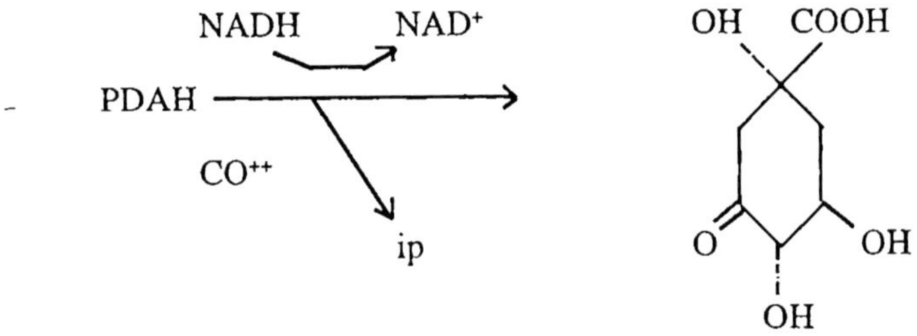

PDAH —(NADH → NAD^+; CO^{++}; → ip)→ [ring structure: OH, COOH, O, OH, OH]

5-dehydroquinic acid

A water molecule is eliminated from dehydroquinic acid to form dehydroshikimic acid which is then reduced with NADPH to yield ultimately shikmic acid.

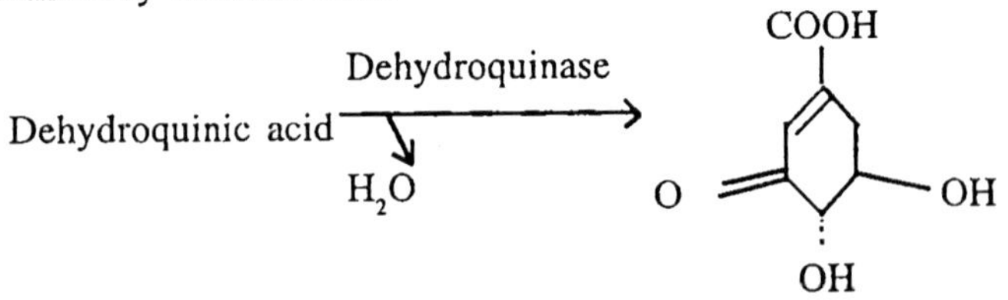

Dehydroquinic acid —(Dehydroquinase; → H_2O)→ [ring structure: COOH, O, OH, OH]

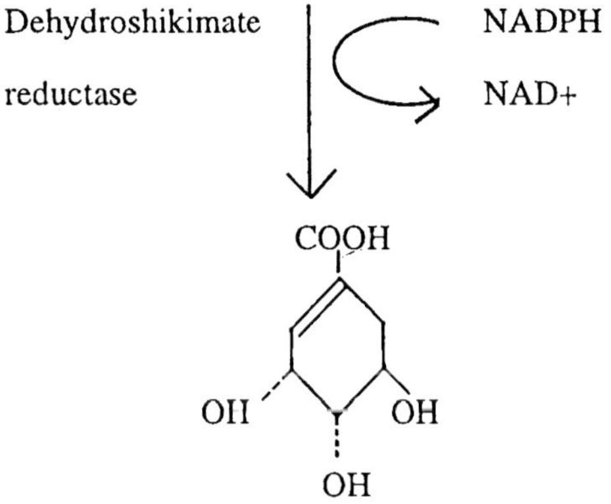

Dehydroshikimate reductase

NADPH → NAD+

COOH

OH OH

OH

Shikimic acid

In the synthesis of chrismic acid from shikimic acid, the later is phosphorylated first to produce phosphoshikimic acid. Phosphoshikimate then reacts with one molecule of phosphoenol pyruvate to yield 3 - enol pyruvylshikimic acid 5 phosphate. It is then dephosphorylated to yield chorismic acid.

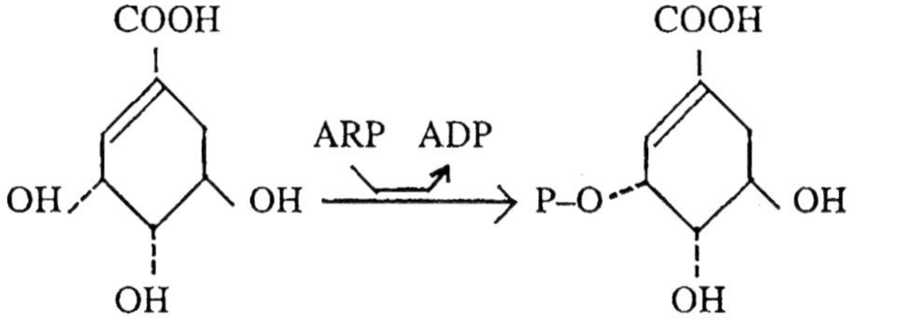

COOH

OH OH

OH

ARP ADP

COOH

P–O OH

OH

shikimic acid

5-phosphoshikimic acid

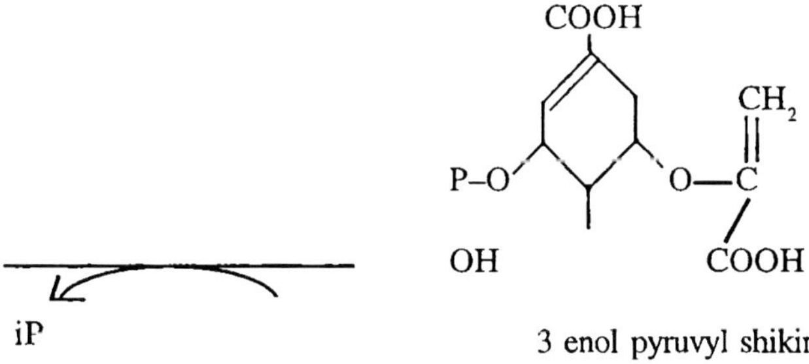

COOH

CH_2

P–O O—C

OH COOH

iP

3 enol pyruvyl shikimic acid 5 phosphate

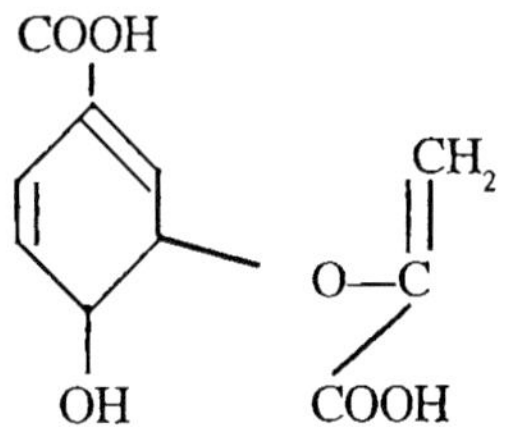

COOH

OH

CH_2

O—C

COOH

chorismic acid

The name chorismic acid is derived from a Greek word meaning separation. This is the product where biosynthetic pathways of amino acids diverge.

PHENYLALANINE AND TYROSINE

Chorismic acid is converted to prephenic acid by the rearrangement of side groups. This reaction is catalysed by the enzyme chorismate mutase.

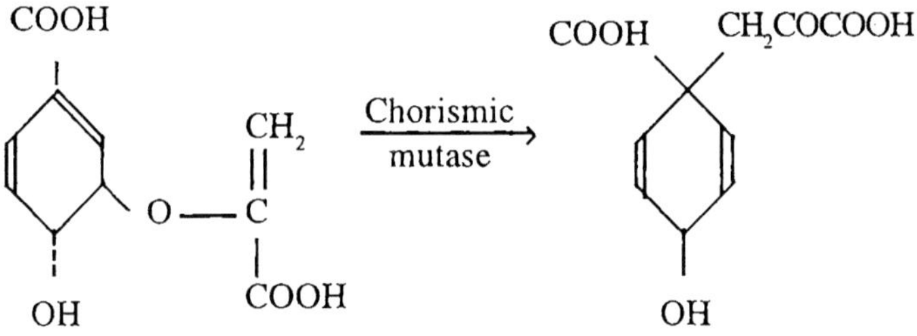

COOH CH_2 O — C COOH OH

Chorismic mutase →

COOH $CH_2COCOOH$ OH

chorismic acid prephenic acid

Prephenic acid is the common precursor for both phenylalanine and tyrosine. In the synthesis of phenylalanine, it is decarboxylated and dehydrated to a compound called phenylpyruvic acid. Phenyl pyruvic acid after transmission produces phenylalanine.

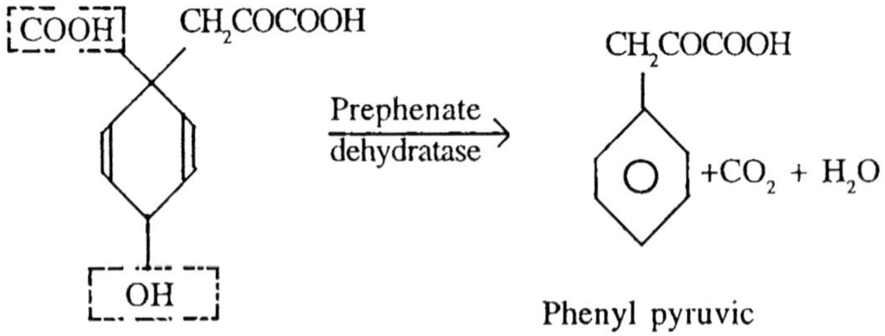

COOH $CH_2COCOOH$ OH

Prephenic acid

Prephenate dehydratase →

$CH_2COCOOH$ $+CO_2 + H_2O$

Phenyl pyruvic acid

For tyrosine synehtsis, prephenic acid undergoes oxidative decarboxylation first in the presence of enzyme prephenate dehydrogenase and hydroxyphenyl pyruvic acid is produced. This compound in then transaminated to tyrosine.

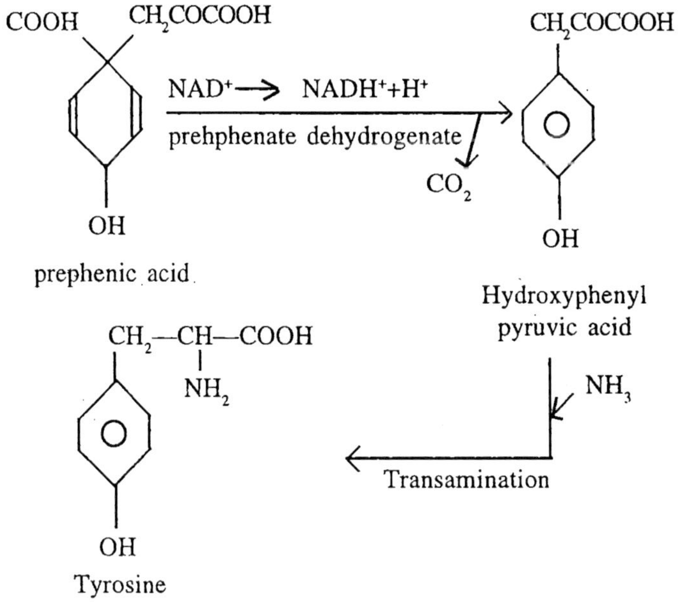

COOH $CH_2COCOOH$

$NAD^+ \rightarrow NADH^+ + H^+$

prehphenate dehydrogenate

CO_2

OH

prephenic acid

$CH_2COCOOH$

OH

Hydroxyphenyl pyruvic acid

NH_3

Transamination

CH_2—CH—COOH

NH_2

OH

Tyrosine

TRYPTOPHAN

Synthesis of tryptophan from chrismate starts with the reaction with glutamine. The product of the reaction is anthranillic acid and the enzyme, anthrinilate synthatase.

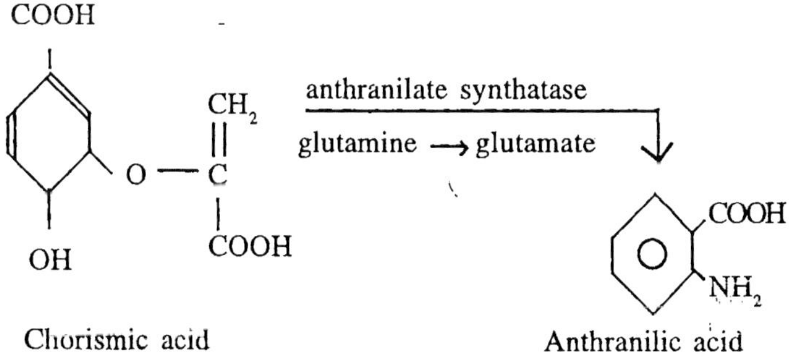

COOH

CH_2

O — C

COOH

OH

Chorismic acid

anthranilate synthatase

glutamine → glutamate

COOH

NH_2

Anthranilic acid

Anthranilic acid reacts with phosphoribosyl pyrosphosphate (PRPP) to produce anthrinilate phosphoribosyl transferase. In the next step it is decaroboxylated with the formation of side ring also to produce indoleglycerol phosphate.

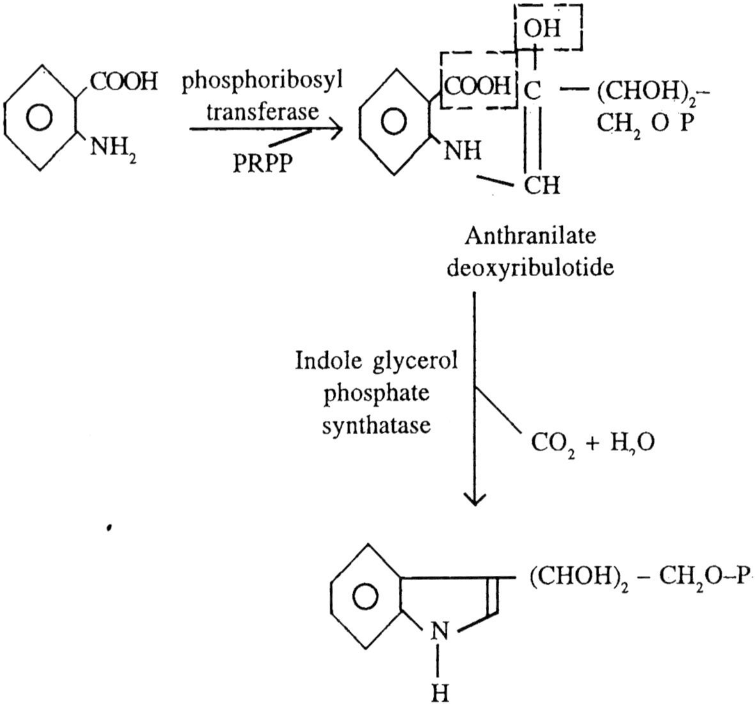

COOH
NH$_2$
phosphoribosyl transferase
PRPP
OH
COOH C — $(CHOH)_2$–
CH_2 O P
NH
CH

Anthranilate deoxyribulotide

Indole glycerol phosphate synthatase

$CO_2 + H_2O$

$(CHOH)_2$ - CH_2O–P
N
H

Indole-glycerol phosphate

Synthesis of tryptophan from indoleglycerol phosphate has been demonstrated in E.coli, Neurospora and higher plants. The enxyme tryptophan synthatase has been crystalized from bacteria and the fungi. Serine condenses with indole glycerol phosphate in the presence of pyridoxal phosphate to produce tryptophan (Fig. 11.4).

BIOSYNTHESIS OF HISTIDINE

Histidine is a heterocyclic amino acid with a unique biosynthetic pathway derived from ATP and phosphoribosyl pyrophosphate (PRPP). The pathway has been worked out in microorganisms but no

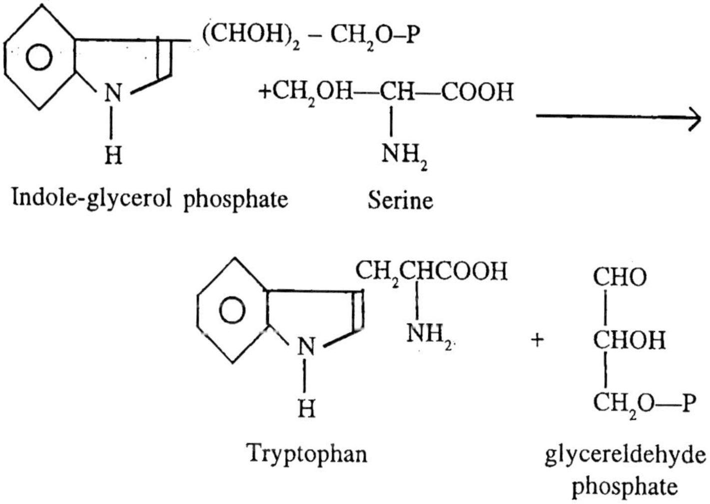

(Indole ring)—$(CHOH)_2 - CH_2O{-}P$ + $CH_2OH{-}CH(NH_2){-}COOH$ ⟶

Indole-glycerol phosphate Serine

(Indole ring)—$CH_2CH(NH_2)COOH$ + $CHO{-}CHOH{-}CH_2O{-}P$

Tryptophan glycereldehyde phosphate

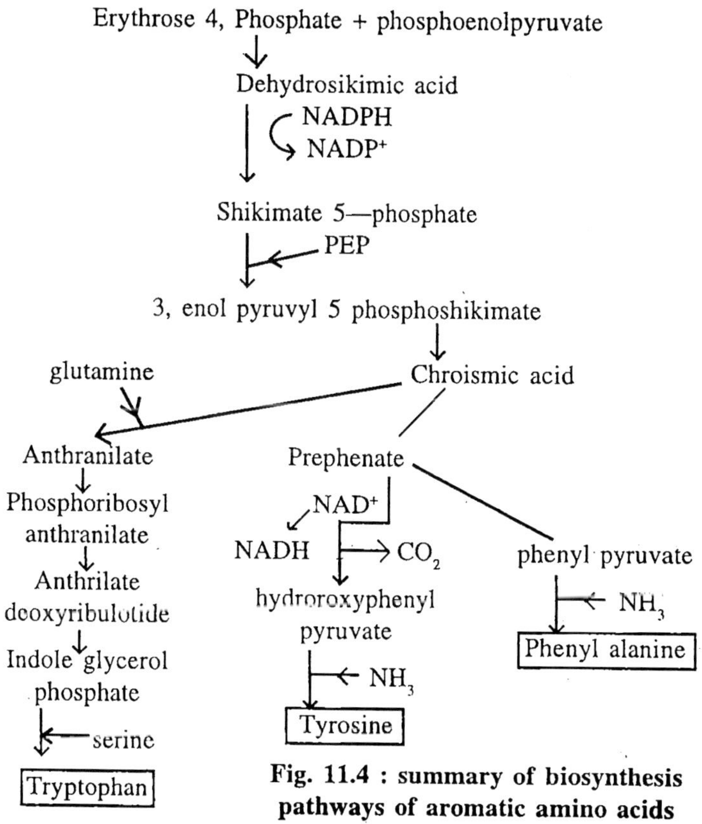

Erythrose 4, Phosphate + phosphoenolpyruvate

↓

Dehydrosikimic acid

↓ (NADPH → $NADP^+$)

Shikimate 5—phosphate

↓ ← PEP

3, enol pyruvyl 5 phosphoshikimate

↓

Chroismic acid

Chroismic acid → (glutamine) → Anthranilate → Phosphoribosyl anthranilate → Anthrilate deoxyribulotide → Indole glycerol phosphate → (← serine) → **Tryptophan**

Chroismic acid → Prephenate → (NAD^+ → NADH; → CO_2) → hydroroxyphenyl pyruvate → (← NH_3) → **Tyrosine**

Prephenate → phenyl pyruvate → (← NH_3) → **Phenyl alanine**

Fig. 11.4 : summary of biosynthesis pathways of aromatic amino acids

information about histidine synthesis in higher plants is available. The pathway (fig. 11.5) is as follows:

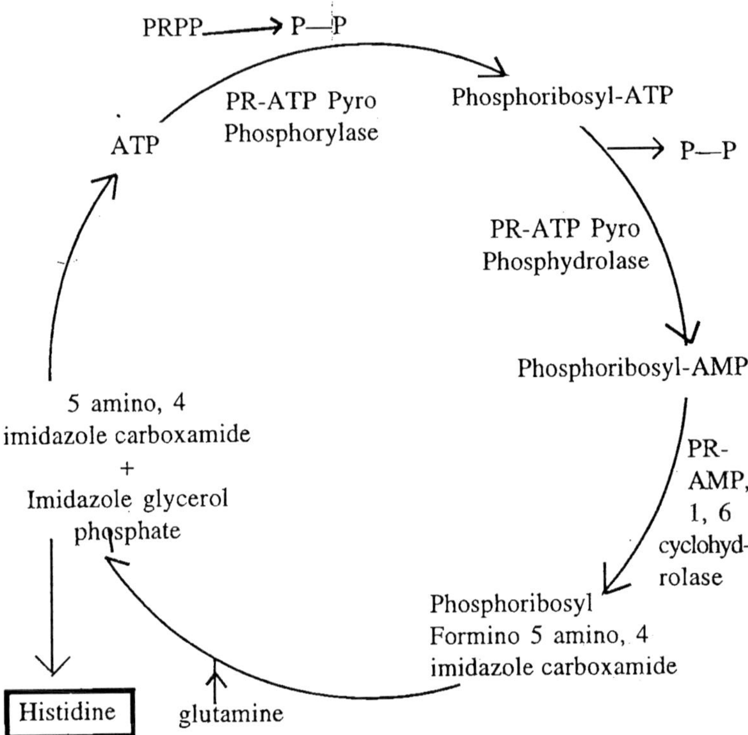

Fig. 11.5 : Histindine synthesis in microorganisms

One of the product in this reaction (5.amino 4 imidazole carboxamide) is precursor of ATP. In this pathway, ATP is recycled during the synthesis of histidine. Synthesis of histidine from imidazole involving imidazole acetol phosphate, histidinol phosphate and histidinol.

CATABOLISM OF AMINO ACIDS

Besides being incorporated into proteins, amino acids also serve as source of carbon and nitrogen when required. The first stage in their catabolism is the removal of —NH_2 group and formation of corresponding ketoacids. Ammonia liberated is quickly converted to urea or it is incorporated in some other aminoacids.

$$R\text{-}CH\text{-}NH_2COOH \longrightarrow RCOCOOH + NH_3$$

$$\downarrow$$

$$\text{Urea}$$

TRANSAMINATION

Transamination is the most important metabolism of the conversion of amino acids into keto acid. Its importance was shown for the first time by russian workers Braunstein and Bych Kou in 1939. In this process, the amino group is transferred to a recipient ketoacid which is converted to amino acid.

$$R\text{-}CH(NH_2)\ COOH + R_2COCOOH \rightleftharpoons R_1COCOOH + R_2CHNH_2COOH$$

The enzymes transaminases or amino transfrases catalyzing this type of reactions are common in biological systems. Coenzyme for each reactions is pyridoxal phosphate, a derivative of vitamin B_6. The mechanism of the reaction has been described elsewhere in this series. Most of the amino acids may act as donor amino acids. The recipients, however, are mostly ketoglutrate, oxalacetate or pyruvate and produce glutamate, asparatate and alanine respectively. Threonine and lysine probably do not participate in transamination.

DEAMINATION

Amino group of the amino acids may be removed either by oxidative or by non oxidative deamination. One example of oxidative deamination is deamination of glutamate, by glutamate dehydrogenase.

$$\text{Glutamate} + NAD^+ + H_2O \rightleftharpoons NH_3 + \alpha\ \text{ketoglutarate} + NADH + H^+$$

This reaction is reversible and the electron is accepted by NAD^+. A group of flavin containing enzymes (amino acid - oxidase) also catalyze oxidative deamination. In this case H_2 is transferred to O_2 and H_2O_2 is produced.

Non oxidative deamination is catalyzed by amino acid ammonia lyase. In this process, NH_3 is eliminated with the formation of a double bond in the remaining carbon skeleton. In bacteria and plants, asparatate is deaminated to fumarate by the enzyme asparatase (asparate ammonia-lyase).

$$\begin{array}{c} COOH \\ | \\ CH[NH_2] \\ | \\ C[H]_2 \\ | \\ COOH \end{array} \rightleftharpoons \begin{array}{c} COOH \\ | \\ CH \\ || \\ CH \\ | \\ COOH \end{array} + NH_3$$

Fumaric acid

Similarly, phenylalanine can be deaminated to cinnamate. Deamination of histidine is more common in mammals.

FATE OF DEAMINATED AMINO ACIDS

Acids produced from deamination or transmission of amino acids are channelled to several metabolic routes. Keto acids are eventually oxodised to CO_2 and H_2O through acetoacetate and acetyl CoA. Acetoacetate is one of the chemical constituents of the ketone bodies appearing in the pathological conditions of urine. Thus, the amino acids which give rise to acetoacetate during their metabolism are called ketogenic. For example, leucine is first deaminated to isovaleryl formic acid which reacts with coenzyme A to yield a series of compounds ultimately producing acetoacetate and acetyl CoA. Fig 11.6 shows catabolic pathway of leucine to acetoacetate.

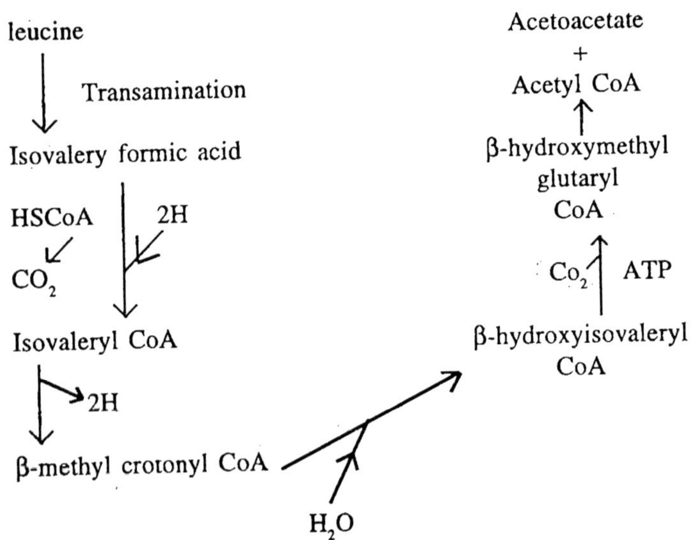

Fig. 11.6 : Catabolism of leucine to acetoacetate

Some amino acids are deaminated to produce acids which are broken down to C_4 dicarboxylic acid or pyruvate and then converted to carbohydrate. For example alanine is deaminated to pyruvate which can be utilized for the synthesis of glucose or glycogen. These amino acids are called glucogenic or antiketogenic amino acids. Examples of other glucogenic amino acids are glycine, serine, aspartate, glutamate, valine, histidine, arginine, proline, threonine, tryptophan, methionine and serine.

DECARBOXYLATION OF AMINO ACIDS

Amino acid decarboxyllases catalyse the removal of CO_2 from carboxyl group of amino acids. All decarboxylases seem to require pyridoxal phosphate as coenzyme. Thus, histidine, may be decarboxylated to histidine by histidine decarboxylase.

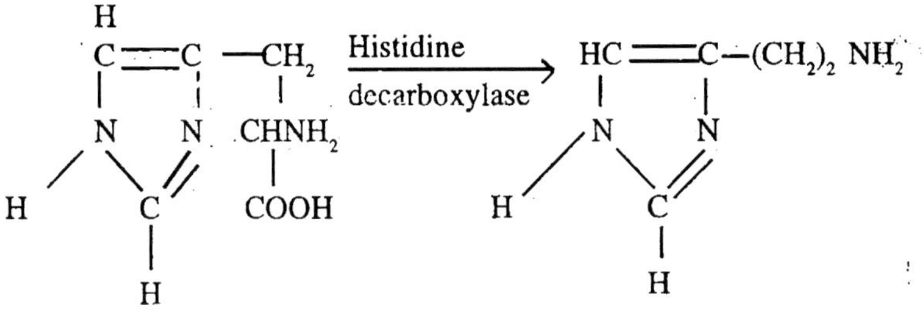

$$\text{HC}{=}\text{C}{-}\text{CH}_2{-}\text{CHNH}_2{-}\text{COOH (imidazole ring: N, N, C–H; N–H)} \xrightarrow[\text{decarboxylase}]{\text{Histidine}} \text{HC}{=}\text{C}{-}(\text{CH}_2)_2\,\text{NH}_2 \text{ (imidazole ring: N, N, C–H; N–H)}$$

Histidine　　　　　　Histamine

Similarly, 3,4-dihydroxyphenylalanine is decarboxylated to dopamine, tryptophane to tryptamine, tryosine to tyramine, glutamate to amino butyrate and so on. This type of amines are called ***Biogenic amines***. Many of these amines have strong pharmacological effects and others are important as precursors of hormones or as coenzymes.

BIOCHEMICAL DESCRIPTION OF PROTEINS

The term protein was suggested by Berzelius & used by Mulder for the first time in his text book in 1840. All the basic functions of life depend upon specific protein. Indeed, we do not know of any life without proteins. They are present in each cell in one form or the other. Proteins account on average about one fourth to one third of the total dry weight of any living organism. Proteins take part in the structural organisation of several cellular components. It has been known for a long time that the proteins are high molecular weight compounds, molecular weigh ranging from 6000 to many million & made of several amino acids.

The techniques of analysis available to us have increased our knowledge about protein structure enormously during the last few decades. Now, methods to isolate & purify proteins are available. The physical configuration & chemical components of several proteins such as haemoglobin, myoglobin, ribonucleous, chymotrypsin, carboxypeptidase, papain, lysozyme etc. are known.

CHEMICAL STRUCTURE

The proteins are big molecules built up from several smallest units. Back bone of all protein molecules is made up of linear chains of polypeptides. These polypeptides are made of amino acids. The number of amino acids in a polypeptide & thus in a protein may vary considerably. Perhaps the smallest protein known is insulin which as 5L amino acids. Examples of some bigger proteins are chymotrypsin (246 amino acids) & human haemoglobin (574 amino acids). Union between amino acids involves amino (-NH_2) & carboxyl (-COOH) groups & releases water. Such bonds (=CO-NH) are called ***peptide bonds*** as presented in fig. 12.1 At one end of the polypeptide chain, there is a free or acylated amino group called N-terminus at the other end a free amidated carboxyl group called **C-Terminus.**

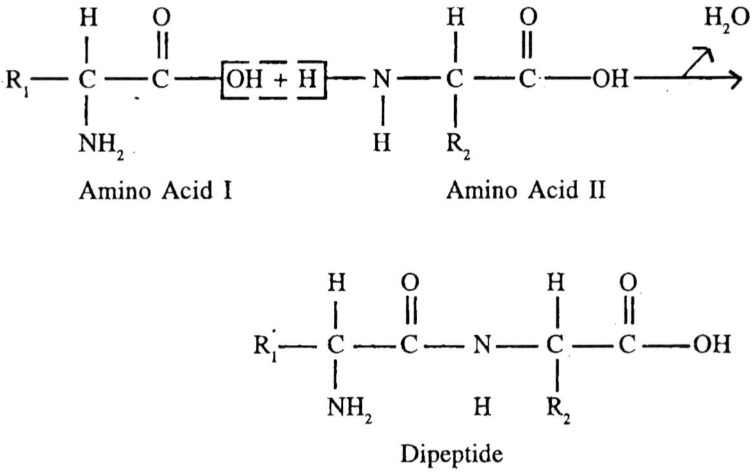

$$R_1-\underset{NH_2}{\underset{|}{\overset{H}{\overset{|}{C}}}}-\overset{O}{\overset{\|}{C}}-\boxed{OH + H}-\underset{H}{\underset{|}{N}}-\underset{R_2}{\underset{|}{\overset{H}{\overset{|}{C}}}}-\overset{O}{\overset{\|}{C}}-OH \xrightarrow{H_2O}$$

Amino Acid I Amino Acid II

$$R_1-\underset{NH_2}{\underset{|}{\overset{H}{\overset{|}{C}}}}-\overset{O}{\overset{\|}{C}}-\underset{H}{N}-\underset{R_2}{\underset{|}{\overset{H}{\overset{|}{C}}}}-\overset{O}{\overset{\|}{C}}-OH$$

Dipeptide

Fig. 12.1 : Peptide bond formation

Although more than 100 amino acids are known, only 20 of them take part in the synthesis of protein. plants are able to synthesize all these amino acids from breakdown products of carbohydrate & reduced nitrogen. Most animal cells however, can make only some of these amino acids. Those amino acids which are not synthesized endogenously & have to be supplied with the diet of the animal are known as essential ammino acids. All amino acids contain a carboxyl group (-COOH) & a amino group attached to the α carbon atom.

Polypeptides are the simplest proteins. Most of the natural proteins, however, are complex polypeptides & have different configurations. Polypeptides are named according to the amino acids constituting them. e.g. if glycine, alanine & serine combine in the following manner, a tripeptide named glycyl ananyl serine will be formed.

```
                                                  OH
                                                  |
                  H             O                 CH2
                  |             ||                |
      CH2         N             O                 CH            OH
     /   \       / \           /  \              /  \          /
NH2       C            CH            N                C
          ||           |             |                ||
          O            CH3           H                O
```

Naming of the polypeptide starts with N-terminal & proceeds one amino acids at a time to the end of chain (C-terminal). Each amino acid except the last one is named as an acyl group.

Very few of the small peptides are important biologically. Glutathione, a tripeptide of amino acids viz. glutamic, cysteine & glycine function in certain oxidation-reduction reactions. Some hormones are cyclic peptides rather than linear one. Oxytocin is a hormone in which amino acids are joined to form a circle. The circle is held in its normal shape by covalent disulphide (S-S) bonds between cysteine molecules. Disulphide bonds are characteristics of other biologically important protein also.

PHYSICAL CONFIGURATION

The linear chains of polypeptides are arranged in space in various ways. e.g. these may stick together to form a sheet or may bunch up randomly into a ball or arrange in a highly ordered screw. This arrangement gives rise to the secondary structure of proteins. The details of the secondary & tertiary structures of proteins were worked out initially by two American scientists, L. Pauling & R.B. Corey with the help of X-ray crystallography.

PRIMARY STRUCTURE

Arrangement of amino acids in a polypeptide chain is the primary structure of the protein. For determining amino acid sequence of of polypeptide chain various methods have been developed during last 25 years. Amino acid sequence of the smallest protein known. That is of insulin, was studied by Fredrick Sanger in 1954. Amino acid sequence of many other proteins have been studied since then. Perhaps the largest protein analysed for amino acids is chymotrypsinogen, which is composed of 246 amino acids. (fig. 12.2) The linear chain is held in a definite shape of 5 disulphide (S-S) bonds.

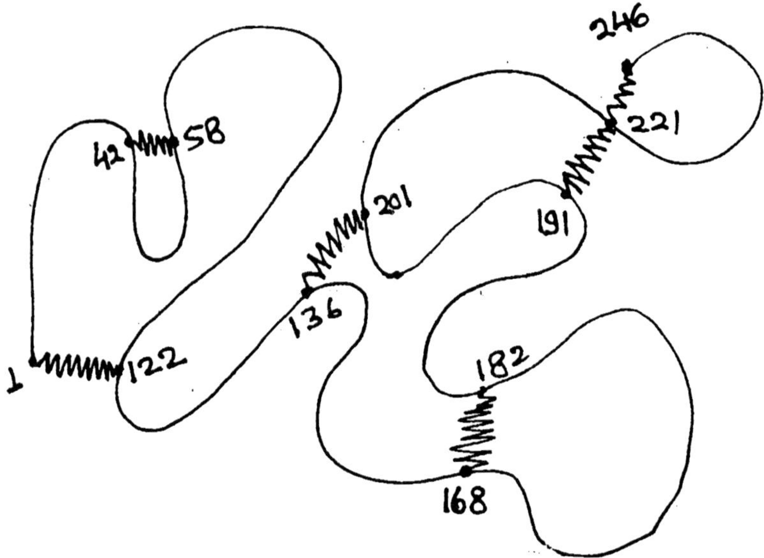

Fig. 12.2 : Disulphide bonds in Chymotrypsinogen

Normal human haemoglobin is composed of 4 polypeptide chains & 4 haeme groups of two identical pairs where each pair consists of an α & a β - chain The β chain is made of 141 amino acids residues & contains relatively more acidic groups than the α chain (with 146 residues). The two chains are held together by ionic & hydrogen bonds. Myoglobin with 153 amino acids contains only one polypeptide chain.

Both myoglobin & haemoglobin contain the prosthetic group haeme, a metal organic compound closely related to porphyrin component of chlorophyll. The oxygen combining properties of myoglobin & haemoglobin are rendered due to the presence of haeme.

The sequence of amino aids in a polypetide chain is not a regular one. Acidic amino acids. however, tend to group themselves together & the basic & aromatic amino acids to the same.

The sequence of amino acid is determined genetically & the species or the individuals are differentiated on this basis. In human being normal haemoglobin differ from sickled cell anaemic person in the sequence of amino acids. (Fig. 12.3) In this only one amino acid, glutamate is replaced by valine first & then by lysine in haemoglobin C.

Normal haemoglobin	Val-His-Leu-Thr-Pro-Glu Glu-Lys
Sickle cell haemoglobin	Val-His-leu-Thr-Pro-Val-Glu-Lys
Haemoglobnin C	Val-His-leu-Thr-Pro-Lys-Gly-Lys....

Fig. 12.3 : Sequence of amino acid in normal sickle celled haemoglobin

SECONDARY STRUCTURE

The polypeptides are held together to give rise a definite shape of the protein molecule by some secondary forces such as hydrogen bonds. Hydrogen bonds are formed between the carbonyl oxygen & imide hydrogen along the polypeptide backbone (N-H ... O=C). The hydrogen &oxygen atoms are separated apart by a distance of 2.8°A. & the energy of the these bonds is about 1400 cal/mole. The second feature of the secondary structure is the bending & folding of polypeptide chain.

The most common type of secondary structure is alpha helix or alpha structure. This is formed by the bending of polypeptide chain to form hydrogen bonds in the same chain. The bedding of the chain is very regular & chain assumes the shape of a helix. It has been revealed by X-ray analysis that large sections of polypeptide backbone of myoglobin are folded into α helices. This type of secondary structure is found in many other proteins of wool, hair & other fibrous structure also. In the α helix of myoglobin & probably in others too, each amino acid is away from the other at a distance of 1.47°A & there is 3.7 amino acids in one turn. Distance of one turn of the helix, which is (3.7×1.47) 5.4°A is the identity period of α helix. Screwing of the helix may be right handed or left handed & side chains interact with each other & the regularity of the chain is disturbed.

Pauling & Corey have considered at alternate type of secondary structure. In this case of polypeptide chains are lead linearly either parallel or antiparalled with respect to one another.

This structure is called Beta structure of β pleated sheet. β structures are quite common in nature & are favoured by the presence of amino acids glycine & alanine.

Silk & certain synthetic fibres such as nylon & orlon are composed of β structures. Fig. 12.4 gives α and β structures & protein.

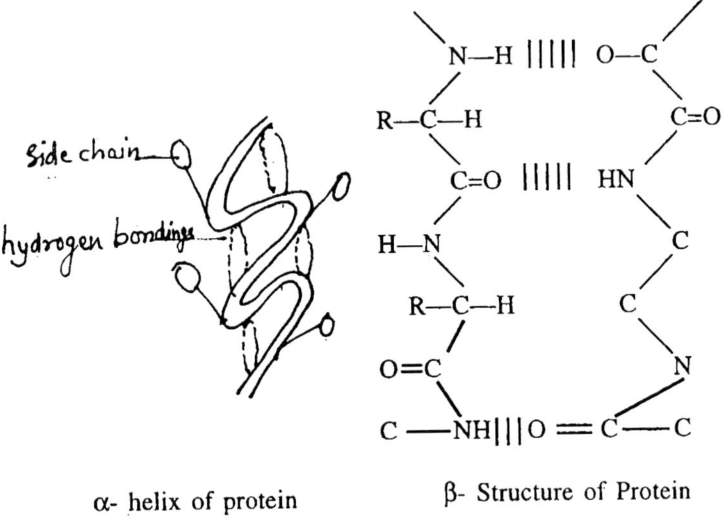

α- helix of protein β- Structure of Protein

Fig. 12.4 : α and β structure of protein

Some synthetic polypeptides assume different types of secondary structures under different conditions. Poly L-lysine is an example of this. At pH values greater than 10.5, this polypeptide forms nearly perfect helices which are stable in aqueous solution below 20°C.

When the temperature is raised upto 35°C, the helix is changed gradually into a random coil. Above 35°C, the polypeptide assumes a β - pleated sheet structure.

TERTIARY STRUCTURE

Three dimensional configuration of proteins of very common in the biological systems. This is the tertiary structure of proteins. The helices & other regular arrangements are not favoured in the nature due to following reasons :

1. Since an amino group is absent from proline, its presence in a chain does not allow normal hydrogen bonding.

2. Disulphide bridges between cysteine molecules join various parts of the polypeptide chain & regular helix is disturbed, &

3. Side chains have different configurations. They can be attached together or with the solvent molecules & cause the chain to fold.

These phenomena result in the tertiary structure of protein which is energetically most stable form & is represented by most of the biologically active proteins.

High temperature & some other unnatural conditions like high acidity & high alkalinity etc. breakdown the native 3-diamensional form to give randomly oriented, biologically inactive proteins. It is denaturation of the protein. When the denatured chains are carefully returned to their original environment, some of them can re assume their native configuration with full biological activity (renaturation). Denaturation is specially reversible in proteins containing some disulphide bonds. Many enzymes which posses tertiary structure can completely restore their catalytic activity even after boiling. Trypsin, ribunuclease & myokinase are examples of this. Presence of certain compounds in the environment prevents renaturation. e.g. when ribunuclease is renatured in the presence of 8M urea, the original structure of the enzyme is restored only in 1% of the cases.

PROPERTIES

Solubility

Earlier the proteins were classified on the basis of their solubility. Albumins are soluble in distilled water whereas globulins are soluble only in dilute salt solutions of neutral pH. Leucosin in the seeds of cereals & ricin in caster oil seeds are examples of plant albumin while legumin from peas & tuprin from potato tubers are examples of globulins.

Some proteins are soluble in acidic or alkaline solutions e.g. glutenin, which is most common protein of cereals. Prolamines is another class of proteins which are soluble in alcohalic solutioins. Giladin of wheat & zein of maize are examples of this class scleroproteins, which are abundant in horns, nails & nairs are relatively insoluble & are hydrolysed only in special conditions. Most of the digestive enzymes are also unable to hydrolyse this protein & that is why it is not digested by most of the animals. Some insects & fungi, however are able to utilise it as food material.

There are some basic proteins also which are rich in basic amino acids arginine & lysine. They are combine with nucleic acids to form nucleoproteins & are thus found in the nucleus of the plant & animal cells. Histones & protamines are examples of this class.

Zwitter ion formation

Like amino acids, proteins are also charged either positively or negatively & thus migrate either to anode or cathode of an electric field. At isolectric point (pl) they are electrically neutral & do not move towards any pole. The isoelectric pH of the proteins depends upon the relative number of acidic or basic groups, which are rendered by amino acids. Serum albumins, which contains many acidic amino acids has an isoelectric pH of 4.7. Gliadin with many basic amino acids has 9.0 isoelectric pH. Isoelectric pH of certain other proteins are given in Table 12.1.

Table 21.1 : Isoelectric pH of some proteins

Protein	*Pl*	*Protein*	*Pl*
Casein	4.6	Cytochrome C	9.8
Avidin	10.0	Tetanus toxin	5.1
Lysozyme	11.0	Serum, globulin	5.4
Pepsin	2.7	Edestin	6.9

DENATURATION

Under certain environmental condition, the spatial arrangement of polypeptide chains within a protein molecules is changed from a native form to disordered arrangement. This has been termed as denaturation only primary structure of the protein is retained. Physical & chemical properties of such proteins are different than the native proteins & they lose most of their biological activities. During denaturation, the soluble globular protein are changed into insoluble fibrous proteins. Commercial fibres are obtained from globular proteins by treating them with various denaturating agents.

There are several chemicals which can denature the proteins. Mineral acids & alkalies, acetone, alcohols, salts such as ammonium sulphate, urea, tungustic, picric & trichloacetic acids are common denaturaing chemicals. Hydrostatic pressure, ultraviolet & ionizing radiations also cause denaturation.

These agents cleave the hydrogen &/or salt linkages & thus polypeptide chain is unfolded.

HYDROLYSIS OF PROTEIN

Proteins can be hydrolysed stepwise to their ultimate building units, the amino acids. This hydrolysis can be brought about either by an acid, a base or hydrolytic enzymes. During stepwise hydrolysis, following products can be obtained from a protein.

Protein ———> denatured protein ———> metaprotein
↓
amino acids <——— dipeptide <——— peptone

Complete hydrolysis with about 20% boiling HCl or H_2SO_4 solution yields free amino acids & their breakdown products. Unless otherwise controlled hydrolysis completely destroys tryptophans and about 10% of the serine & threonine.

Alkali hydrolysis for the protein is carried out with 20% NaOH solution at 120°C in the presence of Nickel. Complete hydrolysis yields a mixture of amino acids & their breakdown products. Theonine, serine, cystine, arginine & lysine may be partially hydrolysed during alkali hydrolysis.

Hydrolysis of proteins by enzymes is quite common in nature. Proteases split proteins stepwise into amino acids. Several proteolytic enzymes are used commercially to prepare amino acids from the digestion of natural proteins.

OXIDATION OF PROTEINS

Proteins are oxidized by putrefaction process. The products are amino nitrogen compounds, CO_2 & water depending upon conditions employed. The bad smell of the dead & decaying bodies is due to putrefaction of the proteins.

BIOLOGICAL SIGNIFICANCE

Several types of proteins have been characterised & isolated from the cellular components of living organisms. Depending upon their chemical & physical structure & location inside the cell, the various proteins perform various functions.

Structural proteins are generally inert to biochemical reactions. They maintain the natural form & position of organs. The cell wall & primary fibrous constituents of the cell have structural proteins.

Collagen is the most abundant known protein in animals forming a major part of the skin, cartilage, ligament, tenden & bone. The scales of fish & reptiles & hairs, feathers, horns, hoofs & claws are made up of the proteins, keratin.

Capacity of motion & flexibility in the organisms is the virtue of certain proteins with great tensile strength. These are contractile proteins. Catalytic proteins are the most active proteins of an organisms. The properties of living cells depend upon the biological processes which are regulated by enzymes. The major portion of an enzyme is made up of proteins. These proteins are mostly spheroidal in shape & are termed globular proteins. Some enzymes are simple proteins containing only amino acids, others are complex proteins. Enzymes catalyze a variety of reactions in the living cells. Certain protein specially in animals are involved in the transport of many essential biological factors to different parts of the organism. Haemoglobin, transporting O_2 from one part of the body to the other is an example of carrier protein

MECHANISM OF PROTEIN SYNTHESIS

Proteins are synthesized in the cytoplasm on ribosomes. The sequence of amino acids in the protein is controlled by DNA through m-RNA. The relation between DNA & protein synthesis is summmarised in the following line diagram:-

DNA ——> RNA ——> Protein

The arrows indicate the direction of the transfer of information from the genetic material.

THE ACTIVATION OF AMINO ACIDS & THE ROLE OF t-RNA

The amino acids to be incorporated into protein are activated by forming their adenylates with ATP. In the synthesis of adenylates the carboxylic group of amino acids is attached by high energy bonding to adenylic acid (AMP) & this reaction is catalysed by the enzyme amino acyl synthatase :

$$\text{AA} + \text{ATP} \xrightarrow{\text{amino acyl synthatase}} \text{AA} \sim \text{AMP} + \text{P–P}$$

Normally, the synthatase is complexed into AA-AMP & this complex collides with the appropriate t-RNA, the enzyme is released

& at the same time amino acid is transferred to the pH group of terminal adenylic acid residue of t-RNA.

$$\text{AA}\sim\text{AMP} + \text{tRNA} \xrightarrow[\text{synthatase}]{\text{aminoacyl}} \text{AA–tRNA} + \text{AMP}$$

AMP & P–P thus released can be used in the formation of ATP again.

AMP + P–P ———> ATP

There are atleast 20 different synthatase recognizing 20 amino acids. In several cases more than one sps. of RNA have been found to be specific for the same amino acid. The synthatase, has atleast two active sites, one for recognizing the amino acid & the other for t-RNA. The maximum frequency at which a wrong amino acid is recognized amino acid is recognized by the enzymjes & is inserted into the protein is about 1 in 1000. Some amino acid analogues, however, readily charged & subsequently incorporated into proteins. Generally, charging of the amino acid is more accurate than the activation e.g. isoleucine t-RNA synthatase converts valine as well as isoleucine to the corresponding amino acyl adenylate but only isoleucine is transferred to the t-RNA.

TRANSCRIPTION OF RNA ON DNA

The genetic information of DNA is transferred to protein by m-RNA molecules. RNA molecules are synthesized on DNA template with the help of enzyme RNA polymerase. Along each gene only one strand of DNA works as a template for the m-RNA. RNA chains synthesized are single stranded linear molecules. They are synthesized in 5' to 3' direction & hence the 5' terminal nucleotide always contains a p-p-p group.

Apparently all normal RNA chains start with either adenine or genuine. Messenger RNA after synthesis diffuses from the nucleus to the cytoplasm & attaches to the ribosomes.

ROLE OF RIBOSOMES

Protein synthesis never occurs in cytoplasmic fluid but only on the surface of ribosomes. Ribosomes are small cellular particles made of r-RNA & protein. There are about 15000 ribosomes in a rapidly growing E.coli Cell. The size of ribosomes & other particles is expressed as S (Svedberg unit). A high Svedberg indicates that particle sediments rapidly because of its bigger size. Ribosomes isolated from E.coli are of 70s type, those isolated from cytoplasm of higher organisms have a size equivalent to 80S. All ribosomes are constructed of two sub units (Fig. 12.5) the larger sub-unit is about twice the size of the smaller unit. Each ribosomes has two holes, the peptidyl (P) & the amino acyl (A). Both the these holes are made partly of smaller unit & partly of larger sub-unit. Most of the proteins in the ribosomes have structural role. Some proteins however, serve in the formation of polypeptide chain also. Role of r-RNA is not known. Three types of r-RNA are found, the smaller sub-unit contains a smaller r-RNA while the bigger sub unit contains a bigger & a very small (SS) r-RNA. The r-RNA are single standard & have unequal amounts of guanine & cytosine & of adenine & uracil. The r-RNA are integral parts of the ribosome & can not be removed without complete disintegration of the ribosome.

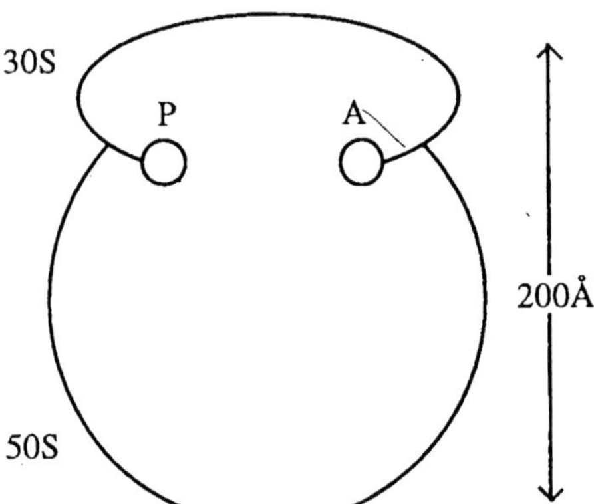

Fig. 12.5 : A Bacterial ribosome

Ribosomes do not take part as whole during protein synthesis but they separate into 50S & 30S subunits. It was evidenced in bacterial system by culturing E.coli on medium containing light (H_2) or heavy (H_3) isotopes. The bacteria are grown for a few generation on medium containing heavy isotopes. When their ribosomes are isolated they are all containing heavy isotopes. When these bacteria are transferred to the light medium heavy 50 S, light 30S & light 50 S & heavy 30 S, light 30 S & light 50 S & heavy 30 S hybrid ribosomes are obtained besides the parental heavy & some daughter light types. (Fig. 12.6)

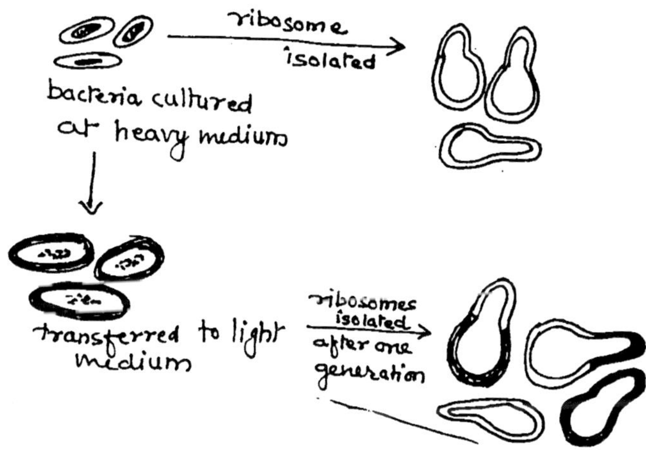

Fig. 12.6 : Role or ribosomes in protein synthesis

In *in vitro* also exchange of heavy & light ribosmes in a protein synthesizing system may be demonstrated when the antibiotic sparsomycin is added to this system exchange of sub units does not occur.

During protein synthesis one m-RNA may be translated at several ribosomes simultaneously & the ribosomes appear to be joined together. Such clusters of ribosomes are called polyribosomes or polysomes.

SYNTHESIS OF POLYPEPTIDE CHAINS

Synthesis of polypeptide chain is initiated when an initiator t-RNA corresponding to a initiator condon joins with ribosomes. Several initiation factors, ribosomal subunits & GTP molecules are as initiation factors required. The initiator t-RNA in bacteria, mitochondria & chloroplasts appear to be formally methionine t-RNA. The formulation of the amino group of methionine occurs after it has attached to t-RNA. The formyl donor is formally textrahydrofolate & the enzyme catalysing this transfer is methionine t-RNA transformylase.

The absence of a free amino group in formyl methionine allows the chain to proceed in only one direction of polysaccharide. The or formyl methionine may be removed & hence many bacterial proteins have methionine, alanine or threonine as N-terminal amino acids. In the cytoplasmic ribosomes of eukaryotic cells it appears that is usually removed from the complete protein, by enzyme amino peptidase. The initiator codons are AUG & GUG for formyl methionine t-RNA & AUG for methionine t-RNA.

At least 3 ribosomal proteins F_1, F_2 & F_3 are necessary for the initiation of protein synthesis. F_3 (m.w. 30,000) is required for the binding of 30S ribosomal unit to m-RNA. The second protein F_1 (m. w. 8000) attaches to F-met-t RNA & helps it bind to 30 S m-RNA initiation complex. It is released when 50 S unit joins the complex. How the third protein F_2 (m.w. 75,000) exactly operates, remain unclear though it is known to bind GTP & also to be needed in the formation of the 30 S initiation complex. This factor is also called translocase.

The requirement of GTP in protein synthesis has been demonstrated experimentally. In *in vitro* systems protein synthesis does not occur in the in two sequential reactions:-

1. in the binding of F-met-t-RNA to the ribosomes &,
2. in the movement of the initiation complex from A to P site of the ribosomes.

Hydrolysis of GTP is probably not required for the first reaction since GMP-PCP an analoque of GTP can also satisfy this requirement.

The series of reactions during the initiation of protein synthesis may be summarized as following:

1. A complex is formed between the smaller 30 S subunit - F met-t-RNA & m-RNA containing initiator codon (complex I). Formation of this complex is catalyzed by factor F_3 & F_1 & also requires a GTP molecule.
2. 50 S subunit attaches to the complex I & complex II is formed which is suppose to be attached to the A site of the ribosomes.
3. F-met-t-RNA is translocated to facilitate this translocation. Hydrolysis of GTP is activated by factor F_2. One GTP molecule is hydrolysed for every translocation.
4. New t-RNA charged with amino acid enters the A site & peptidyl bond formation takes place between F met & the new

aminoacid. The enzyme is peptidyl transfeerase which is present with 50 S subunit.

5. Discharged -RNA is released from the site & a dipeptide is formed.

During protein synthesis the growing pepride chain is esterified to the terminal adenosine residue of t-RNA molécule. When the growing chain is completed this polypeptide t-RNA linkage is cleaved releasing the polypeptide from t-RNA & from ribosomal complex. For release of polypeptide chain the termination codons (UAA, UAG & UGA) & several termination factors have demonstrated that the enzyme peptidyl transferase is itself responsible for the release of polypeptide. During termination process the enzyme catalyzes the transfer of nascent polypeptide to water instead of amino group of a t-RNA.

The inhibition of peptidyl transferase by antibiotics such as chloamamphenicol & puromycin also causes the release of polypeptide.

Release of t-RNA from ribosome is catalysed by another factor, transfer releasing factor (TR) as proposed by H. Ishitsuka et al. This release does not require GTP.

Upon completion of protein synthesis, the ribosomes dissociate into monomeric units & the sub units. RNA released are probably degraded & the nucleotides are reused in their syntheses. Polypeptide chains synthesized fold & refold to assume a definite physical configuration to perform the role assigned to them.

Sometimes, the proteins assume their three dimensional structure while they are polysomes. Diagramatic representation of protein synthesis is given in Fig. 12.7.

CONTROL OF PROTEIN SYNTHESIS

Rate of protein synthesis in a cell is controlled by the genctic make up & the external chemical environment. To accommodate various known facts about the genetic control of protein synthesis, F. Jacob & J. Monod (1961) proposed that genes existed in group of clusters. One cluster consisted of one regulatory gene, an operator gene & one or more structural genes. The structural genes under the control of single operator gene are called operons.

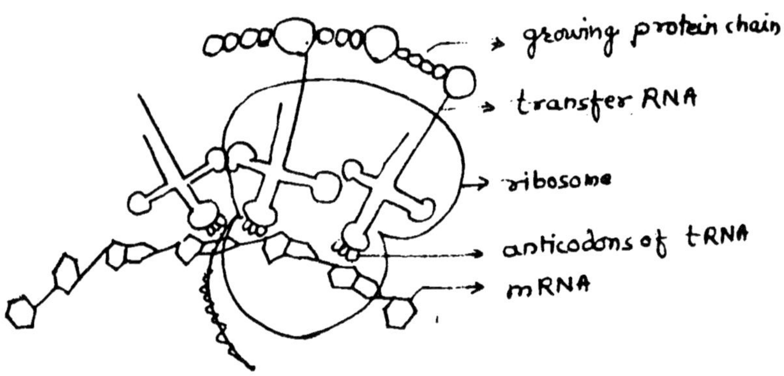

Fig. 12.7 : Protein synthesis - diagrammatic representation

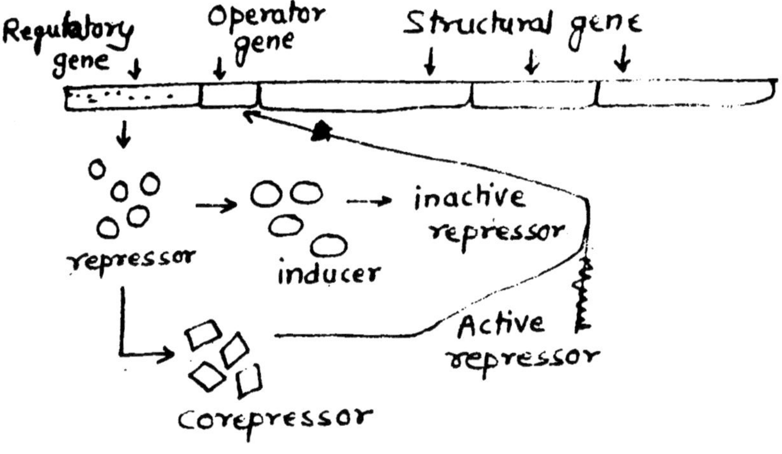

Fig. 12.8 : Protein synthesis - regulatory mechanism based on operon model

The synthesis of enzymes β-galactosidase, β-galactoside permease & galactoside acetylase in E.coli were studied & their regulatory mechanism was consistent with the model.

The regulatory genes code for some special group of molecules called repressors. These repressors molecules are proteins & are present in very small amounts in the cell. Some bacterial cells do not have functional regulatory genes & hence they produce proteins independent of the need. These mutants are called constitutive mutants & these proteins produced in fixed amounts independent of need are called ***constitutive proteins***.

Repressor inhibit protein synthesis by binding to the operators. An operator gene is about 10-12 nucleotide long & the repressor molecule binds with it probably by the formation of hydrogen bonds. The repressors exist in either active or inactive form. When a repressor is combined with an inducer molecule, repressor is inactivated & then does not bind to the operator. e.g. when combined with β-galactoside (inducer) the β-galactosidase repressor cannot bind to its specific operator. Binding of a co-repressor on the other hand, activates the repressor-molecule e.g. several amino acids which inhibit the induction of nitrate reductase may bind with the specific repressor to activate it & stop the synthesis of the enzyme. Binding between a repressor & a corepressor or an inducer is weak & involves hydrogen bonds, salt linkage or van der waals forces.

Thus a repressor is rapidly able to adjust the physiological need of the cell. Binding of repressor to the operator prevent the transcription of adjacent structural gene(s) & hence the synthesis of protein directed by them.

The control of β-galactosidase synthesis by operator gene is negative in character because activity of the repressor molecule prevents protein synthesis. Other operators are under positive control. Arabinose operon is one such case studied in E. coli. Arabinose is converted into L-ribulose, L-ribulose-5-P & D-xylulose 5-P by the successive action of the enzymes isomerase, kinase epimerase.

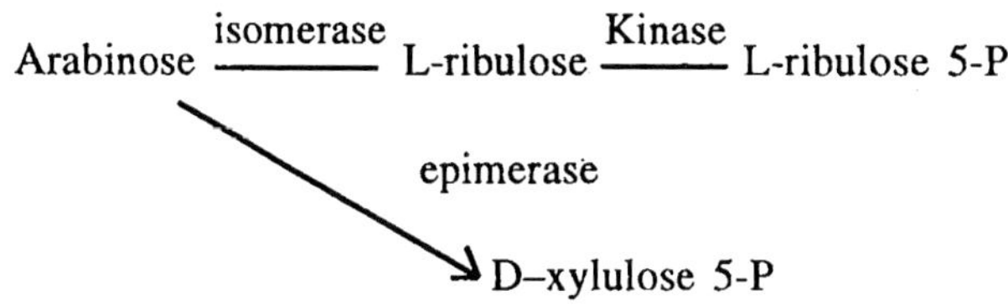

In the absence of arabinose all of these enzymes are present only in small amounts but the addition of arabinose increases the amount of all these three enzymes. The induction is controlled by a fourth gene 'c' which probably has a positive role. A arabinose-c-product complex is necessary for the expression of arabinose operon.

CONTROL AT TRANSLATION LEVEL

The polypeptide chains starts assuming 3-D-Shape while still attached to the polyribosomes. It has been proposed that the altered configuration of the nascent polypeptide may inhibit the further growth of chain by any one of the following three processes:

1. By preventing the further incoming of aa-t-RNA.
2. By preventing the translocation of t-RNA or
3. By preventing the peptide bond formation

First evidence for the translation control came from, the study of haemoglobin synthesis. As mentioned earlier, a haemoglobin molecule is made up of a haememoiety & two α- and two β-polypeptide chains. Two polypeptide chains are the products of two different m-RNA but their syntheses were found to be interdependent. If the structure of one chain was altered the translation of the m-RNA for other chain was affected.

Control of proteins synthesis in T_2 infected E.coli cells at translational level was studied by sueoka & Sueoka (1970). There are 5t-RNA for leucine corresponding to CUG codon of the m-RNA. Messenger tRNA synthesized after T_2 infection carries little or no CUG codon that may code for leucine. Thus after the infection of T_2 leucine t-RNAs are degraded. Regarding the mechanism of degradation, it is proposed that phase T_2 infection induces a specific ribunuclease which nicks Leu-tRNA at the site of translation of E coli m-RNA on ribosomes which in turn leads to the cessation of protein synthesis. The process eliminates half of the leucine t-RNA-1 molecule in the cell.

ROLE OF INTERFERONS

Interferons are known to inhibit the growth of certain viruses. In mammalian cells the synthesis of interferons is induced either by infection with a natural double stranded RNA or by a synthetic double stranded ribunucleotide.

Regarding the mechanism of action it has been observed that ribosomes from interferon injected mammalian cells do not form polysomes with viral RNA, but when they are exposed to hose mammalian RNA, they form polysomes & the protein synthesis is initiated.

This shows that ribosomes can recognize that specific m-RNA & regulate the overall protein synthesis. It is further believed that there are tissue specific factors in the cell which may recognize the RNA from the same tissue but not from the others.

EXTRACTION & ESTIMATION OF PROTEIN FROM BIOLOGICAL ORGANIC MATTER

Reagents

1. 0.2% NaCl solution.
2. 10% C Cl_3.COOH (Trichloroacetic acid)
3. 0.2% NaOH solution.
4. (a) 2% Na_2CO_3 in 0.1% sodium hydroxide solution

   (b) 0.5% $CuSO_4$ in 1% sodium-potassium tartrate.

   Mix 50 ml. of (a) with 1.0 ml of (b).
5. Dilute phenol reagent with equal volume of distilled water.

PROCEDURE

Extraction of protein

Homogenize 5 mg of material with 10 ml. saline solution (0.2% NaCl sol) & add 10 ml. 10% TCA & incubate for 1 hour at 37°C.

Centrifuge at 4000 RPM for 20 minutes. Discard the superatent fluid & dissolve the precipitate in 10 ml. of 0.2 N NaOH.

Estimation of protein (Lowery et al 1951)

Take 0.5 ml. protein sample. Add 5 ml. of reagent (4) & mix well.

Rapidly add 0.5 ml. phenol reagent & allow the mixture to stand for 10 minutes at room temperature & read at 600 nm in a colorimeter.

Standard curve

Dissolve albuin in 0.2 N NaOH & prepare protein solution of different concentration arranging 0.5-20 mg per ml.

Develop the colour & read at 600 nm.

Draw a curve with optical density value against concentration.

Selective extraction of protein

The protein moleculese are electrically charged, their charges being contributed by the constituent amino acids. The net charge of a protein molecule depends upon the amino acids constituents. The net charge of molecule is also variable depending upon the pH.

Protein may also have isoionic or isolectric pH, depending on the constituents of amino acids.

Prepare a homogenate of org. matter in water & lecant the supernatnat. It contain soluble carbohydrate & protein.

Use specific precipitant to precipitate the carbohydrate. The filtrate will contain proteins.

Apply tests to the filtrate containing proteins:-

1. ***Milion's test***

   Take 5 ml. of filtrate in a test tube & add 3-4 drops of Milion's reagent. Mix & heat the mixture till it boils. White ppt. will appear which should turn red when heated further. This is the indication of protein presence.

2. ***Acetate test***

   Take 5 ml. of filtrate containing protein & add 5-6 drops of lead acetate. A ppt will appear. Dissolve the ppt. by adding drop by drop 40% NaOH.

When ppt. is completely dissolved, boil the solution till a brown or black ppt. appear.

This indicate the presence of protein.

3. ***Biuret test***

Take 5 ml. of protein sample (filtrate) & add equal volume of protein sample (filtrate) & add equal volume of 10% NaOH.

Mix them thoroughly & add 0.5% $CuSO_4$ solution drop by drop shake it well till a purple violet or pinkish violet colour appears.

DETERMINATION OF TOTAL NITROGEN BY NESSELERIZATION

REACENT

Nesseler's reagent

10% NaOH solution

10% sodium silicate solution

Concentrated sulphuric acid

Standard solutions of :

$CuSO_4$ $5H_2O$

K_2SO_4

SeO_2

Take 10 ml. 7.5% K_2SO_4 solution in a 100 ml. volumetric flask. Add to ml. $(NH_4)_2SO_4$ solution (94.9 gm/lt.), make up the volume to 100 ml.

Each ml. of this solution is equivalent to 2 mg nitrogen in ammonium form.

PROCEDURE

Digest 100 mg of dry powdered material in a Kjeldahl's flask with 5 ml. concentrated sulphuric acid & 500 mg. potassium sulphate, 50 mg $CuSO_4$. $5H_2O$ & 100 mg SeO_2. Until the fluid becomes colourless.

Transfer the fluid into a 100 ml. Volumetric flask & make up the solution to 100 ml.

Take 5 ml. of aliquot in a test tube. Add 2 ml. of distilled water & 2 ml. of 10% sodium silicate solution. Shake thoroughly & add 1 ml. 10% NaOH solution.

After shaking well, add 1 ml. of Nessler's reagent.

Shake the contents thoroughly & allow it to stand for 10 minutes & read at 470 nm. in a colorimeter.

Develop the colour of standard ammonium sulphate solution in a concentration ranging from 1 mg - 100 mg per ml. & read at 470 nm. Draw standard curve & get the concentration of nitrogen by referring into the standard curve.

CALCULATIONS

Concentration of nitrogen $\frac{N_2(g)}{0.1} \times = \text{O.D.}$

Chapter -13

ENZYMES AND ITS SIGNIFICANCE IN BIOLOGICAL SYSTEM

Enzymes act as catalysts increasing rate of reaction but not itself involved in the reaction.

$$A \xrightarrow{\text{Catalyst}} B$$

The efficiency is measured by the number of moles produced in the reaction by one mole of catalyst. Enzymes are organic catalysts produced within living cells but which act independently of the cell. Efficiency is measured as :-

enzyme efficiency = substrate transfer/unit time.

$$\text{Substrate + Enzyme} \rightleftharpoons \text{substrate - enzyme complex} \rightleftharpoons \text{product Enzyme.}$$

In enzyme system, substrate refers to compound acted upon by enzymes and change produced in substrate occurs after the union of substrate and enzyme. Similarly, products into which substrate is converted are referred to as *end products.* At the end of the reaction Enzymes not used up in reaction are set free to combine again. Most of the enzymatic reactions are reversible.

CHARACTERISTICS OF ENZYMES

Definition : enzymes are thermolabile organic catalysts elaborated by a living cell.

Enzymes are by nature pure proteins or proteins combined with some other material or compound where the major portion is protein.

Enzyme — protein portion = apoenzyme

organic compound = prosthetic

co-enzyme = some prosthetic necessary for activating enzymes

apoenzyme inactive protein high mol.wt. + co-enzyme inactive organic molecule low mol. wt. ⟶ Holo-enzyme active

activators

Some co-enzymes are vitamins. Some co-enzymes are metal ions, e.g. Fe, Mn, Mg, etc. ranges between molecular weight of enzyme = 10000-1000000. Smallest isolated enzyme has about 100 amino acids. About 700 enzymes have been isolated so far.

enzymes have:

1. High catalytic efficiency.
2. High degree of specificity.

Efficiency

1. mole pure enzymes ⟶ 10,000-1000,000 moles substrate/min.

Specificity

1. **Absolute specificity**

   e.g. urease, only urea

2. **Group specificity**

   e.g. Esterase, esters

$$R-\overset{\overset{O}{\|}}{C}-O-R^1 \longrightarrow R-\overset{\overset{O}{\|}}{C}-OH$$

3. **Stereochemistry specificity**

– hydrolize only one stereo group L or D

e.g. L-lactic acid

$$\begin{array}{c} CH_3 \\ \diagdown \\ HOCH \\ \diagup \\ COOH \end{array} \xrightarrow[-2H]{\text{lactic dehydrate}} \begin{array}{c} CH_2 \\ | \\ C=O \\ | \\ COOH \end{array} \qquad -OH$$

$$\text{glucose + yeast} \xrightarrow{\text{many intermediate steps}} \text{ethyl alcohol} + CO_2$$

Enzyme systems involve many intermediate steps.

NOMENCLATURE

A. Based upon name of substrate acted upon -ase

1. Class of substrate

   e.g. Protease, Lipase, Carbohydrase, Ribonuclease

2. Specific compounds

   e.g. Lactase, Sucrase, Urease, Cellulase

B. Based on type of reaction catalyzed.

e.g. Oxidase, Reductase, Decarboxylase, Dehydrogenase

Old common names of enzymes

Rennin ..... coagulates milk in stomach

Pepsin, Trypsin ...... Proteolytic enzymes

Classification Where produced, where acts

1. Intracellular : *endoenzymes* = produced inside cell, retain inside cell, function, inside cell.
2. Extracellular - *exoenzymes* = produced inside cell, act outside.

Hydrolytic enzymes - break large molecular weight compounds to compounds small enough to penetrate cell wall by introducing water.

definition : act by catalyzing introduction of the element of water at a specific bond of the substrate—a reversible reaction.

FACTORS AFFECTING RATE OF ENZYME REACTION

Same as factors affecting growth rate of bacteria :

—Enzymes are more resistant to unfavourable environment conditions

e.g. dried yeast at 100°C for 6 hrs. (oven)

Though dried yeast lost reproductive ability but still can ferment (active).

1. *Effect of Temperature*

— accelerates enzymes activity.

— thermolabile means heat sensitive.

— each has optimal temperature ~ 35°C.

— rate of enzyme activity doubles for each 10 increase.

a. Increase rate of enzyme activity.

b. thermal inactivation beyond optimum (Fig. 13.1).

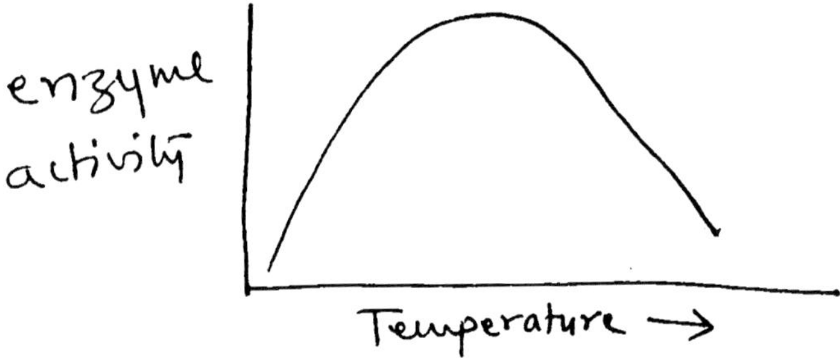

Fig. 13.1 : Effect of temperature on enzyme activity

2. *Effect of pH*

Optimum pH different for different enzymes. Certain types prefer acidic, alkaline or neutral conditions.

Affect of pH on enzymes

a. reversible-slight change in pH

b. irreversible-significant change in pH activity will not be regained if returned to normal conditions (Fig. 13.2).

Fig. 13.2 : Effect of pH on enzyme activity

3. ***Effect of substrate concentration***

Increase of substrate-increase in enzyme activity (Fig. 13.3).

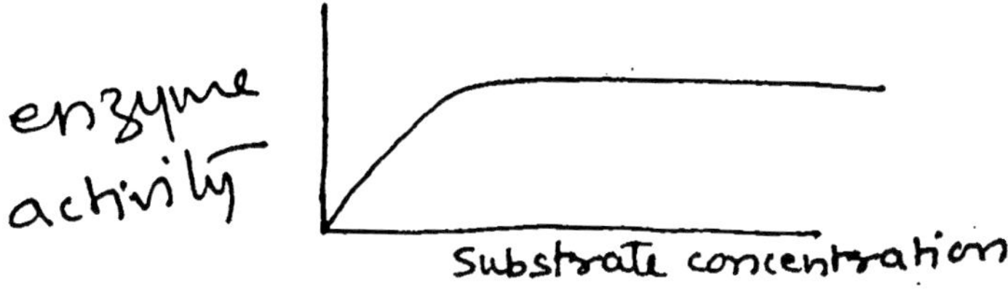

Fig. 13.3 : Effect of substrate concentration on enzyme activity

4. ***Effect of Enzyme concentration***

Fig. 13.4 (a & b) showEffect of enzyme concentration on enzymatic activity for a particular biochemical reaction.

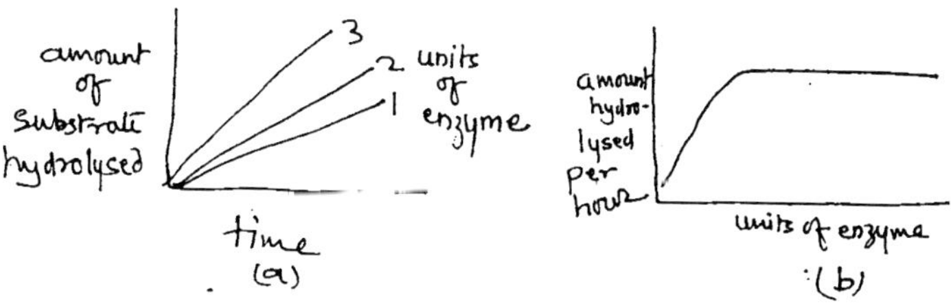

Fig. 13.4 : Effect of enzyme concentration on enzymatic activity

CLASSIFICATION BASIS OF SUBSTRATE AND FORMATION OF ENZYMES

1. Constitutive-always exist inside cell regardless of outside environment, produced by biosynthesis.
2. Adoptive or Induced enzymes

— produced when environmental change creates a need for them

— produced in response to presence of specific substrate enzyme induction (Fig. 13.5 a & b).

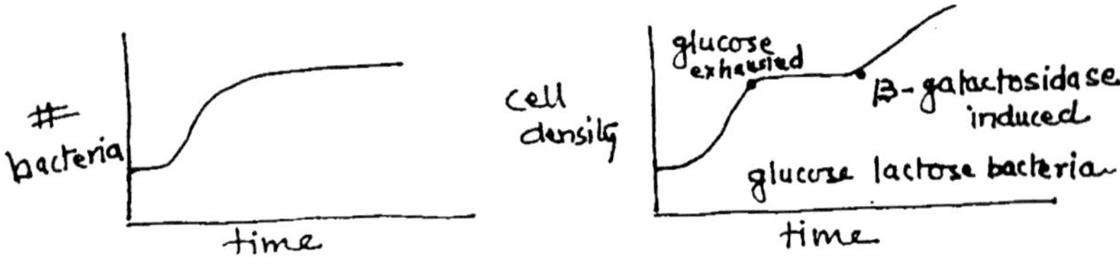

Fig. 13.5 : Microbial enzyme in the presence of inducer

Lag phase time period of enzyme induction e.g. β galactosidase

lactose ⟶ glucose + galactose

Nutrition (affecting biosynthesis and reproduction)

Different organisms:

— differ in source of energy

— differ in nutritive requirements

— differ in end products of metabolism from the same nutritive source as another organism

The largest group is chemi-organotrophic which includes bacteria.

TYPES OF METABOLISM BASED ON FUNCTION

1. Energy producing reactions-breakdown nutrients-catabolism
2. Energy consuming reactions-biosynthesis-anabolism

Link between two types of metabolism provide by ATP, a coenzyme. ATP stores small packets of energy (Fig. 13.6).

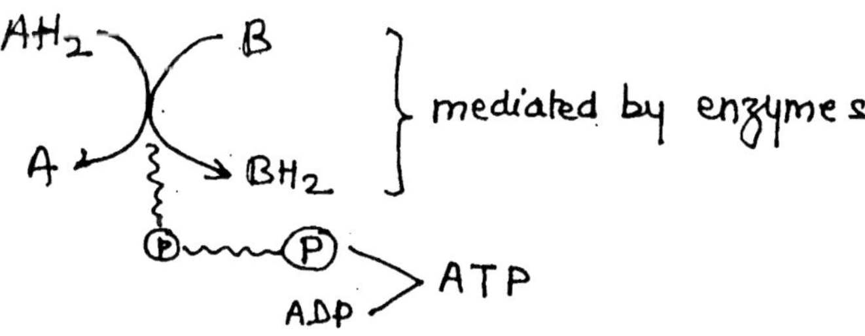

Fig. 13.6 : ATP formation

Oxidizable material provides an energy release in the removal of electrons and removal of H.

— Reaction is mediated by enzymes.

— Energy trapped n Phosphate turns ADP $\longrightarrow$ ATP

— Energy reused turns ATP $\longrightarrow$ ADP.

(aerobic) respiration vs. (anaerobic) fermentation

Both are related in that they refer to the chemical reaction of what is the hydrogen acceptor.

Fermentation is an energy yielding reaction in which organic matter is both the oxizable substrate and the oxidizing agent.

Fermentation is characterized by:

— partial oxidation

— no external source of oxygen there fore the absence of an external H acceptor

— mixtures of end products, e.g. 2 aldehyde ⟶ alcohol + acid + e.

Fermentation defined as absence of external H acceptor.

$C_6H_{12}O_6$ ⟶ $CH_3\ CH_2\ OH$ + CO_2 + Energy (22 Kcal)
glucose ethyl alcohol (reduced) (Oxidized)

ANAEROBIC FERMENTATION

In this type of reaction only a small portion of energy is released. There's still a lot of energy tied up in ethyl alcohol.

— 2 ATP's formed in this type of reaction.

— about 30 x the energy is released in aerobic respiration

— common end products of anarobic reactions (acids, aldehydes and alcohols) hold most of the energy

— it takes more anaerobic reactions to get as much energy as one aerobic reaction.

— more energy is available for biosynthesis in aerobic than in anaerobic reactions (Fig. 13.7).

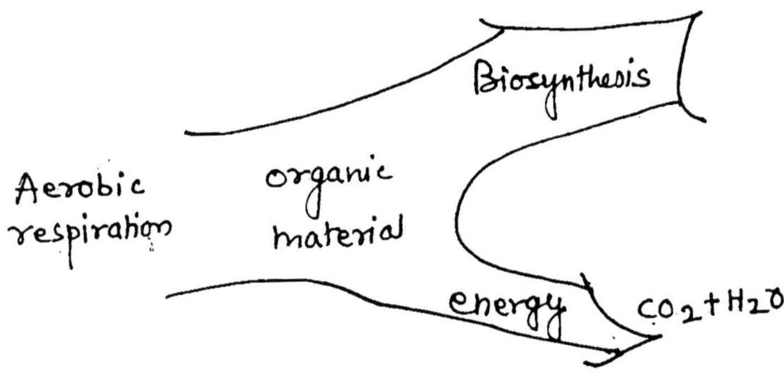

Fig. 13.7 : Aerobic respiration and energy content

— the diagram is just the reverse for anaerobic systems

anaerobic digestion

$4C_3H_6O_2$ + $2H_2O$ ⟶ CO_2 + CH_4 + 20 Kcal

— not the build up of solids as in aerobic

— most of the energy is tied up in end products of acids, alcohols or aldehydes.

— anaerobic digestion usually occurs in two steps :

— methane formers produce 70% CH_4 in a good digester

$$C_3H_6O_2 + O + O_2 \longrightarrow CO_2 + H_2O + 360 \text{ K cal}$$

Propionic acid

In a biological treatment system organic matter is not completely destroyed. Organic matter is transformed into cells, a form that can be removed in the final clarifier

Bulking sludge - won't settle (e.g. : laundry waste)

Rising sludge - happens when Nitrogen gas bubbles up after denitrification and pushes concentrated sludge out of the clarifier.

METABOLISM

— different organisms have different pathways for breaking down a substrate.

— there are many intermediate steps before the final end products.

— pyruvate, is a key compound in many pathways *Oxidation/ Reduction reaction*

$$AH_2 \longrightarrow A + 2H + 2e \quad \text{(oxidation)}$$
$$2H + 2e + B \longrightarrow BH_2 \quad \text{(reduction)}$$
$$AH_2 + B \longrightarrow A + BH_2$$

Respiration defined as reaction with an external H or electron acceptor.

aerobic respiration

$$C_6H_{12}O_6O_2 \text{ external H acceptor} \longrightarrow CO_2 + H_2O + \text{Energy (675 Kcal)}$$

glucose

– This type of reaction yields 38 ATP's from 1 mole glucose.

anaerobic respiration - organic or inorganic compounds are the external electron acceptors.

e.g. using nitrates

$$NO_3 + 2e^- + 2H^- \longrightarrow NO_2^- + H_2O$$

denitrification

In waste water most nitrogen is in the form of NH_3. No oxygen is present.

$$\underbrace{NH_3 \longrightarrow NO_2 \longrightarrow NO_3}_{\text{nitrification}} \; \underbrace{\longrightarrow NO_2 \longrightarrow N_2}_{\text{denitrification}}$$

Nitrogen gas is stripped from the waste as the end product.

Another example of anaerobic respiration is organisms using SO_4 as the external electron acceptor. But, there are fewer sulfate reducing bacteria and the end product is H_2S.

obligate aerobic organisms must have O_2.

obligate anaerobic organisms are "poisoned" by O_2.

Internediate or Facultative anaerobic organisms can function either aerobically or anaerobically. Coliform bacteria of the type that can go either way and in fact are defined with this characteristic.

Yeasts can go either way :

ferment — anaerobic

$CO_2 + H_2O$ —— aerobic

aerobic

$$C_6H_{12}O_6 + 6O_2 \longrightarrow 6CO_2\ 6H_2O + 670 \text{ K Cal}$$

anaerobic (different organisms produce different end products)

$$C_6H_{12}O_6 \xrightarrow{\text{oxidation/reduction}} 2\ CH_3CHOHCOOH + 22\ \text{K cal.}$$

$$C_6H_{12}O_6 \xrightarrow{\text{oxidation/reduction}} 2\ C_2H_5OH + 2CO_2 + 22\ \text{K cal.}$$

$$C_6H_{12}O_6 \xrightarrow{\text{oxidation/reduction}} 3\ CH_5COOH + 15\ \text{k cal.}$$

ethonol —C—C—O[H] ⟶ —C— C(=O)H + $2H^+ + 2e^-$

dehydrogeneation = oxidation

O = $2e^-$ O
H + O OH } simultanaeous
H + OH H_2O

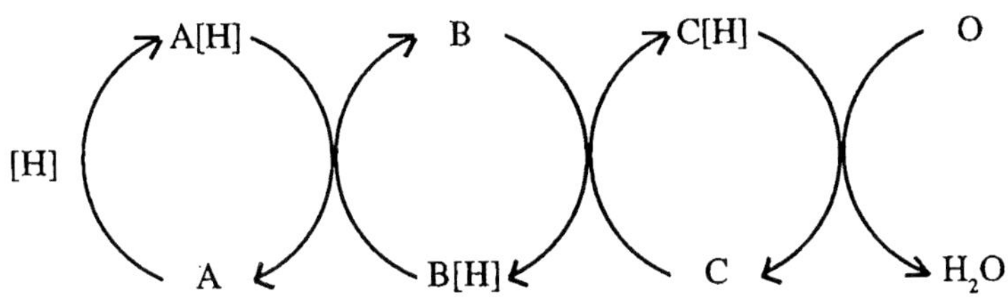

DPN — Coenzyme I
DPN = D–phosphopyridine nucleotide (old name) = NAD

$DPNH_2$ or $NADH_2$ are the reduced forms

$$DPN \underset{-[H_2]}{\overset{+[H_2]}{\rightleftharpoons}} DPNH_2$$

ENZYMES

1. Protein portion —apoenzyme
2. organic portion — prosthetic - transfer of electron coenzyme

   Coenzyme I — NAD (DPN) reduced form NADH ($DPNH_2$)

   Coenzyme II — NADP reduced form $NADPH_2$ ($TPNH_2$)

constitutive enzymes are synthesized regardless of external environmental conditions.

— induced enzymes are synthesized when needed.

Chapter -14

BIOCHEMICAL DESCRIPTION OF ENZYMES AND COENZYMES

Enzymes are the biological catalysts, which accelerate the rate of biochemical reactions. Like catalysts, the enzyme regulate the speed and specificity of the reactions without being used up but unlike catalysts, they are produced by the living cells only. These are the enzymes which run myriads of biochemical reactions in normal cellular environemt. For hydrolysing start in the food for example, high temperature or acidity is required.

In the alimentary canal, however, it is digested to glucose under relatively normal conditions of temperature and pH by the enzymes. Similarly the enzymes are able to hydrolyse other macromolecules such as protein and lipid also in normal physiological conditions.

The study of the enzymatic reactions started with the study of fermentation. In this process, grape juice or sugar solution is converted to alcohol and CO_2 in the presence of bacteria and molds. German writers used the term 'ferment' for the chemical principle capable of fermentation. The existence of biological catalysts was predicted by a Swedish chemist Jon Jakob Berzelius in 1835. He included biological materials in his study of chemical catalysis and he found that degradation of starch was faster with the potato pieces than with the sulphuric acid. Kuhne in 1878, used the term enzyme (meaning in yeast) for the soluble 'ferment' of bacteria or yeast. By that time, it had been believed however, that the living activity of yeast is necessary for the fermentation of sugar. The vitalistic hypothesis was blown off by Buchner brothers in 1897. They prepared yeast extract for some medicinal purposes and added sugar to preserve it. But to their surprise, cell free extract was able to ferment sugar. E. Buchner gave the name 'zymase' to this enzyme which is now known for the first time in 1926, when JB Sumner obtained a reasonably pure crystalline form of urease from jack bean meal. Since then several other enzymes have been purified and crystallized and all have proved to be protein.

CLASSIFICATION AND NOMENCLATURE OF ENZYMES

Mostly enzymes react upon some specific substrates. They are named according to the substrate with suffix ***ase***. Thus, urea hydrolysing enzyme is called ***urease*** and phosphate splitting enzyme is ***phosphatase***. Sometimes the enzymes are named according to the nature of reaction catalysed. e.g enzymes transferring hydrogen are ***dehydrogenases*** and those which transfer chemical groups are called ***transferases***. The complete designation of the enzymes includes both substrate and the nature of the reaction.

The naming of the enzymes was left entirely to the discovers. This resulted in some decriptives names such as pH_5 enzyme. In 1961, however, the commission on enzymes of the International Union of Bio-Chemistry framed certain rules for the nomenclature and classification of enzymes. The commission has recognised 6 classes of enzymes. Each further divided into subclasses and sub-subclasses. Classification of the enzymes upto subclasses are common and systematic names of some ensymes are given in Table 14.1. The enzymes are now named according to their systematic position and given the commission number. First number indicates the class of the enzyme and subsequent numbers indicate subclass and subclass in order. e.g. the enzyme nitrate reductase is written as nitrate oxidoreductase E.C. 1.6.6.6.1 where

EC = enzyme commission

1 = is the class of enzyme, which is oxidoreductase in this case 6 (1-6) is the subclass of enzyme, which is for enzymes acting upon NADH or NADPH as electron donor.

6 (1-6-6) is the sub-subclass of the enzymes who transfer electrons to itrogenous groups 1 (1-6-6-1) indicates the first enzyme of the series, reducing nitrate to nitrite.

Table 14.1 : Classification of enzymes according to I.U.B. System

Enzyme Commission number	*Class, subclass of enzyme and systematic name*	*Common name*
1	2	*3*
1.	Oxidoreductase	
1.1	Acting on the -CHOH group of donors	
1.1.1	With NAD or NADP as acceptor	
1.1.1.1	Alcohol : NAD reductase	alcohol dehydrogenase
1.1.1.8	L-glycerol-3-phosphate: NAD oxidoreductase	glycereldehyde phosphate dehydrogenase
1.1.2	With a cytochrome as an acceptor	
1.1.3	With O_2 as acceptor	
1.1.99	With other acceptors	
1.2	Acting on aldehyde or keto group of donors	
1.3	Acting on = CH=CH group of donors	
1.4	Acting on-$CHNH_2$ group of donors	
1.4.1	With NAD or NADP as acceptor	
1.4.1.2	L-glutamate: NAD oxidoreductase	glutamate dehydrogenase
1.5	Acting on-CNH group of donors	

1	*2*	*3*
1.6	Acting on NAD or NADP as donors	
1.7	Acting on other nitrogenous compounds as donors	
1.8	Acting on sulphur compounds as donors	
1.9	Acting on haeme group as donors	
1.9.3	With O_2 as acceptor	
1.9.3.1	Cytochrome C:O_2 oxidored-uctase	Cytochrome oxidase
1.10	Acting on phenols and related substances as donors	
1.11	Acting on H_2O_2 as acceptor	
2.	Transferring C_1 groups	
2.1.1	Methyl transferase	
2.1.1.2	S-adenosylmethionine: guanidino acetate N-methyl transferase	Guanidinoacetate methyl transferase
2.2	Transferring alkyl or related groups	
2.3	Acyl transferases	
2.4	Glycosyl transferases	
2.5	Transferring alkalyl or related groups	
2.6	Transferring nitrogenous groups	
2.6.1	Aminotransferases	

1	*2*	*3*
2.6.1.2	L-alanine, 2-oxoglutarate amino transferase	Alanine amino transferase
2.7	Transferring phosphorus containing groups	
2.8	Transferring sulphur containing groups	
3.	Hydrolases	
3.1	Cleaving ester linkage	
3.1.1.3	Glycerol ester hydrolase	Lipase
3.2	Cleaving glycosides	
3.2.1	Glycoside hydrolase	
3.2.1.1	α.-1,4 glucan, 4-glycan hydrolase	α amylase
3.4	Cleaving peptide inkage	
3.4.4	Peptide peptidohydrolase	
3.4.4.4	None	Trypsin
3.5	Cleaving C-N bonds other than peptide bonds	
3.6	Cleaving acid-anhydride bond	
3.6.1	In phosphoryl containing anhydrides	
3.6.1.3	ATP-Phosphohydrolase	ATPase
3.7	Cleaving C-C bonds	
3.8	Cleaving halide bonds	
3.9	Cleaving P-N bonds	
4.	Lyases	
4.1	carbon-carbon lyases	

1	*2*	*3*
4.2	Carbon-Oxygen lyases	
4.3.2	Amididine-lyases	
4.3.2.1	L-argininosuccinate arginine lyase	Arginino succinase
4.4	Carbon-Sulphur lyases	
4.5	Carbon-halide lyases	
5.	Isomerases	
5.1	Recemasees and epimerases	
5.2	Cis-trans isomerase	
5.3	Intra-molecular oxidorectases	
5.3.1	Interconverting aldoses and ketoses	
5.3.1.9	D-glycose, 6 phosphate-Ketoisomerases	glucose-phosphate isomerase
5.4	Intra-molecular transferases	
5.5	Intra-molecular lyases	
6.	Ligases	
6.1	Forming C-0 bonds	
6.2	Forming C-S bonds	
6.3	Forming C-N bonds	
6.3.1	Acid-ammonia ligases	
6.3.2.2	L-glutamate ammonia ligase	glutamine synthatase
6.4	Forming - C - C bonds	

PHYSICOCHEMICAL NATURE OF ENZYMES

Most of the enzymes are made up of two portions-a coenzyme and an opoenzyme.

$$\text{Coenzyme (prosthetic group)} + \text{apoenzyme} \rightleftharpoons \text{holoenzyme}$$

Both portions are enzymatically inactive when separated from each other. Specificity of an enzyme to the substrate is due to apoenzyme portions. All chemical analyses have shown that this portion is a protein, composed of I-amino acid units. Enzymes are water soluble and their physco-chemical characteristics are similar to water soluble globular proteins. Heat, acid, alkali or other agents which denaturate proteins destroy activity of enzymes also.

They show colour reactions and ultraviolet absorption spectrum a characteristic of protein. Some enzymes e.g. pepsin are simple globular proteins without any prosthetic groups or coenzymes.

Enzymes either exist as single units or as aggregates of several subunits. Each subunit may have an active centre where substrate molecules bind during reaction. The molecular weight of enzymes ranges from 20,000 to many million. Three dimensional structure and amino acid composition of several enzymes have been worked out.

ENZYME KINETICS

Chemical equilibrium and energy of activation

According to the law of mass action, rate of a chemical reaction is proportional to the active masses of the reactants. The velocity constants, of forward and backward reaction may be written as K_1 and K_2.

$$A + B \underset{K_2}{\overset{K_1}{\rightleftharpoons}} C + D$$

An enzyme increases one of the velocity constants (K_1 or K_2) and reaction proceeds at faster rate to one direction. The equilibrium in a biochemical reaction, therefore, is regulated by the activity of the enzymes catalysing forward and backward reactions. This dynamic equilibrium is different from a true chemical equilibrium which is governed by thermodynamics. In the enzymes catalysed biochemical reaction, energy available for the reaction remains the same but enzyme increases the efficiency of the use of the energy. In other words, the enzyme activates a particular reactant with lower free energy to undergo changes. In fig. 14.1, e.g. the chemical compound A is in a metastable state. Energy E is required as energy of activation to convert the compound in a new activated state of B. The reaction then becomes spontaneous, converting B into C and energy E is recovered. A further increment of energy released according to the difference in enthalpy (ΔH) of A and C. Catalysts increase the rate of conversion of A into C with the lower energy of activation (e).

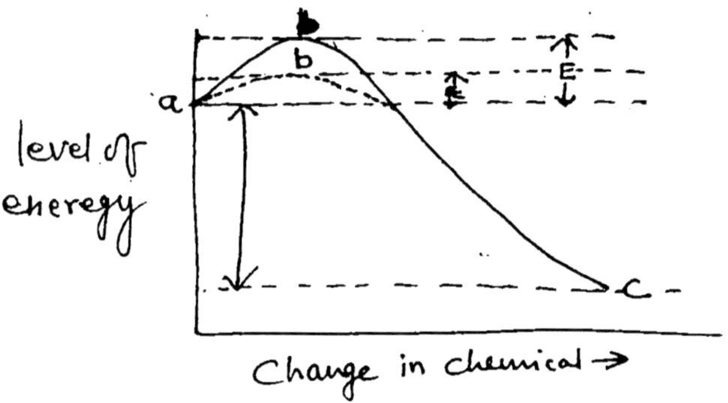

Fig. 14.1 : Existence and absence of a catalyst and corresponding energy of activation of a metastable compound

A catalyst or an enzymes lowers the energy of activation probably by forming a complex with reactants, which in turn can decompose liberating the catalyst and the products. Formation of enzyme/Catalyst-substrate complex requires less free energy than the activation of a reactant (substrate) in the absence of catalyst. Further, decomposition of a complex is spontaneous because it produces additional strains on the bonds in reactants, which have to be broken during the reaction e.g. the energy of activation for decomposition of H_2O_2 is 10 Kcal per mole, but in the presence of enzyme catalase it is only 6.4 K. cal per mole. The energy of activation in this case is reduced to about one third in the presence of enzyme.

Michaelis constant

In 1913, L. Michaelis and M. Menten postulated that enzyme forms a complex with the substrate during reaction. The complex then dissociates into the product (p) and the enzyme. The rates of these reactions are governed by constants K + 1 (for the formation of complex), K-1 (for the formation of complex), K - 1 (for the dissociation of complex), and K + 2 (for the formation of product) according to the following equation:

$$K + S \underset{K-1}{\overset{K+1}{\rightleftharpoons}} ES \xrightarrow{K+2} E + P \qquad ...1$$

Function of the rate constants of three different steps can be written by one constant Km the Michaelis constant-

$$Km \neq \frac{(K + 2) + (K - 1)}{(K + 1)} \qquad ...2$$

The rate of the conversion of the substrate to products will be proportional to the enzyme, substrate and complex. Substituting the concentration of enzyme, substrate and complex in equation 2, we get-

$$Km = \frac{(e - P)\ S}{P} \qquad ...3$$

or the rate of the reaction :

$$V = K + 2\ (ES) = K + 2\ (P) \qquad ...4$$

Where e is the total concentration of enzyme, both in free form and in complex from, p is the concentration of Es concentration of free substrate and then e-p is the concentration of enzymes in free form.

By re-arranging equation 3, we derive at-

$$\frac{Km}{s} = \frac{E-P}{P} = \frac{(E)}{(ES)} \qquad ...5$$

This equation can be rearranged to obtain the value of p:

$$P = \frac{es}{Km+s} = \frac{e}{Km/s+1} \qquad ...6$$

By substituting this value of p in equation (4),

$$V = K+2p - \frac{K+2e}{Km/s+1} \qquad ...7$$

At very high concentrations of substrate, virtually all the enzyme will be in the form of complex with the substrate. Under these condition s p = e and v will reach maximum value of V.

Therefore $V = K + 2e$ ...8

Substituting this value for K + 2e in equation (7),

$$V = \frac{V}{(Km/s) + 1} = \frac{Vs}{Km + s} \quad ...9$$

This is the final equation for the calculation of Michaelis constant (Km) in an enzymatic reaction.

The experimental values of Michaelis constant for many enzymes have been found to be that concentration of substrate at which the rate of reaction is at half of the maximum rate (fig. 14.2).

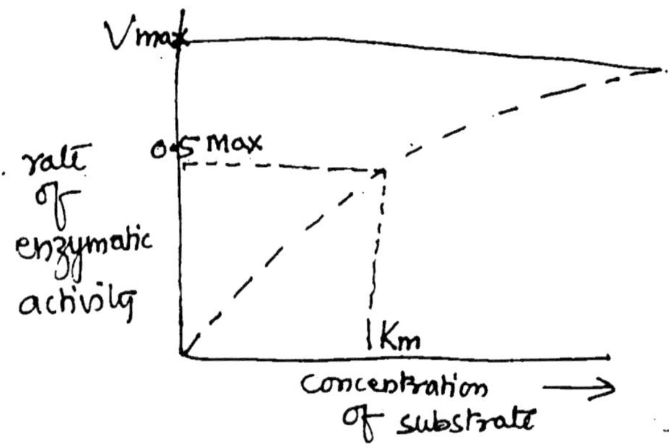

Fig. 14.2 : Determination of Michaelis Constant (Km)

Thus if v = 1/2V then equation (9) can be written as -

$$1/2V \quad \text{or} \quad \frac{V}{2} = \frac{V_s}{Km + S}$$

$$\text{or} \quad Km + S = 2s$$

$$\text{or} \quad Km = s$$

Thus, the experimental definition can be achieved through mathematical derivation also.

The Michaelis constant for most of the enzymes range between 10^{-2} to 10^{-5} M substrate concentration. The value of Michaelis constant is inversely proportional to the enzyme activity. A larger Km means that a high substrate concentration is required to achieve half of the maximum rate. It means that the enzyme has lower affinity to the substrate.

For experimental measurement of the Michaelis constants reciprocal of enzyme activity and substrate concentrate are plotted to yield a straight line (fig. 14.3). This kind of double reciprocal plot is the line Beaver-Burk plot. The value of Km from this plot is determined by the extension of line towards the abcissa.

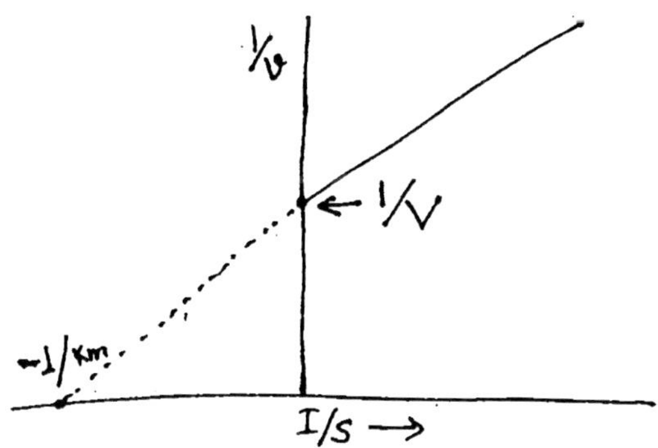

Fig. 14.3 : Deaver-Burk Plot

MECHANISM OF ENZYME ACTION

The mechanism of conversion of a substrate to a product by the enzyme appears to be the same as for a catalyst, except that it is very specific . It is known that enzyme combines with substrate to form a complex

$$E + S \rightleftharpoons ES$$

The evidence for the formation of enzyme-substrate complex was provided by the spectrophotometric analysis of the reactants. The substrate may have a specific colour or absorption spectrum. After adding the enzyme, a change in colour or absorption spectrum of the substrate indicates the formation of complex. This type of observation was first made using horse radish peroxidase and its substrate, hydrogen peroxide by B. Chance in 1940. Enzyme-substrate complex formation is further evidenced by enzyme-blocking phenomenon. When a compound similar to the substrate.

chemical structure is mixed with the enzyme, the enzyme is inactivated for the original substrate e.g. the activity of succinic dehydrogenase is inhibited by malonate, the structure of which is similar to succinate and which forms a complex with the enzyme. The binding of the substrate involves only active centre of the enzyme. The exact mechanism of the union of the enzyme with substrate may vary with the enzyme, however, that only active centre is involved in the process of evidence from the fact that the molecules of certain enzymes can be trimmed to smaller sizes without losing their catalytic activity. Thus, papain, may be trimmed to 60 from 180 amino acid residues without losing its activity.

The formation of enzyme-substrate complex involves a number of weak bonds and it causes some kind of stress on substrate molecule. Because of this stress, substrate molecule becomes reactive even under normal conditions of temperature etc.

Some specific bond(s) of the substrate molecule is broken or made under this condition and new compound is formed. After the reaction, product molecules leave the enzyme surface and the enzyme is freed to combine with the other substate molecules.

$$ES \Longrightarrow PE$$

After going through above para, it is apparent that the mechanism of action of enzymes is similar to the catalysts. The mechanism in enzyme action, however, is more precise only the specific bonds of substrate molecules are put out to stress and broken. The mechanism also explains the specificity of enzymes. Enzymes have a definite three dimensional configuration with a precise active centre which can accommodate only specific substrate molecules.

Khosland's induced fit model

D.E. Khosland et. al. (1966) postulated that the enzyme molecule has atleast 2 sites which are able to combine with the substrate. Combination of substrate at one site induces a conformational change in the enzyme which results in a greater affinity of the substrate to the other site. In this way, there is a co-operative effects of substrate molecules.

FACTORS AFFECTING ENZYME ACTIVITY

Substrate concentration

Enzyme activity increases with increase in concentration of the substrate to a maximum and then it levels off. Probably at this concentration of substrate amount of enzyme becomes limiting. Some enzymes are inhibited even by high concentrations of substrate. The phenomenon is called substrate inhibition.

Two types of graphs are obtained with substrate concentration and enzymes activity plot. Straight line or Michaelis type of curve is obtained with the enzymes which are made up of one unit (fig. 14.1). The enzymes which are made up of several subunits and show interaction amongst sub-units (allosteric ensymes), exhibit sigmoidal types of curve (fig. 14.4).

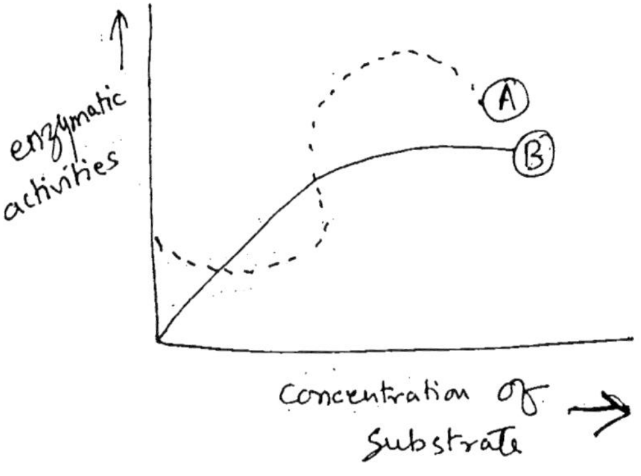

Fig. 14.4 : Substract concentration and respective enzymatic activities (A) Sigmoidal curve (B) Michaelis-Mentern type curve

Enzyme concentration

Increase in enzyme concentration increases enzyme activity. If an excess of substrate is present, increasing the enzyme by two times usually doubles the rate of formation of the product. With the concentration of the enzyme also, a point may be reached where all the substrate molecules are bound to the enzyme and further increase in enzyme has no effect on the formation of the product (fig. 14.5).

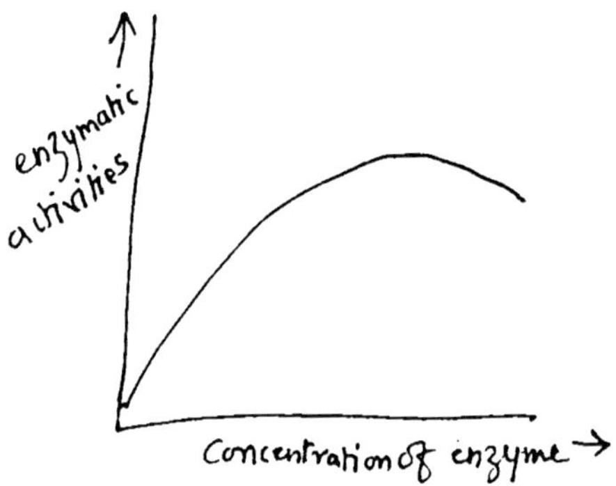

Fig. 14.5 : Enzyme concentration and corresponding enzymatic activities

pH

Because of their protein nature, enzymes are very sensitive to change in pH. A graph of enzyme activity against pH is usually bell shaped (fig. 14.6). There is a optimum pH, which varies with the enzyme. However, some enzymes have two optimum of pH or no sharp pH but a range of pH. Generally the optimum pH lies between 4 and 9, which depends upon the isoelectric pH of the enzyme and the substrate upon which they react.

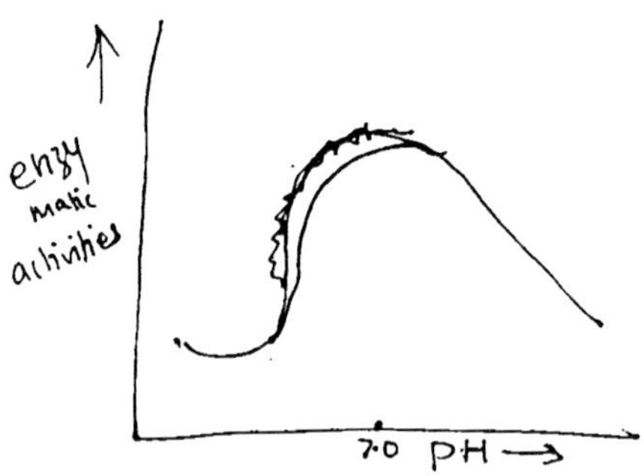

Fig. 14.6 : Effect of pH on enzymatic activites

Other ions

Many enzymes are inactive unless a cation are loosely bound to the enzyme, while in other cases they bind with the substrate. Anions are also effective kin enhancing the enzyme activity, as chloride enhances the activity of salivary amylase. Concentration of ions also effects the enzyme activity.

Temperature

For most enzymatic reactions, rise of 10°C in temperature (Q_{10}) doubles the rate of reaction between 5° to 40°C. At higher temperatures Q_{10} drops to 1° or lower.

Enzymes are damaged irreversibly at 50° to 60°C, because of their proteinaceous nature. A few enzyme continue to be active above 60°C also. The enzymes of dry tissues such as seeds and spores are most resistant to high temperature. A general curve for the relationship of enzyme activity and temperature in given in fig. 14.7.

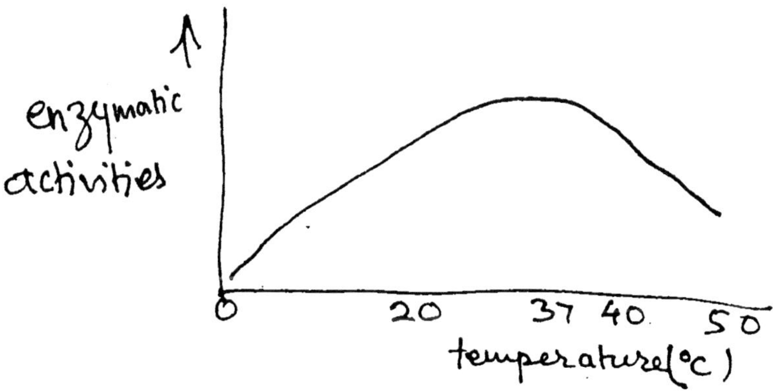

Fig. 14.7 : Temperature effect and corresponding enzymatic activities

Redox potential

Enzymes are sensitive to the redox potential of the cell also. Oxidising and reducing enzymes are involved in changing the redox component of a cell and hence they are influenced by the redox potential. Many other enzymes are affected by redox potential because of the presence of readily oxidisable-SH group in their molecule.

Enzyme inhibitors

The activity of enzymes can be inhibited by a great variety of substances. Some inhibitors resemble the substrate molecules in the structure. They unite with the enzyme at a place where substrate is to bond with. This leaves less or no enzyme molecules available for binding with the substrate. This kind of inhibition is called competitive inhibition. In competitive inhibition, if substrate concentration is increased the inhibitor is displaced from the enzyme and the rate of enzyme activity for the normal substrate increases. Inhibition of succinic dehydrogenase by malonic acid, as mentioned earlier is an example of competitive ihibition. Malonate complete with succinate since bath the structurally similar.

$$\begin{array}{c} COOH \\ | \\ CH_2 \\ | \\ COOH \end{array} \qquad\qquad \begin{array}{l} CH_2 —COOH \\ | \\ CH_2—COOH \end{array}$$

Malonic acid   Succinic acid

Many antibacterial drugs work on the principle of competitive inhibitors of the enzymes. Sulfa-drugs are believed to be competitive inhibitors of the enzymes metabolising p-amino benzoic acid (as a coenzyme. Para amino-benzoic acid is an essential vitamin for many bacteria and is structurally similar to sulphanilamide. Sulphanilamide inhibits the growth of those bacteria which require p-amino benzoic acid.

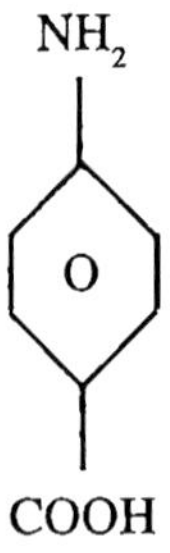

NH_2 — (benzene ring, O) — $COOH$

p-amino benzoic acid

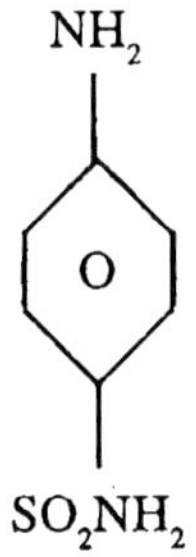

NH_2 — (benzene ring, O) — SO_2NH_2

Sulfanilamide

Non-competitive inhibitors are not specific and combine irreversibly with the enzyme.

Many-SH containing enzymes are inhibited by Hg and As. Similarly many Cu and Fe containing enzymes are inhibited by Hg and As. Similarly many Cu and Fe containing enzymes are inhibited by CN. Cyanide combines with these element and prevents them from performing their role in enzyme action. The non competitive inhibitors are different from other factors or inhibitors which destroy the enzyme integrity.

CO-ENZYMES

Although some enzymes such as lysozyme, chymotrypsin and trypsin are made up exclusively of protein, most of the enzymes have a non-proteinaceous portion also. ***The non-proteinaceous part of enzyme, which can be reversibily removed from the enzyme by dialysis is called coenzyme.*** Those groups which are tightly bound to the enzyme and can not be removed by dialysis are prosthetic groups. The coenzymes are also called co-substrates. Coenzymes and prosthetic groups particulate actively in the reaction and are altered chemically. Usually a second reaction involving another enzyme restores the original structure of coenzyme and prosthetic groups.

STRUCTURE AND CLASSIFICATION OF COENZYMES

Metallicions, vitamins and other types of organic compounds function as prosthetic groups or coenzymes. An essential component of most of coenzymes is phosphate in the form of nucleotides. Nicotinamide adenine dinucleotides, commonly known as co-enzyme I (NAD) or co-enzyme II (NADP) are coenzymes for dehydrogenases. Iron containing cytochromes transfer electrons from flavoproteins to cytochrome oxidase. Pyruvate decarboxylase, an enzyme which splits pyruvate to CO_2 and acetaldehyde in yeast cells contains thiamine (vitamin B_1) as a part of its coenzyme.

Coenzymes are classified according to the reactions they participate in. Such a classification with their corresponding vitamins is given in Table. 14.2.

Table 14.2 : Classification of major coenzymes

Class	*Abbreviation transferred*	*Group*	*Corresponding Vitamins*
1	*2*	*3*	*4*
1. Hydrogen transferring coenzymes			
Nicotinamide Adenine dinucleotide	NAD/NADP	Hydrogen/ Electrons	Nirotinamide
Glavomono nucleotide	FMN Electrons	Hydrogen/	Riboflavin
Flavin adenine dinucleotide	FAD	Hydrogen/ Electrons	Riboflavin
Ubiquinone	Q Electrons	Hydrogen/	Riboflavin
Lipoic acid	Lip (S_2) Electrons and acyl group	Hydrogen/	-
Cytochromes	-	electrons	-
2. Group transferring coenzymes			
Adenosine triphosphate	ATP	phosphate AMP	-
Adenosyl methionine	"	methyl group	methionine
Phosphoadenyl sulphate	PAPS	Sulphate group	-
Pyridoxal phosphate	PALP	Amino Group	Pyridoxine
Cytidine diphosphate	CDP choline	Phosphoryl	-
Uridine diphosphate	UDP acid	Sugar uronic	-
Tetrahydrofolate	COF group	Formyl Folate	
Biotin	- group	Carboxyl Biotin	
Coenzyme A	CoA	Acyl group	Pentothenate

1	*2*	*3*	*4*
Thiamine pyrophosphate	RPP	C_2 aldehyde group	Thiamine
3. Isomerase and lyase coenzymes			
Uridinediphosphate	UDP	sugar isomerisation	-
Pyridoxal phosphate	PALP	decarboxylation	pyridoxine
Thiamine pyrophosphate	TPP	"	Thiamine
B_{12} coenzyme	-	carboxyl displacement	Cobalamin
Glutathione	-	Isomerisation of olifins	-

MECHANISM OF ACTION OF COENZYMES

Some important coenzymes are discussed in following paragraphs with special emphasis on chemical structure and mechanism of action:-

Nicotinamide Adenine Dinucleotide

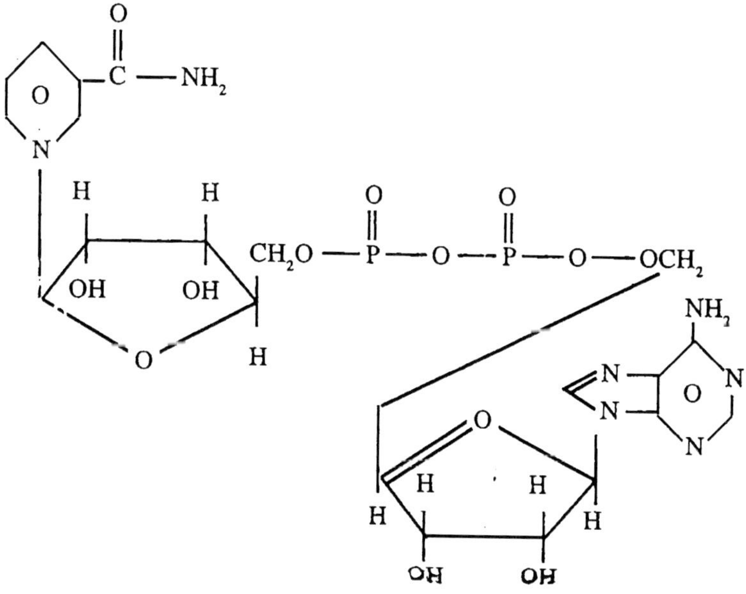

In the process H is added to a substrate from C-4 atoms of the pyridine nucleus. The pyridine is reduced retaining only two double bonds.

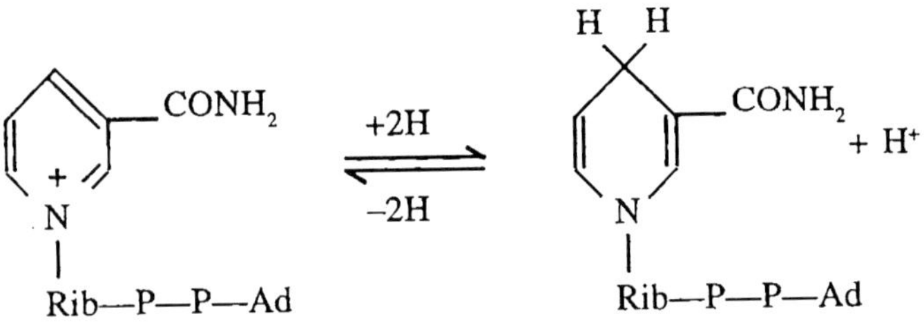

$$NAD^+ \underset{-2H}{\overset{+2H\cdot}{\rightleftharpoons}} NADH + H^+$$

NADH absorbs light at 340 nm but NAD^+ does not absorb at that wave length. Thus, if NADH is produced during a reaction, absorption of light at 34" nm increases. The increase in absorption is proportion in the enzyme activity. This property has been used extensively in the bioassay of dehydrogenases and other oxidoreductases which utilize NAD^+/NADH as coenzymes.

ADENOSINE TRI PHOSPHATE

Adenosine tri phosphte has three phosphoric acid residues with the adenoisine ribose nucleotide.

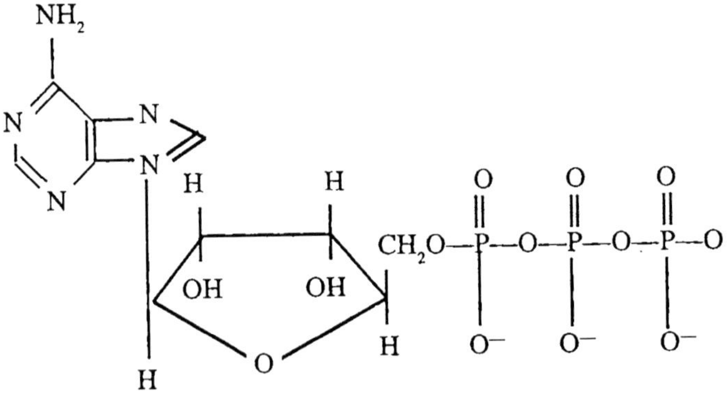

Adenosine tri phosphate (ATP)

ATP contains two energy rich bonds and can transfer following types of groups:

1. Transfer of orthophosphate group and release of ADP.
2. Transfer of diphosphate and release of AMP.
3. Transfer of adenosyl monophosphate release of pyrophosphate
4. Transfer of adenosyl group and release of both ortho and pyrophosphate.

Transfer of orthophosphate groups is the most common reaction of ATP. ATPase transfer this orthophosphate to water molecules. Kinases transfer it to hydroxyl, acyl of amide groups.

PYRIDOXAL PHOSPHATE

Several transmines, decarboxylases and synthatase use pyridoxal phosphate as coenzyme. Some of these reactions are catalysed by PALP even in the absence of enzyme although at a slower rate.

O
||
HO—P—OCH$_2$ CHO OH
|
OH O
N CH$_3$

Pyridoxal phosphate (PALP)

The mechanism of catalysis by this coenzyme involves intra molecular chemical reactions via series of steps. In these reactions, a series of electron pairing and proton transfer processes occur. Pyridoxal phosphate forms an intermediate compounds (Shiff's base), which reacts further in various ways to form different product.

THIAMINE PYROPHOSPHATE

This coenzyme is involved in the transfer of acyl and carboxyl groups. It is derived from the vitamin, thiamin (Vit. B_1).

CHO
P—O—CH$_2$ X
—OH + R—CH $\xrightleftharpoons{-H_2O}$
—CH$_3$ NH$_2$
N

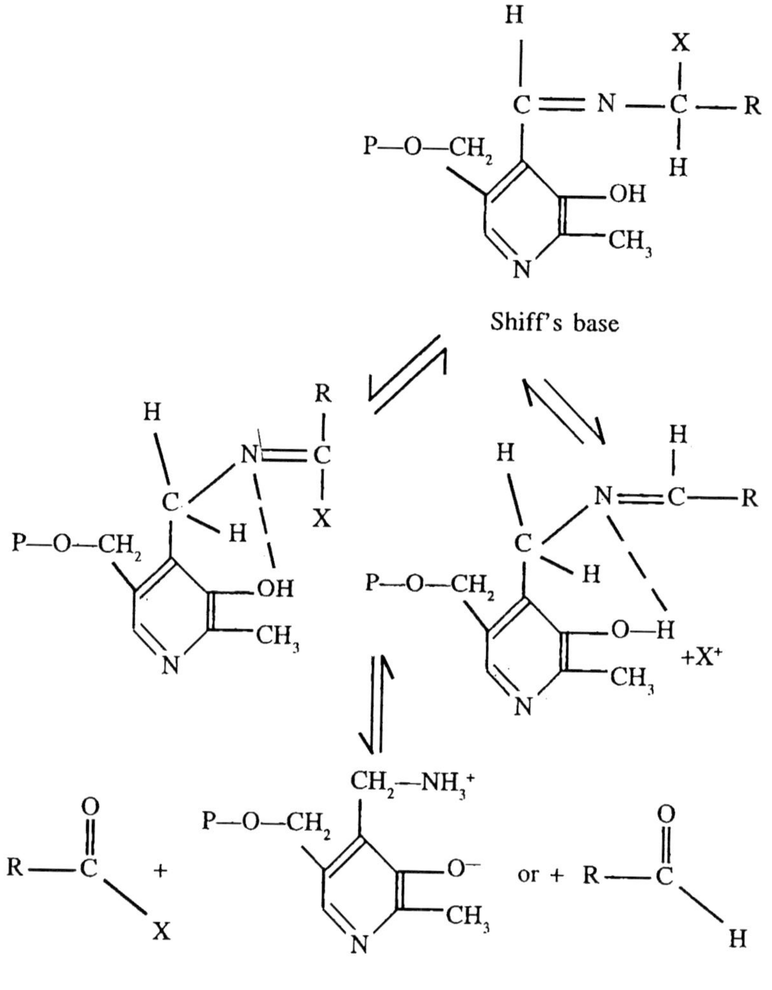

Shiff's base

Pyridoxamine phosphate

THIAMINE PYROPHOSPHATE (TTP)

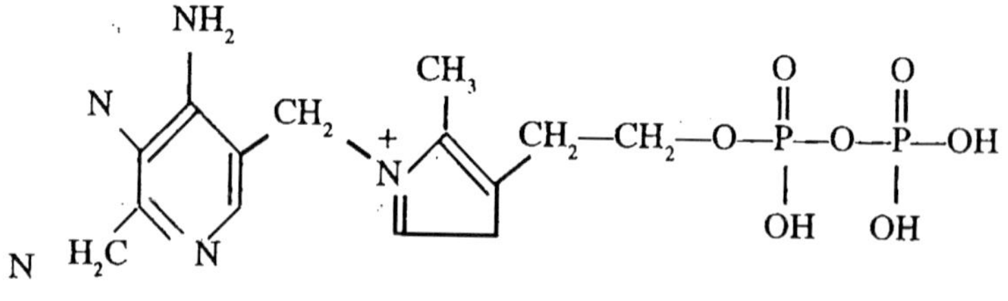

Thiamine pyrophosphate (TTP)

Thiamine pyrophosphate is a prosthetic group, since it remains attached to the enzyme protein. The functional part of TPP molecule is the sulphur containing thizolium ring. The hydrogen atom of C_2 of this ring is exchanged in aqueous solution and takes part in the reaction.

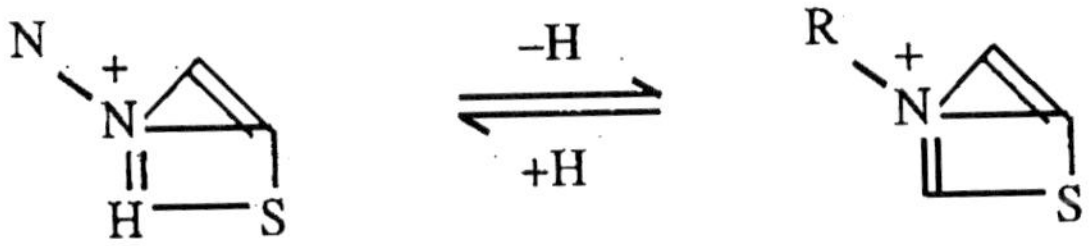

-H

+H

The most important reaction catalysed by TPP is the decarboxylation of pyruvic acid to acetaldehyde. The reaction mechanism is as follow:

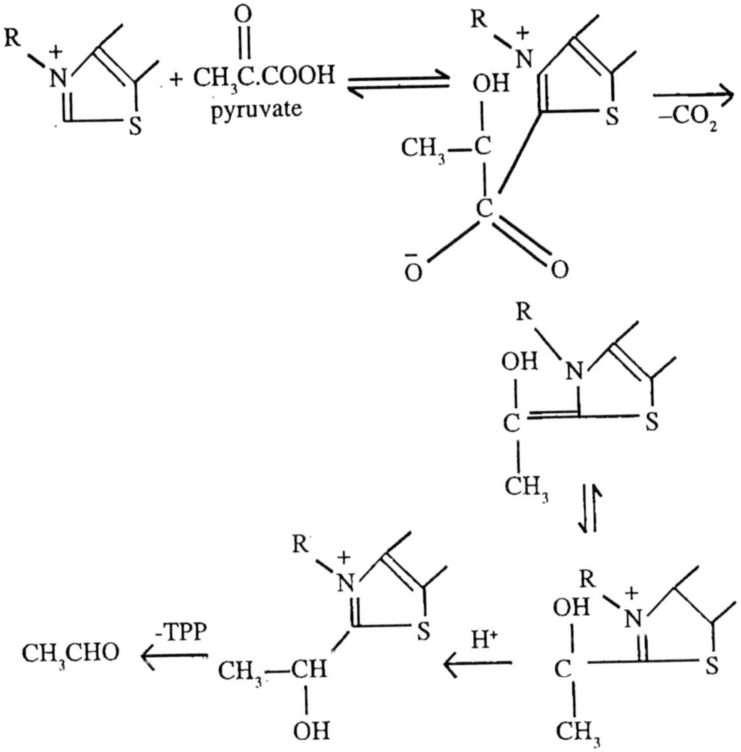

+ $CH_3C.COOH$
pyruvate

$-CO_2$

H+

-TPP

CH_3CHO

The decarboxylation of pyruvic acid by this mechanism is a non enzymatic process and it involves transfer of electron pairs. In all other thiamine catalysed enzymic reactions also, the mechanism is almost similar. Thiamine cleaves the carbonyl-x bond, where x is usually -H or - COO^- group. A substrate - thiamine complex is later on cleaved at a-carbon atom to release the coenzyme.

COENZYME A

Coenzyme A is involved in several acyl group transfer reactions. A common example is the formation of CoA from acetate.

Coenzyme A is a complex compound. The most important CoA compound is acetyl CoA (activated acetate), which participates in several acyl transferring reactions. In this, the acyl residue is bound to free-SH groups, forming a very reactive thioester. The formation of thioester and acetoacetyl CoA, occurs through the following pathway.

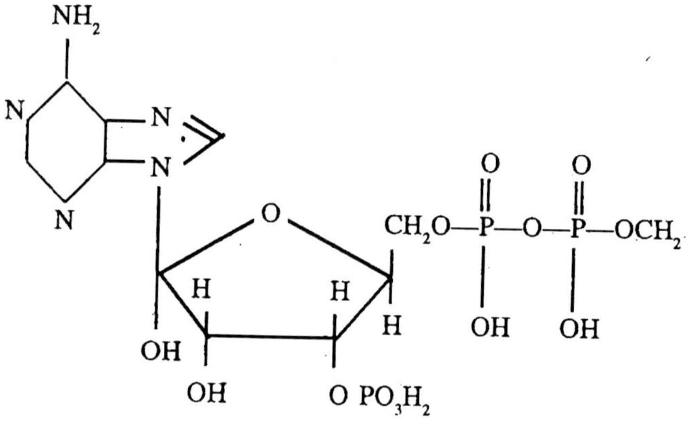

Adenosine triphosphate (ATP)

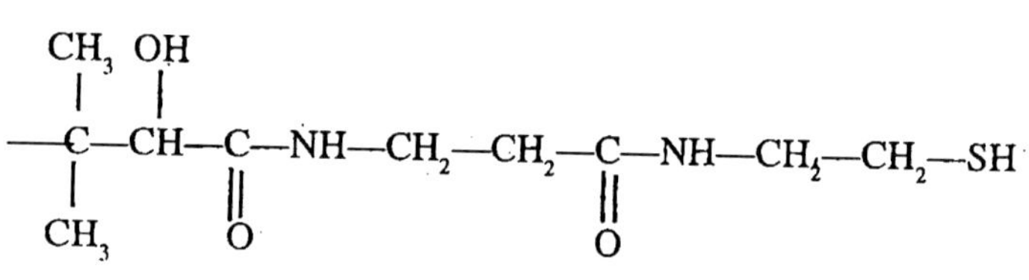

Coenzyme A

$$\underset{\text{acetaldehyde}}{CH_3CHO} + \underset{\text{coenzyme}}{CoA—SH} \xrightarrow{NAD^+}$$

$$\underset{\text{Acetyl CoA}}{H_3—\overset{\overset{O}{||}}{C}—S—CoA} + NAD^H + H^+$$

$$\searrow$$

$$H_3C—\underset{\underset{O}{||}}{C}—CH_2—\overset{\overset{O}{||}}{C}—SCoA + HSCOA$$

H_2O (↙) → $CH_3\overset{\overset{O}{||}}{C}.CH_2COO^-$ + CoASH + H^+

(↘) Further biosynthesis

The formation of acetoacetyl CoA from acetate thioester involves nucleophillic displacement at the carbon and one more acetate group is added at the place of coenzyme A. Besides acylation, coenzyme A can undergo phosphorylation also :

$$CoASH + X—O—\underset{\underset{OH}{|}}{\overset{\overset{O}{||}}{P}}—OH \longrightarrow CoA—S—\underset{\underset{OH}{|}}{\overset{\overset{O}{||}}{P}}—OX + H_2O$$

In the reaction catalysed by phosphotrans acetylase, thiokinase and phosphokinase, phosphosylation of CoA takes place. Thiokinases catalyse the following reaction through CoA.

$$\text{acetyl CoA + adenylic acid} + 2H_2\,PO_3 \longrightarrow \text{ATP + acetate + CoA—SH}$$

Coenzyme A is involved in the elimination reaction also, e.g. in the formation of unsaturated fatty acids from saturated fatty acid. Another coenzyme FAD cooperate in this reaction.

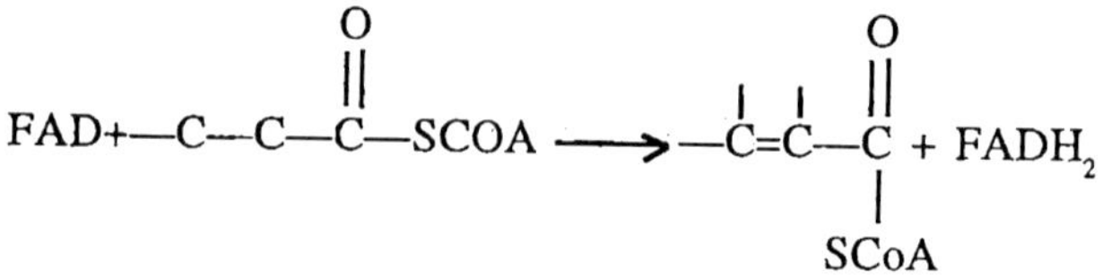

$$FAD^{+} - C - C - \overset{O}{\overset{\|}{C}} - SCOA \longrightarrow -\overset{|}{C}=\overset{|}{C}-\underset{SCoA}{\underset{|}{\overset{O}{\overset{\|}{C}}}} + FADH_2$$

FLAVIN NUCLEOTIDES

Flavin nucleotides exist in two forms: mono nucleotide (FMN) and dinucleotide (FAD) These nucleotides are involved in oxidation reduction reaction. The common substrates are reduced pyridine and saturated carbon compounds.

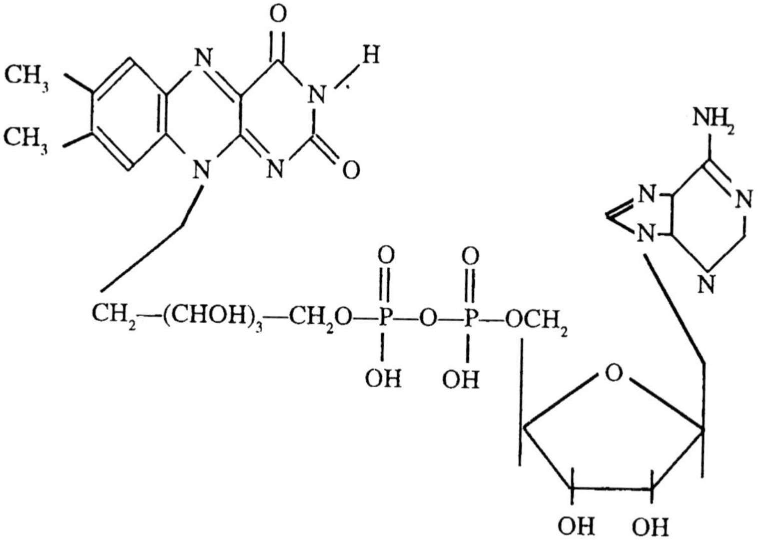

CH_3, CH_3, N, O, H, N, N, N, O, NH_2, N, N, N, N, O, O

CH_2—$(CHOH)_3$—CH_2O—P—O—P—OCH_2

OH OH O

OH OH

Flavin Adenine Dinucleotide

The nucleotides are in either reduced or in oxidized form:-

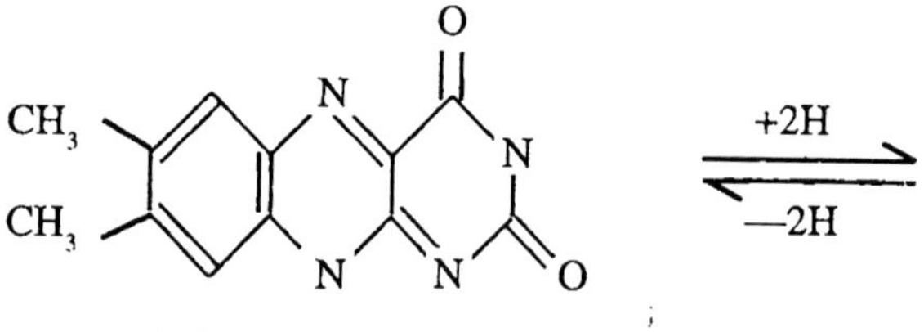

CH_3, CH_3, O, N, N, N, N, O

+2H ⇌ —2H

Oxidized form

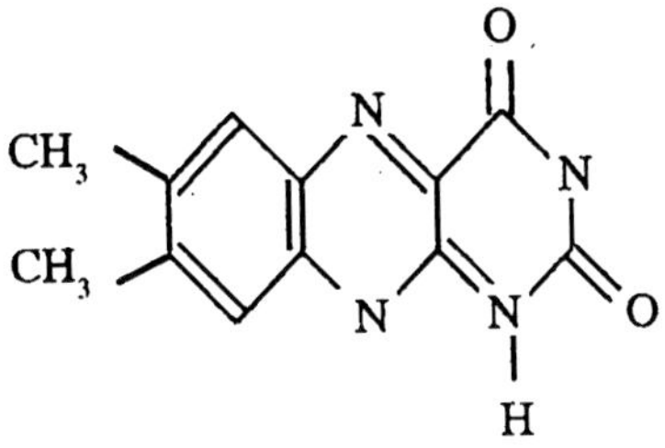

reduced form

Hydrogen is transferred to the prosthetic group through the action of enzyme. The prosthetic group is later on released as usual by another enzyme.

Chapter -15

NUCLEIC ACIDS IN BIOLOGICAL SYSTEM

Nucleic acids are considered as fundamental in life processes and important in carrying genetic transformations from one generation to another. There are types of nucleic acids are knwon. These are RNA (Ribonucleic acid having molecular weight 20,000-1,000,000 and DNA (Deoxyoibonucleic acid) Molecular weight 1000,000,000+. Generally, nuclic acids are composed of :

1. sugar,
2. phosphoric acid and,
3. base.

1. sugar- ribose- +5 carbon atom compound

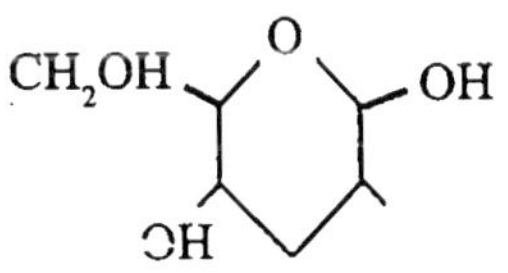

CH_2OH O OH OH

β-D-2 Deoxyribose

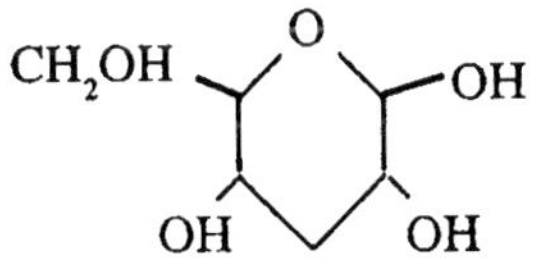

CH_2OH O OH OH OH

β-D-2 ribose

2. Phosphoric acid

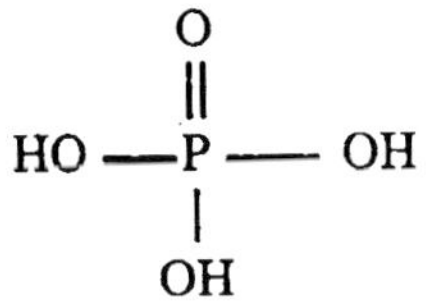

$$HO-\overset{\overset{\displaystyle O}{\|}}{\underset{\underset{\displaystyle OH}{|}}{P}}-OH$$

3. Base

a. heterocyclic *purines*

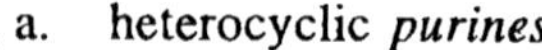

adenine

guanine

b. pyrimidines

cytosine

uracil (uradine)

thymine

- RNA bases = guanine, adenine, uracil, cytosine
- DNA bases = guanine, adenine, thymine, cytosine

Nucleic acids like protein occur in all living cells. They derive their name because of their primary occurrence in the nucleus & acidic nature. Deoxyribose nucleic acid (DNA), one of the two nucleic acids is a major component of chromosomes. Small amount of DNA is also found associated with chloroplast & mitochondria. Certain animal viruses & many bacteriophases are DNA-protein particles. Ribonucleic acid (RNA), the other nucleic acid is found mostly in the cytoplasm & in the plant viruses. Very little of DNA & RNA occur in free forms. Nuclear DNA is associated with arginine & lysine rich proteins.

CHEMICAL COMPOSITION

Elements taking part in the constitution of nucleic acids are C,H,O,N & P. In almost all nucleic acids, there are approximately 15 to 16% nitrogen & 9-12% phosphorus.

Nucleic acids are made up of monomeric units of nucleotides which are analogous to amino acids of proteins. But the number of nucleotides taking part in the synthesis of nucleic acids is fewer. Nucleotides are not stable units & can be further hydrolysed to purpine pyrimidine bases, ribose & deoxyribose sugars & phosphoric acid. After partial hydrolysis, nucleotides & phosphoric acid are obtained. Thus, stepwise degradation of nucleic acids will yield different compounds in the following manner:-

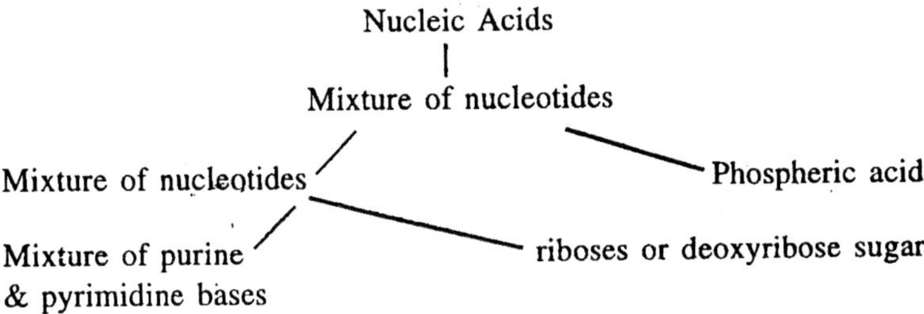

PURINES

The parent substance purine contains 6 membered pyrimidine ring fused to a 5 membered imidazole ring. Adenine & genuine are the major purines that have been isolated from nucleic acids. Adenine is 6-amino derivative of parent purine, while guinine is 2-amino-6-oxypurine.

Adenine

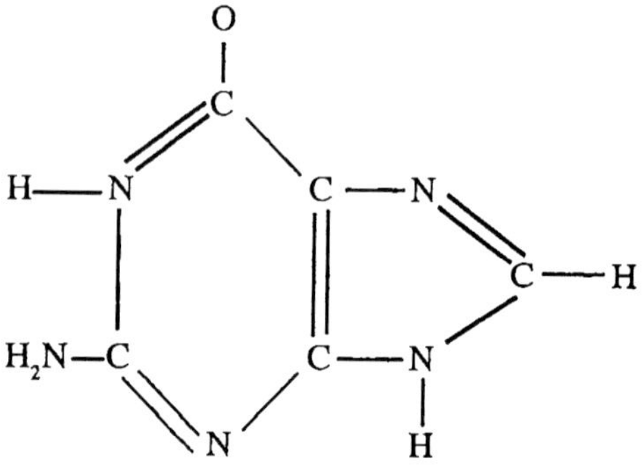

Guanine

PYRIMIDINES

Pyrimidine is a 6 membered heterocylic compound. Pyrimidine itself has not been isolated from natural products. Pyrimidines prevalent in nucleic acids are Uracil, thymine, cystosine & of more limited distribution, 5 methyl cytosine.

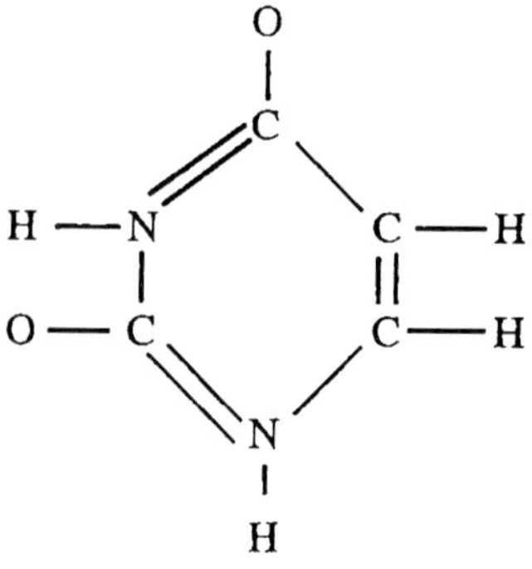

Uracil

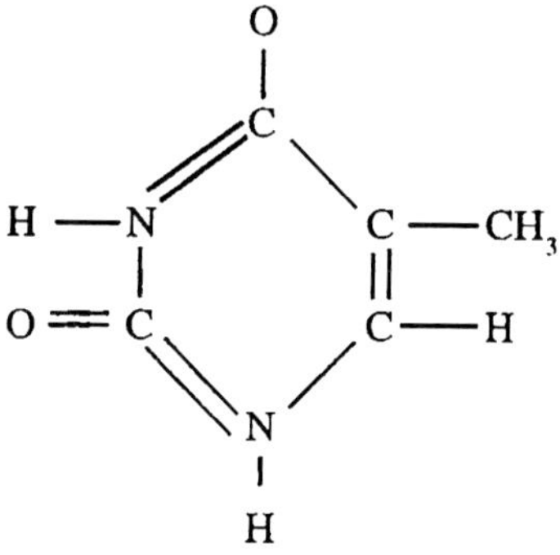

Thymine

Cytosine **Methyl Cytosine**

SUGARS

Sugar components of the nucleic acids are ribose & deoxyribos sugars which are penta aldoses in pyranose ring form.

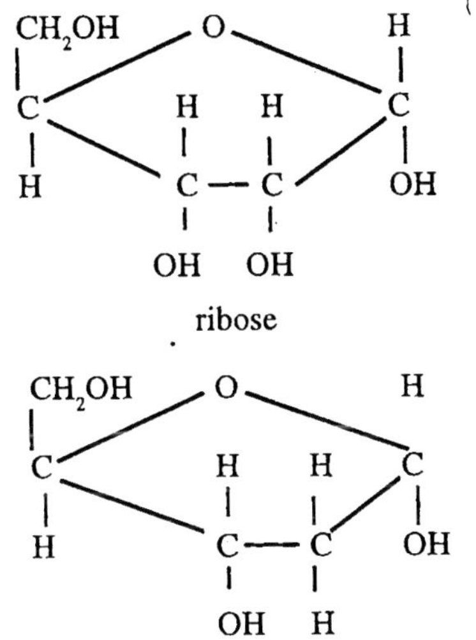

ribose

Deoxyribose

NUCLEOSIDES & NUCLEOTIDES

Ribose or deoxyribose attached to a purine or pyrimidine molecule is called a **nucleoside**. Adenine linked to ribose or deoxyribose is called adenosine & the genuine nucleoside is called

guanosine. The purimidine nucleosides are uridine, cytidine & thymidine. Deoxyribose sugar also forms nucleosides in a similar way.

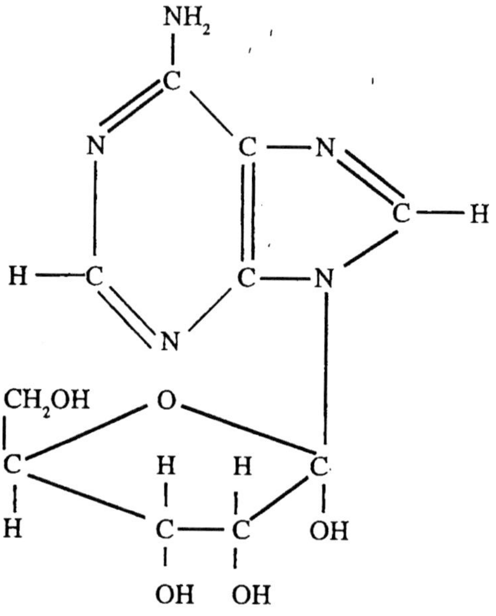

Ribose adenosine

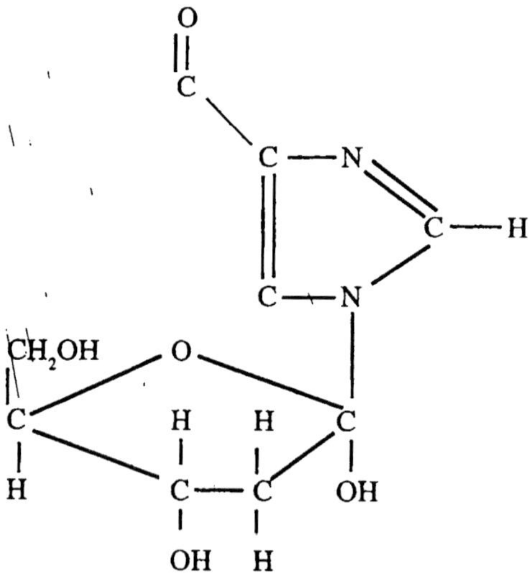

Deoxyribose Guanosine

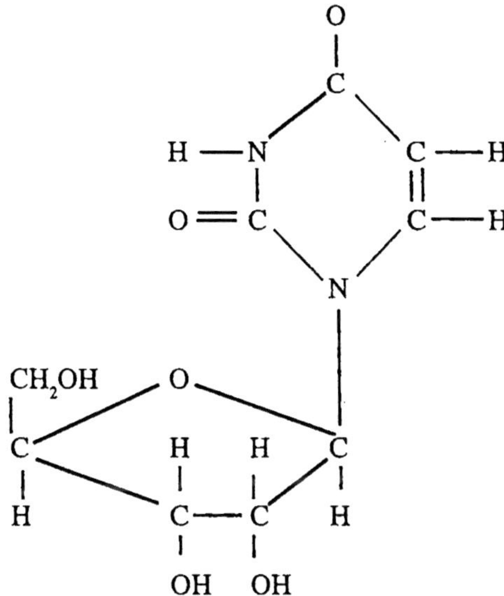

Ribose Uridine

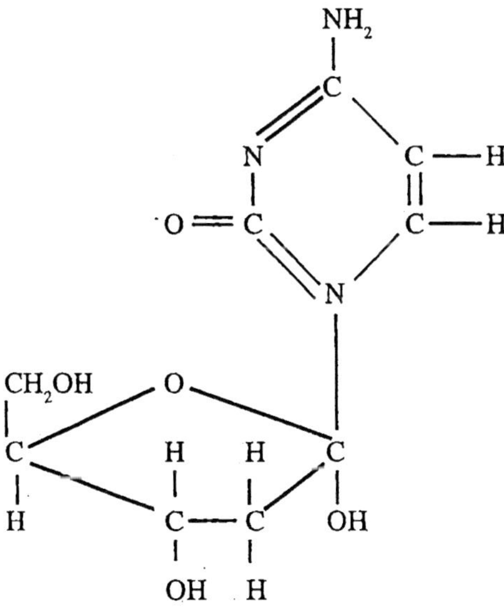

Deoxyribose Cytidine

Nucleotides are phosphate esters of nucleosides & are strongly acidic. They are thus called acid. They are also named as adenosine monophosphate (AMP) and similarly GMP, TMP, CMP & UMP according according to the purine or pyrimidine base present. The naming of nucleosides & nucleotides is given in Table- 15.1.

Table 15.1 : Nomenclature of ribose nucleosides & nucleotides

Base	*Nucleosides*	*Nucleotide*
Adenine	Adenosine Adenine riboside	Adenosine monophosphate Adenyclic acid
Guanine	Guanosine Guanine riboside	Guanosine monophosphate Guanylic acid
Cytosine	Cytidine cytosine riboside	Cytidine monophosphate cytidylic acid
Uracil	Uridine Uracil riboside	Uridine monophosphate Uridylic acid

Thymine is associated with deoxyribose sugar only & the corresponding nucleosides & nucleotides are thymine deoxyriboside & thymidine monophosphate or thymidylic acid.

The nucleotides are monomeric units of nucleic acids. That is why the primary structure of nucleic acids is designated as polynucleotide. The two nucleic acids DNA & RNA differ in their sugar & pyrimidine base components (Table 15.2).

Table 15.2 : Chemical differences between DNA & RNA

Component	*DNA*	*RNA*
Sugar	Deoxyribose	Ribose
Pyrimidine bases	Thymine	Uracil
	Cytosine	Cytosine

PHYSICAL CONFIGURATION

DEOXYRIBOSE NUCLEIC ACID

The first satisfactory model of DNA structure was proposed by J.D. Watson & F.H. Crick in 1953. This model with minor modification is still a valid representation of structure of DNA. A DNA molecule consists of two long chains twisted in the form of a regular double helix (Fig. 15.1) helical structure of DNA differs from that of protein where amino acids are arranged in a sing. chain. The distance between two chains of the helix is about 20Å & the helix makes a turn after every 24Å. Each turn of the chain contains about 10 nucleotides. the adjacent nucleotides in the same chain are joined together through phosphate group attached to the sugar part of the nucleotide. The joining together of the sugar & phosphate is very regular, alway involving the same group.

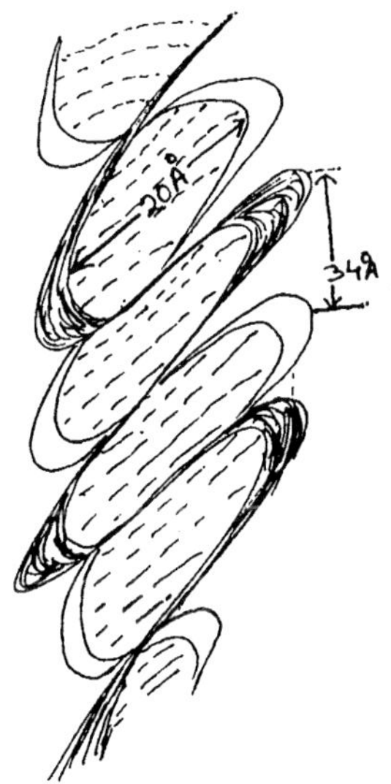

Fig. 15.1 : Watson Crick model of DNA, molecule

Purine or pyrimidine bases of two chains are joined together by hydrogen bonds. The pairing of bases in the opposite chains is so regular that they leave space for only one hydrogen atom. Adenine is always paired with thymine & guinine with cytosine. A purine occupies more space than a pyramidine so to make a regular double helix with a pyrimidine. Thus, the two chains of DNA are complimentary e.g. if the sequence of nucleotides, in one chain is ATGTC (adenine, thymine guanine, thymine, cytosine), the opposite chain must have TACAG (thymine, adenine, cytosine, adenine, genuine). The arrangement of nucleotides in a chain is very specific & is characteristic of the living system.

Hydrogen bonds in the DNA molecules are generally formed at precise locations. In the hydrogen bond, - C =O group of the one base faces opposite the -NH or NH_2 group of the other base with a proper distance (fig. 15.2).

Average molecular weight of natural DNA is very high. The m.w. ranges between 6×10^6 to 10×10^6.

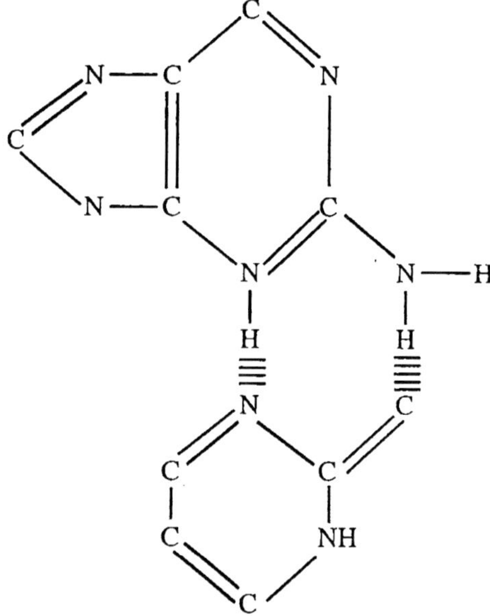

Fig. 15.2 : Hydrogen bonding between the nucleotides

In bacteria & viruses, DNA of lower m.w. are found. DNA is denatured by alkali, acid, heat or low ionic strength. After denaturation, it is converted to a single stranded flexible structure. Ultraviolet absorption, optical rotation & viscosity of denatured DNA is different than the natural one. Denaturation of DNA by heat occurs at a characteristic temperature called the ***melting temperature*** (Tm). During heating or other means of denaturations, the hydrogen bonds of DNA break to separate the two strands. Sometimes, when denatured DNA is cooled or brought back to original condition slowly, a part of the DNA molecule resumes double stranded structure. This is called **Renaturation.**

SINGLE STRANDED AND CIRCULAR DNA

Although most of the organisms have regular double helical DNA, some bacteriophages & animal viruses contain single stranded DNA. Further, DNA of most phages is not linear but is in the form of a circle. The DNA øx 174 phase & of the polyoma virus, which causes cancer in monkeys, are circular. DNA from the mitochondria of higher organisms is also circular.

The *in vitro* origin of circular DNA in the phase has been studied. DNA isolated from some viruses is in fact linear. When it is heated, the two strands of DNA separate & it is denatured. Controlled recoiling brings the two strands together & some normal double stranded DNA is obtained. In this recoiling some portion of DNA may assume circular form also (fig. 15.3) probably join to form a circle. By observing under electron microscope, it has been established that about 20% of the renatured DNA in polyoma virus assumes circular shape during renaturation.

RIBOSE NUCLEIC ACID

There are several types of RNAs which differ in the composition, size, function & properties. Most of the RNAs are long unbranched chain containing 4 types of ribonucleotides linked together by phosphate bonds. Unlike double helical DNA, many of these do not have a regular structure.

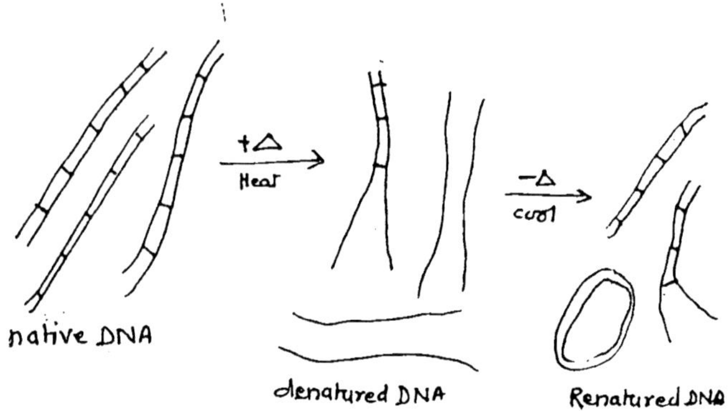

Fig. 15.3 : In ivtro origin of circular DNA in bacteriophages

THREE TYPES OF RNAs

SOLUBLE OR TRANSFER RNA

It makes about 10 to 20% of the cellular RNA & is composed of relatively smaller molecules (m.w. 25000 to 30000). It acts as mediator in the synthesis of polypeptide. The number of nucleotides in a RNA ranges from 75-85. Many unusual bases like inosine, pseudo-uridine, dihydrouridine etc. are present in transfer RNA. The function of these unusual bases is not known. Two dimensional structure of t-RNA was studies & a clover leaf model was proposed by G.R. Phillips (1969). As shown in fig. 15.4 end of the RNA is made of ACC & a variable fourth nucleotide. The amino acid to be transported to the ribosomes for protein synthesis attaches to this end. The clover leaf model of t-RNA envisages three loops in the structure. First loop is made up of 7 unpaired. nucleotides. This region is probably involved in the binding with the ribosomes. Extra arm or the lump of t-RNA consists of several nucleotides. Second loop consists of 7 unpaired nucleotides & contains anticodon of 3 nucleotides which binds covalently with the codon of messenger RNA during translation. The third loop contains & to 10 unpaired nucleotides & is rich in dihydrouracil nucleotides. It is probably involved in the binding of the enzyme amino acyl synthatase. The stem of the model consists of about 10 paired nucleotides.

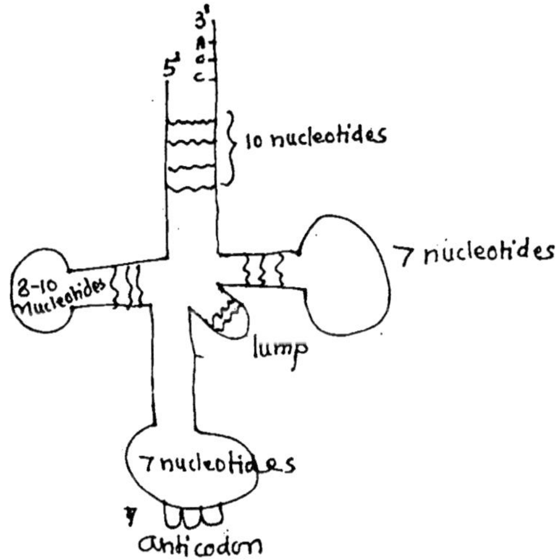

Fig. 15.4 : Clover leaf model tRNA

RIBOSOMAL RNA

Ribosomes contain a large portion of the RNA of the cell, representing as much as 80 percent of the total in some bacteria. Associated with the two sub units of ribosomes, r-RNA are of two sizes 23S & 16 S. A third species of 5S has also been described from E.Coli & mammalian cells. Although most t-RNA are single stranded, many assume a hair pin structure by hydrogen bonding in the same chain (fig. 15.5). The r-RNA has unequal amounts of guinine & cytosine & of adenine & uracil.

It has been possible to reconstitute the ribosomes from RNA & protein. The exact function of r-RNA, however is not known.

MESSENGER RNA

The base composition of this type of RNA is similar to that of DNA except that thymine is replaced by uracil.

It is present in cell in small quaintly, less than 5 percent of the total RNA. Molecular wt. varies but is in the order of 2 to 5 $\infty 10^5$. In E.coli, the average size of m-RNA is 900-1500 nucleotides.

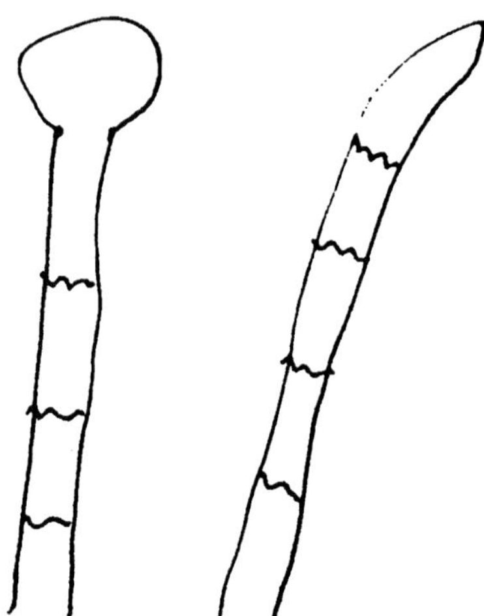

Fig. 15.5 : Hair pin models of rRNA

Messenger RNA molecules carry genetic information from DNA to the protein molecules. Although; m-RNA in most organisms is single stranded chain, it may assume secondary folding in many viruses.

BIOLOGICAL SIGNIFICANCE

DNA is found in all biological systems, with the exception of certain viruses. In higher organisms, it is confined to the nucleus, chloroplasts & mitochondria. DNA is genetically active component of chromosomes & functions as basic carrier of genetic information. Many of the enzymes present in mitochondria & chloroplasts are synthesized under the direction of DNA present in those organelles. In some plant viruses, the genetic function is performed by RNA. DNA also works as a template for the synthesis of various RNAs present in the cell. RNAs in turn perform other functions like structural organisation of ribosomes, transfer of amino acids from cytoplasm to the ribosomes & carrying the message for protein synthesis. Besides this, many nucleotides and bases are important metabolites. Vitamin in B_1 (thiamine) is a pyrimidine derivative. In recent years, certain pyrimidine derivatives have been found to be of important biological applications, Transfer RNA from yeasts, bacteria, mammals & higher plants when degraded by enzymes or acids, yields active cytokinins.

Isopentenyl adenosine is found in the rRNA coding for serine & tyrosine in the yeast. Zeatin riboside is found in the t-RNA of many higher plants.

BIOSYNTHESIS OF NUCLEOTIDES

PYRIMIDINE NUCLEOTIDES

The first step in the synthesis of pyrimidine is the formation of carbamyl phosphate from ammonia or amide, CO_2 & ATP.

$$NH_3+CO_2+ATP \xrightarrow[\text{Kinase}]{\text{Carbamate}} \underset{\text{carbomyl phosphate}}{NH_2\text{-}COO\text{-}PO_3H_2} + ADP$$

Carbamyl phosphate condenses with aspartic acid in the presence of enzyme aspartate transcarbamylase. The phosphate is released & carbamyl asparate is synthesized -

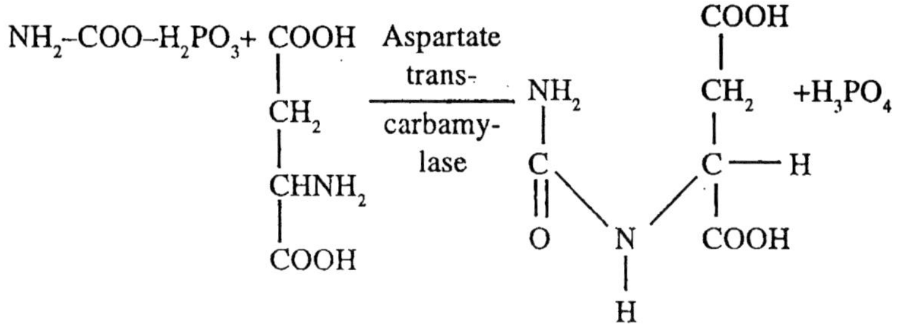

$$NH_2\text{-}COO\text{-}H_2PO_3 + \begin{array}{c} COOH \\ | \\ CH_2 \\ | \\ CHNH_2 \\ | \\ COOH \end{array} \xrightarrow[\text{carbamy-lase}]{\text{Aspartate trans-}} NH_2\text{-}\underset{\underset{O}{\|}}{C}\text{-}\underset{\underset{H}{|}}{N}\text{-}\underset{\underset{COOH}{|}}{\overset{\overset{\overset{COOH}{|}}{CH_2}}{\underset{|}{C}}}\text{-}H + H_3PO_4$$

Carbamyl aspartate

Cyclization of carbamyl aspartate, which is also called ureidosuccinae is catalysed by dihydro-orotase, with the elimination of water molecule.

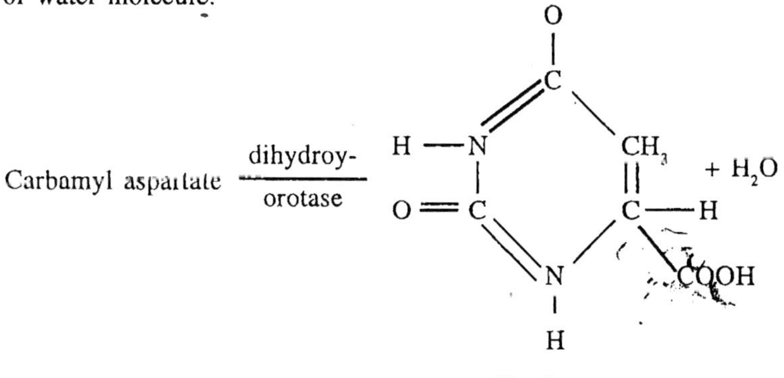

$$\text{Carbamyl aspartate} \xrightarrow[\text{orotase}]{\text{dihydroy-}} \text{(ring: } O{=}C,\ H\text{—}N,\ O{=}C,\ N\text{—}H,\ C\text{—}H\ (COOH),\ CH_2\text{)} + H_2O$$

Dihydro-orotate

Dihydro-orotate is dehydrogenated to orotate or uracil 4-carboxylate. The hydrogen acceptor may either be FAD or NAD^+.

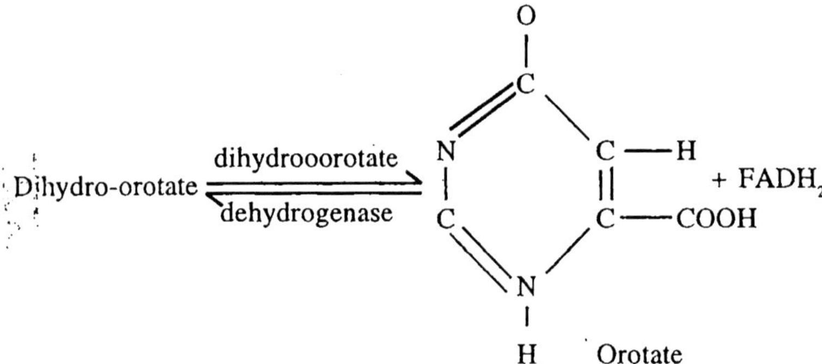

Orotate reacts with 5-phosphoribosyl 1 pyrophosphate (PRPP) to produce oritidine 5-phosphate, which is converted by a specific decarboxylase to uridine 5-phosphate (UMP).

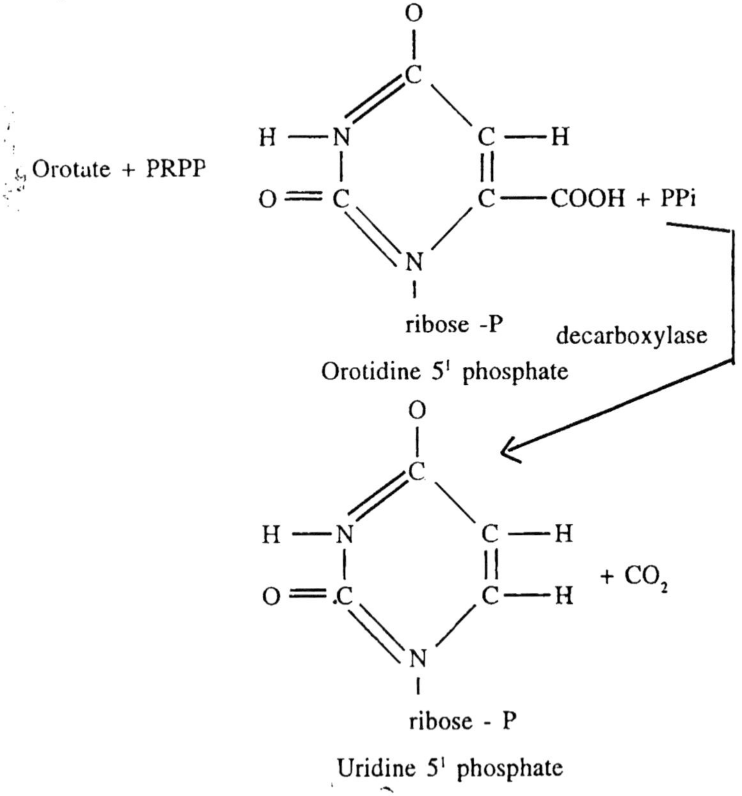

Orotate + PRPP

O
|
C
H — N   C — H
O = C   C — COOH + PPi
N
|
ribose -P

decarboxylase

Orotidine 5' phosphate

O
|
C
H — N   C — H
O = C   C — H   + CO_2
N
|
ribose - P

Uridine 5' phosphate

UMP thus synthesized is phosphorylated to UDP & UTP by specific kinase.

UMP + ATP ———> UDP + ADP

UDP + ATP ———> UTP + ADP

Cytidine triphosphate (CTP) & other pyrimidine nucleotide is synthesized from UTP, Cummonia & ATP.

$$UTP + NH_3 + ATP \xrightarrow{\text{UTP kinase}} CTP + ADP + H_3PO_4$$

In the synthesis of DNA nucleotides, sugar moiety has to be reduced. Synthesis of deoxy CTP has been studied in E.coli. Cytidine diphosphate is enzymatically reduced to deoxycytidine diphosphate in the presence of ATP + Mg^{++}.

$$CDP \xrightarrow[\text{CDP reductase system}]{ATP + Mg^{++}} dCDP$$

CDP reductase system is composed of more than one enzyme. The hydrogen donor for the system is NADPH. Phosphorylation of d CDP by a specific kinase produces d CTP.

$$dCDP + ATP \xrightarrow{\text{kinase}} dCTP + ADP.$$

Condensation of carbamyl phosphate with aspartic acid is rate limiting reaction in the synthesis of pyrimidine nucleotides. The enzyme asparatate franscarbamylase which catalyses this reaction, has been crystallised from E. coli & its properties have been studied extensively. The enzyme has a molecular wt. of 3,10,000. It is an allosteric enzyme made up of two catalytic & four regulatory subunits. The enzyme saturation curve with respect to the substrate aspartate is sigmoid. Cytidine triphosphate the end product of the reaction is a specific inhibitor of the enzyme. The inhibition of enzyme by CTP is also sigmoid (fig. 15.6). The inhibited enzyme is activated by ATP. It is proposed that ATP induces some structural changes in the enzyme & then the enzyme follows a Michaelis. Menten Curve in the presence of by CMP, d-CMP, CDP, d-CTP, GTP and d-GTP also. Cytosine, however, has no effect on the enzyme activity.

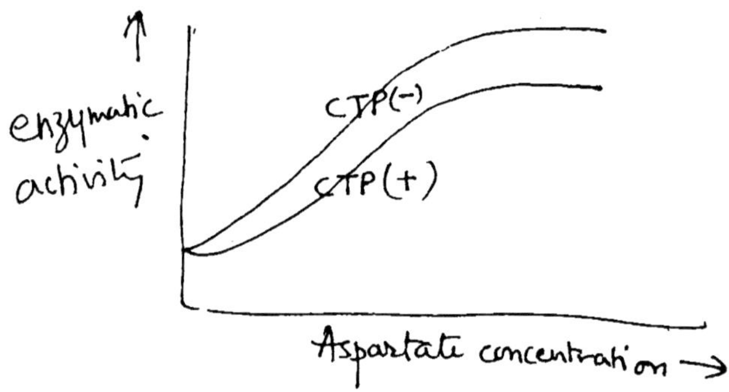

Fig. 15.6 : Effect of substrate concentration on enzymatic activity (CTP as an example)

PURINE NUCLEOTIDES

From labelling experiments, it has been observed that the precursors of purine are CO_2, formate, glycine, aspartate & glutamine. The starting material, however, is ribose-5-phosphate, which is phosphorylated by ATP to yield 5-phosphoribosyl 1 pyrophosphate (PRPP).

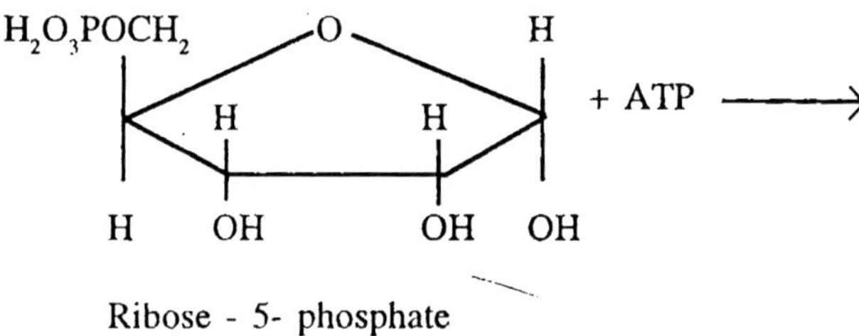

Ribose - 5- phosphate

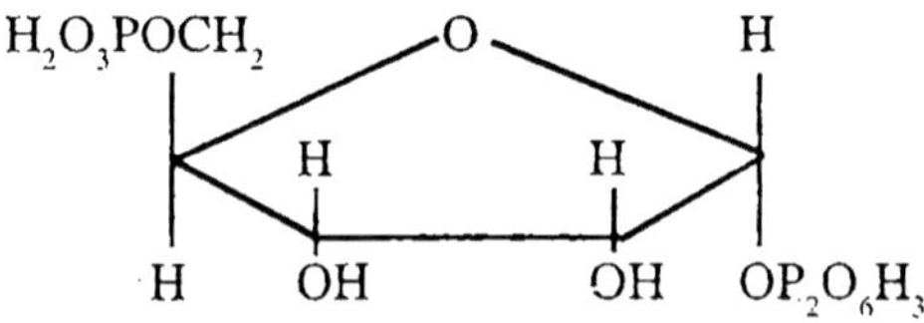

5- phospho-ribosyl 1-pyrophosphate

Glutamine donates its y-amino group to PRPP & 5-phospho-ribosyl amine (PRA) is produced with the loss of pyrophosphoryl residue.

5-phospho-ribosyl 1 pyrophosphate + Glutamine ⟶

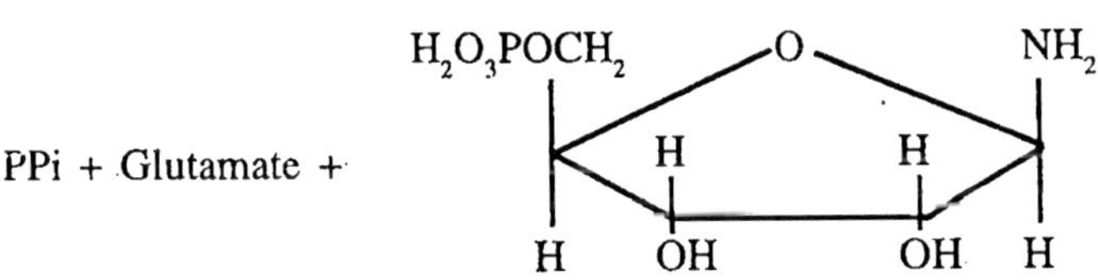

5-phospho ribosyl amine

Glycine then reacts with PRA to produce glycinamide ribonucleotide (GAR). ATP provides energy for the reaction.

5-phosphoribosyl amine + Glycine

↓ ATP ⟶ ADP + Pi

C—N
H_2N—C
N—CH
|
Phosphate ribose

Amino-imidazole ribonucleotide(AIR)

Carboxylation of AIR gives rise to 5-amino 4 imidazole carboxylic acid ribonucleotide (carboxy - AIR)

Amino imidazole ribonucleotide + CO_2

↓

COOH

C—N

H_2N—C

N—CH

Phosphate ribose

5 Amino 4-imidazole carboxylic acid ribonucleotide

This is followed by the addition of a molecule of aspartic acid in the presence of ATP to yield 5 amino 4 imidazole carboxymide after elimination of fumaric acid.

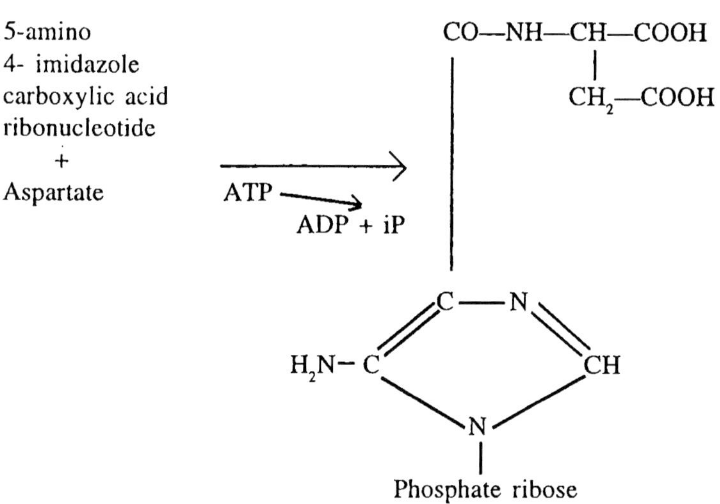

5-amino, 4-imidazole N-succinyl carboxamidae ribonucleotide

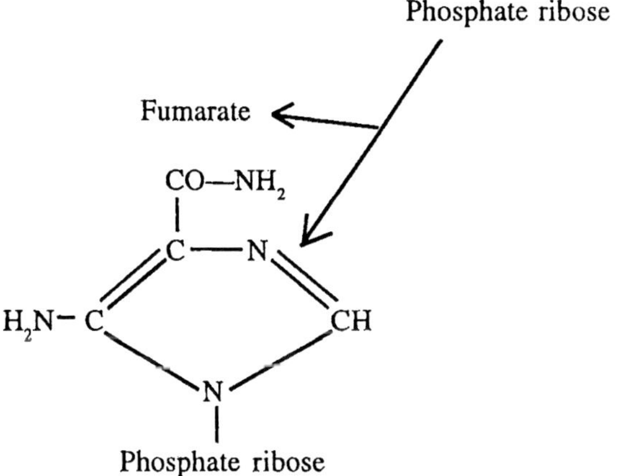

5-amino, 4-imidazole carboxamide ribonucleotide (AICAR)

AICAR is formylated again N-formyl tetrohydrofolic acid to 5-formamido 4 imidazole carboxamide ribonucleotide (FAICAR). FAICAR then loses one water molecule to form cyclic ionisinic acid, where the purine ring appears for the first time.

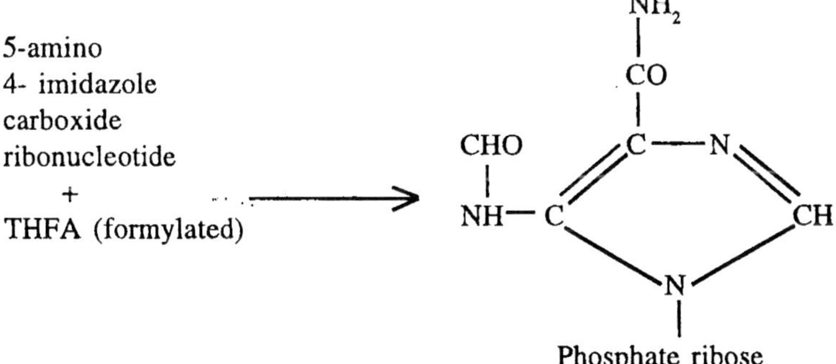

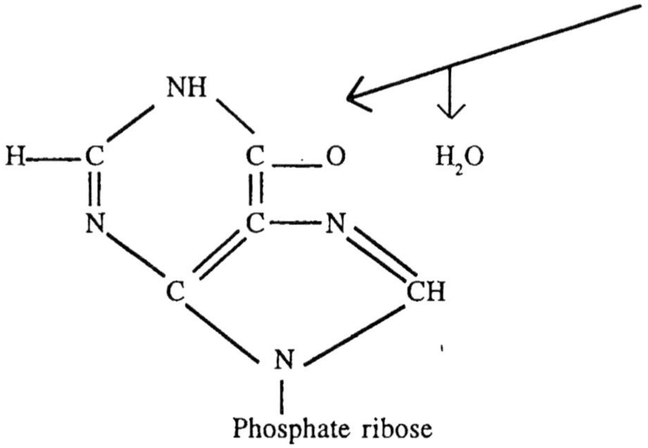

NH
H—C C—O H_2O
N C—N
C CH
N
Phosphate ribose
Inosintic acid (IMP)

Inosinic acid is the precursor for two of the purine nucleotides, guanylic and adenylic acid.

Inosinic acid $\xrightarrow{\text{oxidation}}$ Xanthylic acid $\xrightarrow{\text{amination}}$ Guanylic acid

Inosinic acid $\xrightarrow{\text{Aspartate}}$ Adenylosuccinic acid

GTP → GDP + iP

Fumarate

Adenylic acid

The two purine nucleotide AMP & GMP are phosphorylated by specific kinases as in the case of pyrimidine nucleotides, to produce ATP & GTP respectively

REPLICATION OF DNA

Synthesis of DNA from monomeric units of deoxyribose nudeotides on a DNA template is known as ***replication***. Three mechanisms have been proposed for replication.

1. Dispersive

It is proposed that the original DNA is broken into many pieces, which are more or less uniformly distributed among the progenies. This postulates, however, can not explain the regular arrangement of nucleotides. Among the progenies there many be great variation in the nucleotide sequence of DNA.

2. Conservative

According to this mechanism, the original DNA is copied & the copy is transmitted to the future generation as such. Thus both primary and secondary structure of DNA are conserved. There is neither breakage nor unwinding of the parental DNA during replication. Though, this mechanism of DNA replication is possible structurally and chemically, it is unknown that none of the progenics contain whole set of parental DNA.

3. Semiconservative

The mechanism is based on watsoncrick model of DNA & it is the accepted mechanism of DNA replication. The replication of DNA involves the unwinding of double helix of DNA but no breakage of the separated strands. Thus the primary but not the secondary structures of parental DNA is conserved, during the replication process.

Idea of semiconservative mechanism of DNA replication came out of the careful studies of the double helical model of DNA (Fig. 15.7). The complimentary strand structure is an indication of the possible strand separation & formation of complimentary molecules on each of free single strands. During semiconservative replication, the hydrogen bonds between the complimentary base pairs dissociate & the two separated parts of the strand start replicating. However, the two stands do not completely separate before the new strands are synthesized. That is why in the process of DNA replication, some Y shaped regions are observed. Each strand acts as a model or template for the formation of new complimentary chain. The nucleotides are put together into short pieces of DNA by the enzyme DNA polymerase.

These short fragments of newly synthesized DNA are known as Okazaki fragments. The complimentality of the strands is maintained because of the selective pairing between purine & pyrimidine bases. Adenine Always pairs with thymine & quanine with Cytosine. The small okazaki fragments are joined together by the enzyme polynucleotide ligase to give rise to complete DNA strands. In experiments, where this enzyme is selectively inhibited for short times, accumulation of okazaki fragments occur.

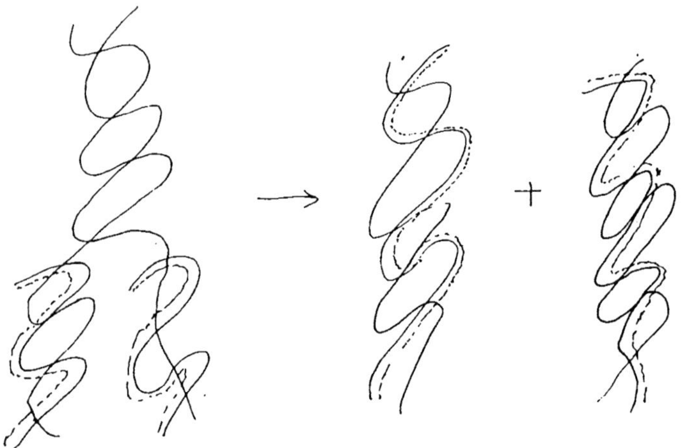

Fig. 15.7 : RNA replication in semionservation manner

Evidence for the semiconservative model of DNA replication came from the experiment of Messelson & statil (1958). These workers allowed E.coli to grow in a medium containing heavy nitrogen (N^{15}). After reproduction, the DNA of bacteria became labelled with N^{15}. This heavy DNA could be separated from light DNA (Containing N^{14}) by centrifuging the mixture with high concentration of heavy DNA were transferred to light medium (containing N^{14}) & allowed to reproduce. As the precursor nucleotide available for use in DNA synthesis were light, the new DNA synthesized were light (Fig. 15.8). The proportion of heavy & light DNA was measured after first & second generation of the bacterial growth in the light medium. After one generation in the light medium, the ratio was half heavy DNA & half light DNA. After the second generation, the ratio was 3 light to 1 heavy. In early replication, the new strand incorporated N^{14} but N^{15} was retained in the original strand. These studies indicated that original strand was maintained during each replication.

REPLICATION OF SINGLE STRANDED DNA

As mentioned earlier, many bacterial viruses have single stranded DNA. The replication of single stranded linear molecules of

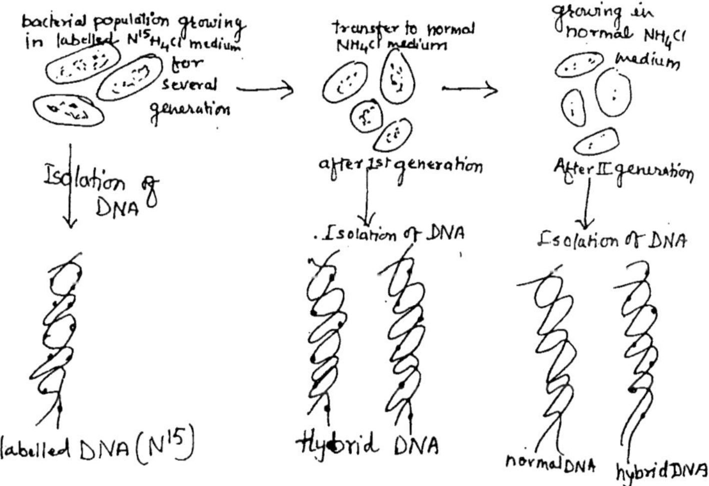

Fig. 15.8 : In vitro semiconservative type DNA replication

DNA also occurs through base pairing on the parental strand. The parental DNA strand is commonly designated as plus (+) strand of the virus. As soon as it enters a host cell, it forms a complimentary minus (-) strand (fig. 9) & thus a double helical intermediate DNA is formed. This is the replicating form of DNA because now it serves as a template for new + strands. The + strands become ensheathed by protein molecules to give rise to progeny viruses. Thus fundamental mechanism of the replication of single stranded virus DNA is the same as double stranded DNA. The only difference is that in single stranded viruses only one (strand) of the two strands in the double helix of replicating form functions as a template for the daughter DNA molecules.

REPLICATION OF CIRCULAR DNA

Replication of circular DNA is a bit complicated than that of linear DNA chains. The first event in the replication of circular DNA is the enzymatic breakage of one of the two parental strands. Enzymes involved in this type of breakage are called endonucleases. Enzymes start working at very specific region of the DNA molecule. This is called the initiation point. Regarding further mechanism of replication of circular DNA, two models have been proposed.

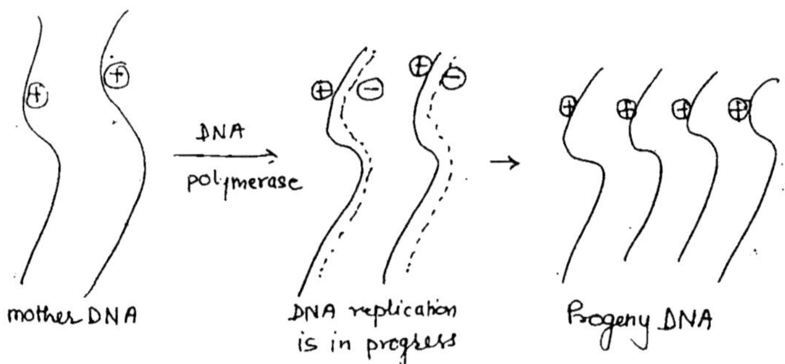

Fig. 15.9 : Replication of single stranded viral DNA

1. Rolling circle model

One of the two strands in the circle is cut open by the enzyme endonuclease. The two free ends are chemically different. The nucleotides are then added to free hydroxyl group end of the strand by the activity of DNA polymerise (fig. 15.10). As the nucleotides are added to this end the other end is pulled out of the circle. This end also replicates in segments. Later on, the segments, of new DNA are joined together by the enzyme polynucleotide ligase.

2. Opposing rolling circle model

In some bacteria during DNA replication, shaped intermediates of DNA are observed. To explain this observation, opposing rolling circle model for the replication of circular DNA was proposed. It is essentially the replication two circles together in opposite directions. Besides the growth of one strand of the circle in a rolling circle model, a second nuclease cuts open the other strand also. The -OH end of this strands grows & the strand with phosphate end is displaced as another tail. This tail is complimentary to the tail of other circle and eventually both join together by hydrogen bonds to give rise to a Ø shaped structure. The ultimate product of replication by opposing circle model is a double circle with increased diameter. Later on these circles are cleaved into linear DNA by specific nucleuses.

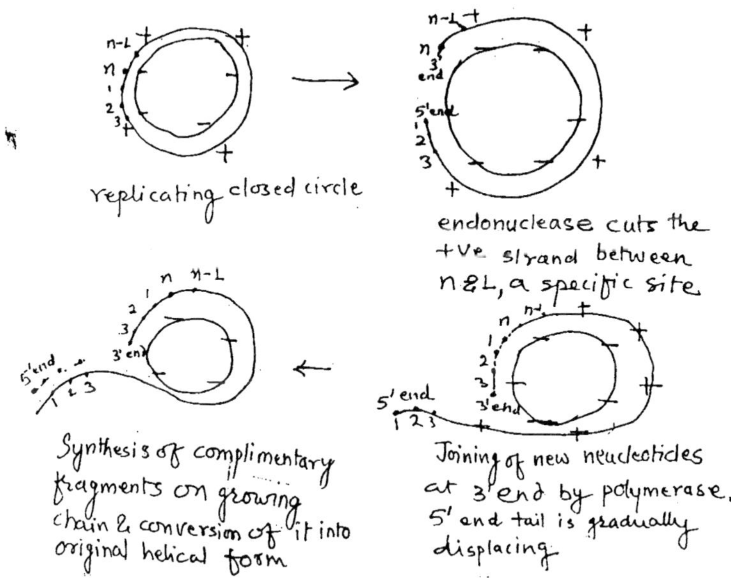

Fig. 15.10 : Replication of circular DNa - rolling circle model

DNA POLYMERISE

Since this enzyme is the pivot in the replication of DNA, whether in vitro or in vivo, it deserves a special account. This enzyme was discovered & first studied by Arthur Kornberg et al. in the E.coli. Since then, it has been isolated and studied from many other organisms also. This enzyme is made up of single polypeptide chain of m.w. about 109×10^9 and contains about 1000 amino acid residues. The molecule possesses at least one free sulphydryl group and the N-end amino acid residue is methionine. It is believed that the enzyme molecules is generally folded in the form of a sphere with the diameter of 60Å (fig. 15,11).

Kornberg et. al. studied the properties of DNA polymerase & described the following nature of the enzyme.

1. It joins only the monomeric units of deoxynucleotides.

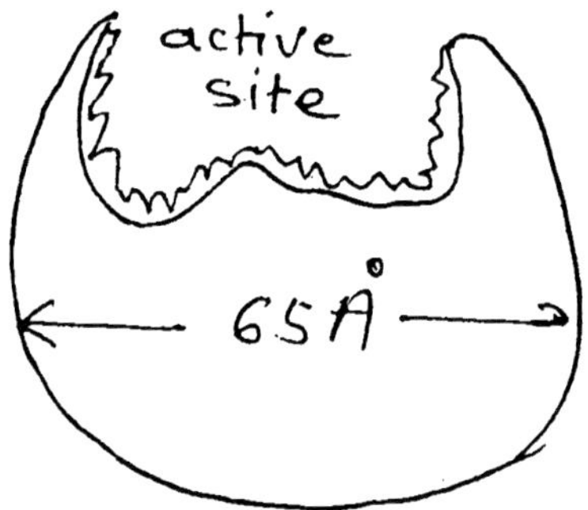

Fig. 15.11 : Molecule of DNA polymerase

2. It requires a single stranded DNA templates.
3. The new nucleotides are joined together in a sequence which is determined by the sequence of nucleotides in the template.
4. The rate of synthesis is about 1000 nucleotides per min. per mole DNA polymerise.

Besides the original polymerise enzyme, which is now called DNA polymerase I polymerase II & III are also identified and characterised. The three enzymes can be separated on the basis of their ion sensitivity. Increase in concentration of $(NH_4)SO_4$ has different effect on the three enzyme activities (fig. 15.12).

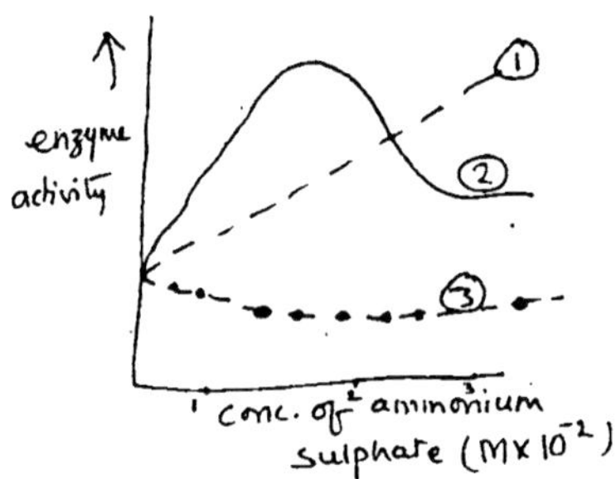

Fig. 15.12 : Polymerase sensitivity to various salt concentration

It is now believed that real enzyme involved in the replication of DNA is DNA polymerase III. DNA polymerase I is used in the filling of gaps between small DNA segments. No specific role has been found by polymerase II.

REGULATION OF REPLICATION OF DNA

DNA replication is controlled by a complex system composed of many parts. In absence of some proteins, DNA is not replicated. If there is no protein synrthesis in the cel, DNA replication stops after first round. To Understand the regulation of DNA, a model called 'replicon' has been proposed. A genetic element such as bacterial chromosome, an episome or a phage is considered as a replicating unit or a replicon. The replican always replicates entirely. Replication of one part of a genetic system but not of the other indicates that the system is composed of 2 or more replicons.

Initiation of DNA replication occurs when the initiator molecule starts functioning. The initiator molecule is synthesized by one of the genes of replicon & it is believed to be an integral component of DNA polymerise. Its structure is probably like a wedge and contributes to the correct 3-D structure of the native enzyme. The synthesis of initiator commits the cell to replicates its DNA & it divide. Several other growth stimulators & cell transforming agents describes in the literature may be equivalent to the initiator molecule. The initiation can be recognized by only a specific sequence on the replicon called operator of replicator or replicator (fig. 15.13). After the initiator is attached to replicator, the replicator, the replication of DNA starts & continues till the whole replicon is copied. The next round of replication is not possible till the cycle is complete & then the new initiator is made. The replicaiton of DNA occurs in a fixed direction, mostly clockwise.

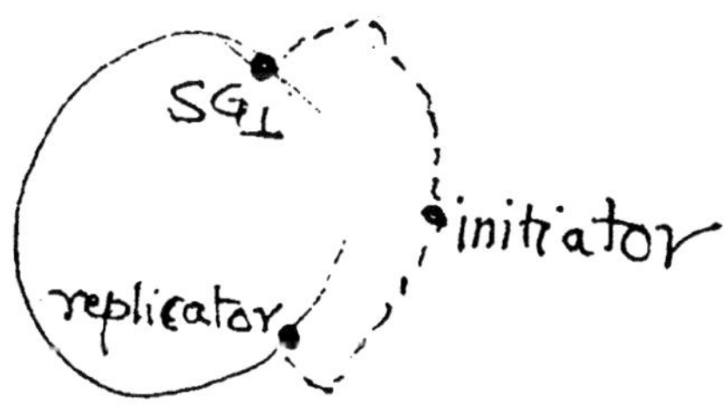

Fig. 15.13 : Circular replicon : diagrammatic

The initiator and replicator are very specific. An initiator produced by a phage genome will attach to replicator of phage only.

Transcription of DNA, synthesis of RNA on DNA template:

The fundamental mechanism of the synthesis of RNA of DNA template is very much similar to DNA replication. All the three types of RNAs are transcribed on DNA template. The DNA strands during transcription separate or melt at a fixed point & function as template on which complimentary ribonucleotides are joined together by the enzyme RNA polymerase, (fig. 15. 14) The sequence of bases in RNA is thus determined by the template/DNA.

$$\begin{matrix} n\ ATP \\ n\ GTP \\ n\ CTP \\ n\ UTP \end{matrix} \xrightarrow[\text{DNA template } Mg^{++},\ Mn^{++}]{\text{RNA ploymerase}} (Ap)_n, (Gp)_n, (Cp)_n\ (Up)_n + 4n\ ppi$$

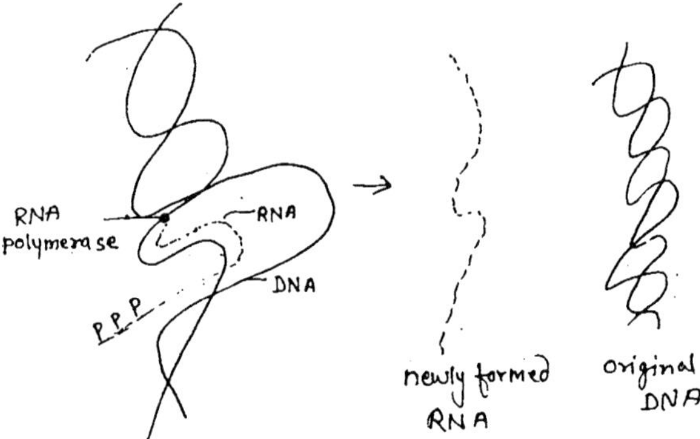

Fig. 15.14 : Transcription of RNA on DNA

As DNA polymerase moves along the DNA template the growing RNA chain is detached & again complimentary DNA chains join together by hydrogen bonds. The direction of enzymatic transcription is always in a fixed direction either 3' to 5' or 5' to 3' of the triphosphate, the later being more common.

The transcription starts at a fixed site on DNA. In E. coli, there are sites rich in cytosine & thymine nucleotides. Therefore, all normal RNa chains start wtih either Adenine or guinine. During transcription, the enzyme RNA polymerase is complexed with DNA : RNA hybrid. This complex has been isolated from E. coli. The rate of transcription in E.coli is about 43 nucleotides per second at 37°C *in vivo. In vitro,* it is only 7-12 nucleotides per second.

The enzyme RNA polymerise was isolated & characterised by M. chamberlain & Paul Berg (1962) . It sediments at 15S with m.w. of about 4,95,000 & is madc up of 6 several sub units.

The sigma (δ) factor is an essential component of the enzyme. It is probably this factor which recognizes the DNA for initiation of transcription & forms an initiation complex. The factor is released when RNA chain has been synthesized. The released factor can be roused for further initiation of transcription. In the absence of sigma factor, the RNA polymerise (core enzyme) of bacteriophage Fd can transcribe DNA but looses the ability to select the appropriate initiation sites on the template & both strands are transcribed.

REPLICATION OF RNA

When RNA viruses enter the host cells, they multiply by replicating their RNA. The mechanism is similar to the replication of single stranded parental RNA (+strand) is replicated to give rise a double stranded replicating form (+-). In the next step, minus strands work as templates for plus strands (fig. 15.15). The plus strands become part of the progeny virus particles.

The enzyme catalysing the replication of RNA in RNA synthatase or replicase. Like both RNA polymerase & DNA polymerase, RNA synthatase catalyses the formation of a complimentary strand upon a single stranded template. The RNA synthatase from a RNA virus QB, is made up of 4 sub units. The synthesis of only one of these sub units is coaded by the Viral genome. In vitro studics with replication of the virus have demonstrated that Mg^{++} & two other macromecular factors are required for replication. The synthesis of new RNA chains from ribonucleotides occurs in 3' to 5' direction.

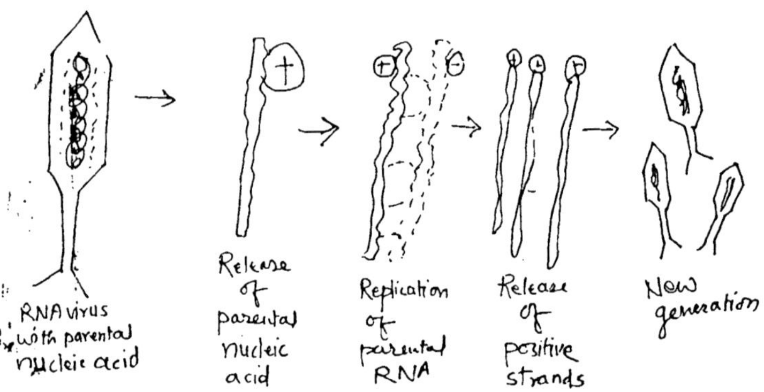

Fig. 15.15 : RNA replication in viruses

SELECTION EXTRACTION OF NUCLEIC ACIDS FROM TISSUES:

Following methods can be used :

1. ***Strong salt extraction method***

   BY this method, DNA & RNA can be extracted from many tissues. Homogenise the tissue in strong salt solution, a mixture of NACl & Sodium acetate & add excess of ethyl alcohol till the precipitatate appears.

   The precipitate contains the nucleic acids.

2. ***Phenol extraction method***

   This technique involved extraction of tissues homogenate with phenol. The phenol denatures the proteins & releases the nucleic acid which enter the aqueous phase.

   The tissue can be processed in the following manner in order to separate different constituents of the matter separately:

Take 5 gms. organic matters & homogenise in 50 ml. cold acid alcohol (80% alcohol in 0.1% per chloric acid).

(ii) Centrifuge the homogenate & separate the supernatant conains free amino acids, soluble sugars & nucleotides.

(iii) Remove the sedimented material (pallet) from centrifuge tube & heat it on 5% 20 ml. TCA at 80°C for 20-30 minutes.

(iv) Decant the supernatant. It contains nucleic acids in hydrolysed state.

(v) Treat the pallet (sedimented mat) successively with 50 ml. of 70, 80 & 90% alcohol in the centrifuge.

(vi) The supernatant contain pigments & lipids.

(d) Treat the pallet of (c) sediment with 50 ml. 2% Na_2CO_3 in 1.0 NaOH solution at 30°C for 30 minutes.

(viii) Centrifuge & separate the supernatant. It contains denatured protein.

(ix) The pallet (sediment) contains cell wall material.

Chapter -16

LIPIDS IN BIOLOGICAL SYSTEM

Lipids are a heterogenous group of substances, widely distributed throughout the plant and animal kingdom. In plants, they are present in the seeds, nuts & fruits while in the animals they are stored in adipose tissues, bone marrow and nervous tissues. Lipids are relatively insoluble in water & readily soluble in organic solvents such as ether, chloroform, carbon disulphide, benzene, hot alcohol etc. Bloor (1943) defines Lipids as ***naturally occurring compounds which are insoluble in water and soluble in one or more organic solvents such as benzene, chloroform, ether and acetone the so-called fat solvent & on hydrolysis yield fatty acids which are utilized by the living organisms.***

Lipids are generally classified into simple lipids, compound lipids & derived lipids.

SIMPLE LIPIDS

Simple lipids are the esters of fatty acids with various alcohols. Examples of this class are natural fats & waxes. Fats are triesters of glycerol & fatty acids, while waxes are esters of fatty acids with long chain alcohols.

FATS

Chemically, fats are triglycerides, since 3 molecules fatty acids condense with 1 molecule of glycerol to yield fat e.g. if 3 molecules of butyric acids are linked to glycerol, the product is tributryin, a fat. Similarly, palmitic acid condenses to produce tripalmition.

$$\begin{array}{l} CH_2OH \\ | \\ CHOH \\ | \\ CH_2OH \end{array} + \begin{array}{l} COOH-CH_2-CH_2-CH_3 \\ | \\ COOH-CH_2-CH_2-CH_3 \\ | \\ COOH-CH_2-CH_2-CH_3 \end{array} \longrightarrow$$

$$\begin{array}{l} CH_2O-CO-CH_2-CH_2-CH_3 \\ | \\ CHO-CO-CH_2-CH_2-CH_3 \\ | \\ CH_2O-CO-CH-CH_2-CH_3 \end{array}$$

Tributyrin

The general formula of the triglycerides may be written as -

$$\begin{array}{l} CH_2-O-CO-R_1 \\ | \\ CH-O-CO-R_3 \\ | \\ CH_2-O-CO-R_3 \end{array}$$

When all the fatty acid acyl groups (R_11, R_2, R_3) are the same, the product is a simple glyceride. When two or more of these groups are different, it is mixed. Natural fats are largely composed of mixed glycerides. Since these glycerides have no free acid or basic groups, they are often called ***neutral fats***.

Fats are solid or liquid (oil) at room temperature and are lighter than water (sp. gravity 0.86). The melting point of the fat depends upon chain length and degree of saturation of fatty acids. Saturated fatty acids containing more than & carbon atoms are solids. Fatty acids with shorter chains and more unsaturation are liquids. The melting points of fats are always higher than their solidification points e.g. tristear in melts at 72°C but solidifices and cooling at 52°C.

FATTY ACIDS

Since all the fats contain glycerol, their properties differ according to the nature of fatty acids present in them. Some of the fatty acids are saturated with hydrogen, whereas others are partially unsaturated. A few have branched chain or container hydroxyl groups or have a cyclic chain at the end. All the fatty acids, however have one carboxylic group and all the fatty acids have even number of carbon atoms.

General formula for saturated fatty acid is R-COOH, where R is CH_3 or $(CH)_n$, n varying from zero in acetic acid to 86 in mycolic acid. The most abundant saturated fatty acids in nature are palmitic (C_{18}) & stearic (C_{16}) acids. Some examples of saturated fatty acids occurring in plants and animals are listed in Table 16.1.

Unsaturated fatty acids have one or more double bonds. The general formula for there fatty acid is R-CH=CH (CH_2) n COOH. Generally the double bonds range between 1 to 6 and occur after 9,12,15,18 etc. carbon atoms. Presence of double bonds in the fatty acids lower their melting point considerably. Most plant fats contain unsaturated fatty acids as oleic & linoleic acid & hence they are liquid (oil) at room temperature. Contrary to this, animal fats have more of saturated fatty acids & hence solid at room temperature. Some common unsatured fatty acids are listed in Table 16.2.

Table 16.1 : Major saturated fatty acids found in plants & animals

Name	*Formula*	*Common source*
Butyric	$CH_3\ (CH_2)_2$ COOH	Butter
Caproic	$CH_3\ (CH_2)_4$ COOH	Butter
Caprylic	$CH_3\ (CH_2)_6$ COOH	Coconut oil, palm oil
Lauric	$CH_3\ (CH_2)_{10}$ COOH	laurel kernel oil
Myristic	$CH_3\ (CH_2)_{12}$ COOH	nutmeg oil
Palmitic	$CH_3\ (CH_2)_{14}$ COOH	palm oil, animal fat
Stearic	$CH_3\ (CH_2)_{16}$ COOH	cocoa, butter
Arachidic	$CH_3\ (CH_2)_{18}$ COOH	peanut oil

Table 16.2 : Major unsaturated fatty acids in plants and animals

Name	*Formula*	*Source*
Palmitoleic	$CH_3(CH_2)_5$ CH = CH $(CH_2)_7$ COOH	Milk fat, sardine oil
Oleic	CH_3 $(CH_2)_7$ CH= CH $(CH_2)_7$ COOH	Olive oil, park fat
Linoleic	CH_3 $(CH_2)_4$ CH = CH CH_2 CH = CH $(CH_2)_7$ COOH	linseed oil soyabean
Linolenic	CH_3 CH_2 CH = CH CH_2 CH = CH $(CH_2)_7$ COOH	linseed hemp seed
Erucic	CH_3 $(CH_2)_7$ CH = CH $(CH_2)_{11}$ COOH	mustard oil

In some animal fats and in bacteria, branched chain fatty acids are also found. They generally have odd number of carbon atoms (C_{13} to C_{17}).Tuberculostearic acid is an example, which is found in ***Bacillus tuberculosis***.

$$CH_3\ (CH_2)_6\ \underset{\displaystyle CH_3}{\underset{|}{CH}}\ -\ (CH_2)_8\ -\ COOH$$

Tuberculo - stearic acid

Fatty acids containing cyclic systems ate also found in nature. Chaulmoogric acid obtained from chaulmogra (Hydnocarpus wightinia) is an example.

```
CH = CH
|      \
|       >CH — (CH)12 COOH
|      /
CH2 — CH
```

Chaulomoogric acid

Linoleic, linolenic and arachinodic acids are unsaturated acids which are also found in small amount in animal fat. There three are sometimes called essential fatty acids since they can not be synthesized in animal body & must be supplemented with the diet.

CHEMICAL PROPERTIES

a) Hydrolysis

Hydrolysis of fats with akali or enzyme lipase yields fatty acids & glycerol. Lipases split fats in steps, from triglyceride to deglyceride to monoglyceride & finally to glycerol & fatty acids. Lipases may work in the temperature range of O° to 40°C.

$$\begin{array}{l} R_1COOCH_2 \\ | \\ R_2COOCH \\ | \\ R_3COOCH_2 \end{array} \xrightarrow[R_3COOH]{H_2O \text{ lipase}} \begin{array}{l} R_1COOCH_2 \\ | \\ R_2COOCH \\ | \\ CH_2OH \end{array} \xrightarrow[R_2COOH]{H_2O \text{ lipase}} \begin{array}{l} RCOOCH_2 \\ | \\ CHOH \\ | \\ CH_2OH \end{array} \xrightarrow[H_2O]{\text{lipase}} \begin{array}{l} CH_2OH \\ | \\ CHOH \\ | \\ CH_2OH \end{array}$$

triglyceride → diglyceride → monoglyceride → glycerol

When the fats are hydrolysed with alkali, the free fatty acids react with alkali to form metallic acid. These salts are soaps & this process is called **saponification.**

$$\begin{array}{l} RCOOCH_2 \\ | \\ CHCOOR \\ | \\ CHCOOR \end{array} \xrightarrow[40°\text{-}120°C]{3NaOH} \begin{array}{l} CH_2OH \\ | \\ CHOH \\ | \\ CH_2OH \end{array} + 3\ RCOONA$$

Soap

The soaps of sodium & potassium are water soluble & used for cleaning purposes. Lead soaps and other insoluble soaps and used as lubricants in grease formulations.

OXIDATION

Oxidation of fats in the air is often accompanied with hydrolysis. Oxidation occurs at double bonds of the fatty acid, producing short chain acids, aldehydes etc. This together with the fatty acids produced during hydrolysis gives a less palatable taste and odour to the fat. It is called ***randicity***. Oxidation can be prevented by using small amounts of certain organic substances. These organic substances are termed ***antioxidants***. Commercially employed antioxidants are non dihydroguiretic acid, tertiary butyl-para-cresol. Tocopherol is a natural antioxidants occurring in vegetable oil.

HYDROGENATION

Unsaturated plant fats are converted into more saturated & solid fats by catalytic hydrogenation. This is usually done over finally divided nickel. In the production of margarine & vegetable shortening, this property is exploited commercially.

$$\text{Vegetable oil} \xrightarrow[\text{Ni Powder, Pressure}]{H_2 \text{ gas}} \text{Solid fat}$$

$$\begin{array}{l} CH_2OCOC_{17}H_{33} \\ | \\ CHO\ COC_{17}H_{33} \\ | \\ CH_2OCO_{47}H_{33} \end{array} \quad + 6H \longrightarrow \quad \begin{array}{l} CH_2OCOC_{17}H_{35} \\ | \\ CHOCOC_{17}H_{35} \\ | \\ CH_2O.COC_{17}H_{35} \end{array}$$

Triolein (oil) Tristearin (fat)

In commercial production of margarine, hydrogenation is stopped at a semisolid fat stage-complete hydrogenated fats are solids brittle & unpalatable. Such fats are used in making candles.

WAXES

Waxes are another class of simple liquids containing one molecule of fatty acid & one high molecular weight alcohol. The constituent acids & alcohols have usually 24 to 36 carbon atoms. Carnauba wax consists of myricyl alcohol (C_{30}) & cerotate (C_{26}). In bee wax, the fatty acid constituent is a smaller chain acid. palmitic acid (C_{16}) and alcohol is myricyl. Amberetolide found in the seeds of Abelmoschus esculentus is a hydroxy acid & is responsible for the characteristic smell of the seed.

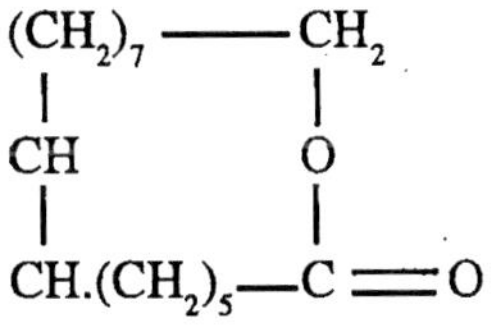

```
(CH2)7 ——— CH2
 |          |
CH          O
 |          |
CH.(CH2)5—C ═ O
```

Ambretolide

Waxes are insoluble in water and very resistant to atmospheric oxidation. They also have melting point. Because of these properties, they are used in furniture polishing. They are chemically inert & not digested by the fat splitting enzymes. They can be split slowly with hot alcohalic KOH, however.

COMPOUND LIPIDS

Compound lipids contain some additional groups or elements besides fatty acids and alcohol. The additional group may contain phosphorus, nitrogen, sulphur or it may be protein.

Phospho-lipids

This class includes lecithins, cephalins & plasmalogens :

Lecithins

It is widely distributed in nature on hydrolysis, they yield glycerol, fatty acids, phosphoric acid & choline. It is also called ***phosphatidyl choline***.

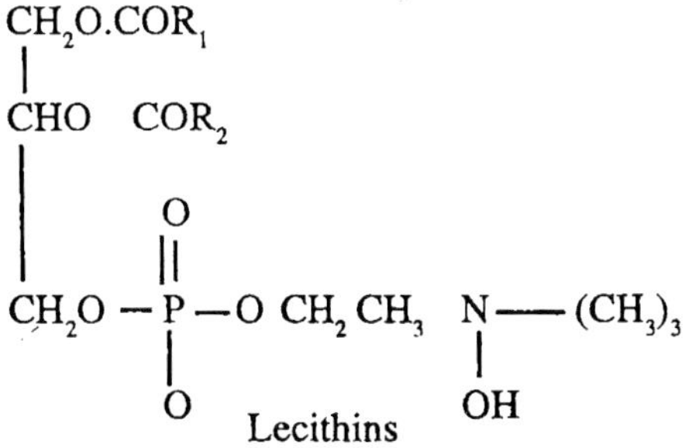

```
CH2O.COR1
|
CHO   COR2
|
|         O
|         ||
CH2O — P — O CH2 CH3  N —— (CH3)3
          |            |
          O            OH
```

Lecithins

The fatty acids commonly found in lecithin are palmitic, stearic, olic, linolinic and archidonic acids.

Lecithins and other phospholipids are yellowish greasy solids, soluble in ether & alcohol but insoluble in acetone. On exposure to air, they rapidly darken in colour and absorb water, forming dark greasy mass. Lecithins are broken down by enzyme lecithinase to lysolecithin. The enzyme lecithinase is present in venoms of bee and cobra. When injected into the blood, lysolecithins cause rapid haemolysis of the red blood cells.

$$
\text{Lecithin} \xrightarrow{\text{lecithinase}} \begin{array}{l} CH_2OCOR_1 \\ | \\ CHOH \\ | \\ CH_2OPO_3\ (CH_2)\ N\ (CH_3)_3\ OH \end{array}
$$

Lysolecithin

Cephalins

It is found in animal tissues in close association with lecithins. They are also found in soyabean oil. The fundamental difference between the lecithins & cephalins is the nature of nitrogenous base. Cephalins contâin ethano-lamine, colamine & sometimes serine in place of choline. The fatty acid components of cephalins are usually stearic oleic, linoleic and arachidonic acids.

Plasmalogens

$$
\begin{array}{l} CH_2OCOR_1 \\ | \\ CHOCOR_2 \\ | \\ CH_2OPO_3(CH_2)\ NH_2 \end{array}
$$

Cephalin (phosphatidyl ethanolamine)

These are abundant in brain & muscles. They are also found in seeds of higher plants. In this group of phospholipids a complex oldehyde is attached to the α-carbon atom of glycerol. Rest of the molecule is similar to lecithin or cephalin. Plasmalogens are soluble in all the lipids solvents.

$CH_2CH{=}CHR_1$
|
CH_2COR_2
|
$CH_2O\ PO_2\ (CH_2)_2\ NH_2$

Plasmalogen (or choline)

Phosphoinositides

Phosphoinositides contain hexahydric alcohol inositol. They can be either mono- or disphosphoinositides. The name lipoinositol was also proposed for these substances.

CH_2OCOR_1
|
$CHOCOR_2$
|
CH_2OPO_3 —H— [inositol ring: H, OH, H, H, OH, OH, OH, OH]

Monophosphoinositides

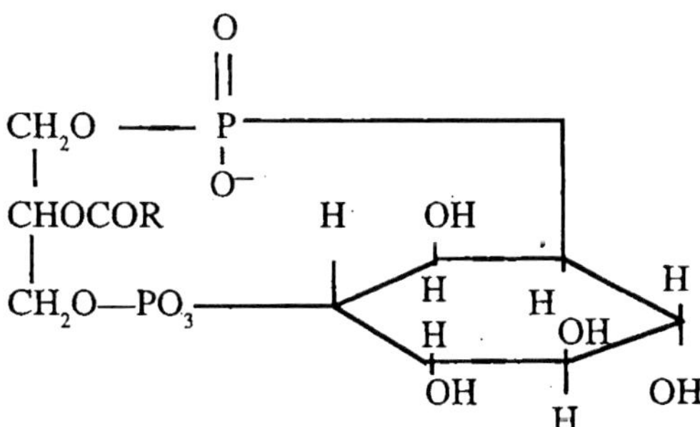

Diphosphoinositides

Monophosphoino sotides occur widely in animals and plants but diphosphoinositdes have been reported from brain tissues only.

Phosphosphingosides

In there lipids, glycerol is replaced by either sphingisone or phytosphingosine. These lipids are abundant in nervous tissues and apparently lacking in plants and microorganisms.

$$CH_3—(CH_2)_{13}—(CHOH)_2—CHNH_2—CH_2OH$$

Phytosphingosine

$$CH_3\ (CH_2)_{12}—(CH_2)_2—CHOH—CHNH_2—CH_2OH$$

Sphingosine

Phisphosphingosides found in plants are mostly associated with carbohydrates & they are called ***glycosphingolipids*** or ***glycolipids***. Galactose, arabinose & fucose are common carbohydrate units of glycolipids occurring in the seeds.

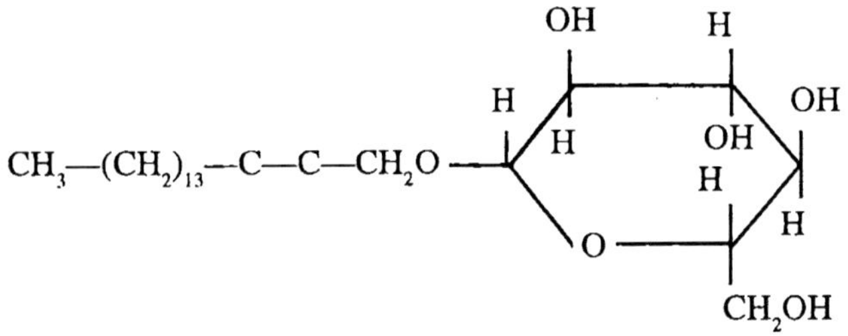

OH H

H OH

H OH

$CH_3—(CH_2)_{13}—C—C—CH_2O$ H

H

O

CH_2OH

Glycolipid

DERIVED LIPIDS

The group, derived lipids, includes hydrolytic products of lipids & in addition other lipids like compounds such as sterols, caratenoids, essential oils, aldehydes, ketones, alcohols, hydrocarbon etc. The sterols are solids wax like substances. Chemically, they are alcohols & occur either as esters of fatty acids. They are highly soluble in fat solvents. They all contain a cyclopentano-perhydro-phenanthrene (stearance) nucleus.

Sterols are widely distributed in plants, animals & microoganisms. They are found in cell membranes & other cellular components containing lipids. Unlike other lipids, sterols can not be saponified & by this process they can be separated from other lipids. The best known animal sterol is ***cholestrol***, which is present in relatively high concentration in nervous tissues & in bile. Cholesterol is a crystalline solid with rhombic crystals & its solution in chloroform is levo-rotatory.

Ergosterol is present in food in small amounts. It has been isolated from parasitic fungus claviceps purpurea (ergot) growing on rye plants. It differs from cholestrol only in the side chain attached to the sterol nucleus.

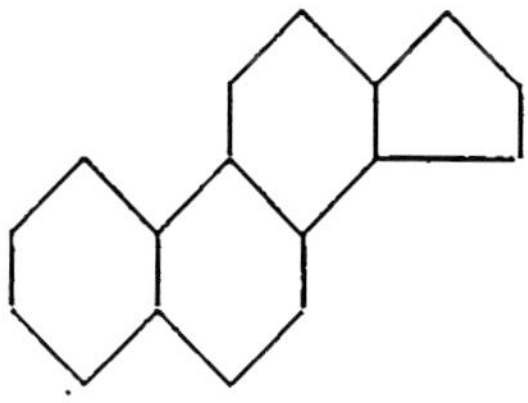

Sterane

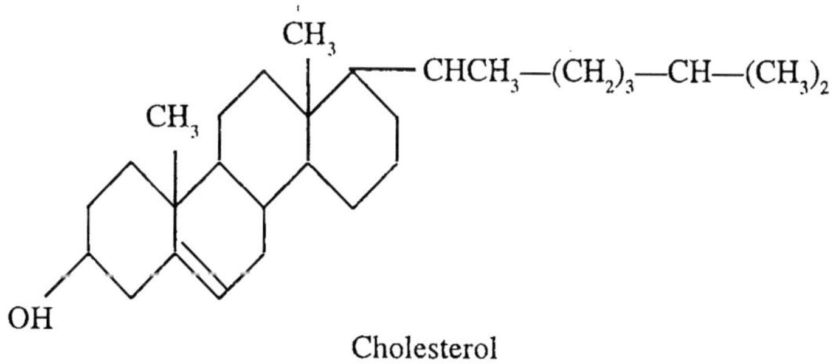

CH$_3$ CHCH$_3$—(CH$_2$)$_3$—CH—(CH$_3$)$_2$

CH$_3$

OH

Cholesterol

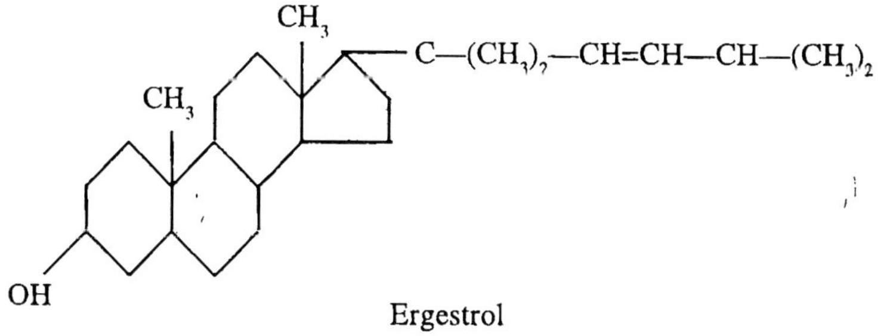

CH$_3$ C—(CH$_3$)$_2$—CH=CH—CH—(CH$_3$)$_2$

CH$_3$

OH

Ergestrol

Exposure to ultraviolet radiation converts ergosterol to vit. D_2 (Calciferol).

Other plants sterols are spinasterol obtained from spinach and cabbage stigmasterol, found in coconut & soyabeans & sitosterol from cereal seeds. The male & female sex hormones of vertebrates are also sterols. They are secreted by the adrenal cortex and testes and ovaries. The mixture of adrenal cortex hormone contains cortisterone, desoxycortisterone, a drenosterone and some other sterols. The testes secrets testesterone, androsterone, adrenosterone. Estaradiol, estrone, estriol, equilin, equilinin & progestrone are secreted by the ovaries.

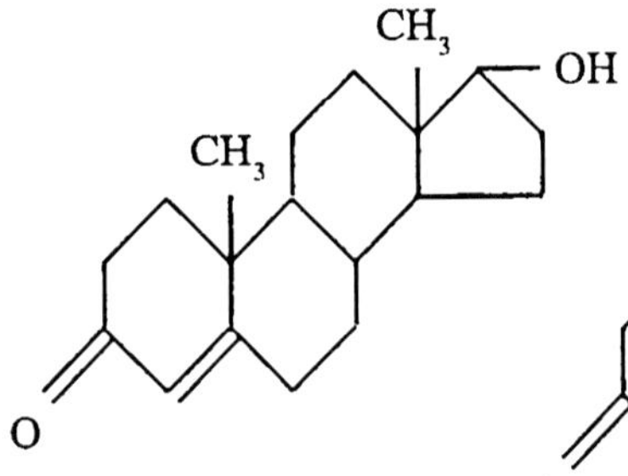

Testosterone 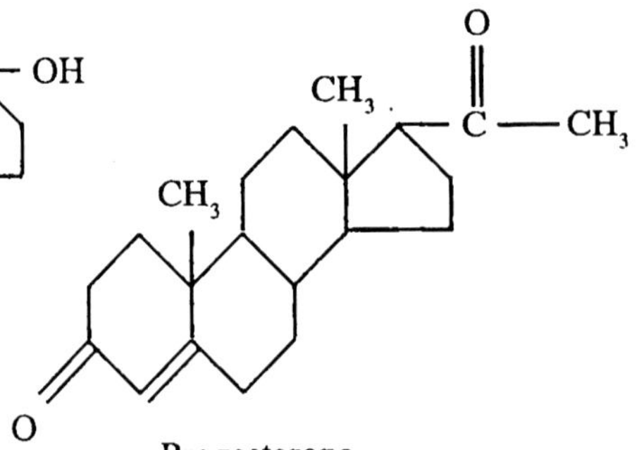 Progesterone

Carotenoids (carotene and xanthophyll) are also included in lipids because of their solubility in fat solvents. Carotenes consist of carbon & hydrogen only, whereas xanthophylls contain oxygen in additoj. Vitamin A is derived from carotene.

Essential oils are also grouped under lipids because of their solubility and natural occurrence. Most of the essential oils are either terepenes related to isoprene or isopentene or benzene derivatives. Some are straight chain carbon compounds without any side branches. The essential oils are usually obtained from plants such as pine, peppermint, rose, lemon, eucalyptus etc.

Other derived lipids such as fatty acids, glycerols, aldehydes etc. Occur only in small amounts as intermediates of metabolic conversion.

BIOLOGICAL SIGNIFICANCE

The main function of fats is that they serve as a source of energy. Complete combustion of 1 g of fat releases 9.0 K. cal. of energy. Most of the foodstuff taken in beyond normal requirement in the animals in converted into fat as principal reserve product.

Lipids are involved in the building of cellular components also. Phospholipids and steroids are present in biological membranes. Since phospholipids posses both acidic and basic groups, they behave as zwitterions. They serve an important role in binding both type of compounds, hydrophilic and hydrophobic, in the membrane. Lecithin is the most important phospholipid found in membrane. Waxes generally form protective covering in plants and animals. They protect living surfaces against bacteria & insect invasion.

In animals, fat is often deposited subcutaneously & protect against extreme environmental conditions. Fatty tissues around many vital organs protect against mechanical injury also.

FAT METABOLISM

Fat oxidation

Fats in the natural systems are hydrolysed by glycerol and fatty acids by the enzyme lipase before oxidation.

$$\text{Fat} + H_2O \xrightarrow{\text{Lipase}} \text{glycerol} + \text{fatty acid.}$$

Glycerol reacts with ATP to form glycerol phosphate which is oxidized to glycereldehyde 3-phosphate. Glycogen or glucose and fructose may be synthesized from this by the reversal of glycolytic process. Alternatively, it is converted to pyruvate & then oxidized by TCA cycle.

Oxidation of fatty acids is a sequential process for which the following two pathways have been suggested.

1. β-oxidation

The principles of β–oxidation were discovered in 1907 by F. Knoop. Although this process was discovered & illucidated in animal system, the enzymes involved in the process have been discovered in plants also. In the first step of oxidation, fatty acids are activated by their conversion to thioesters with the help of coenzyme A. The energy for this process is derived from ATP. The enzyme thiokinase catalyses this reaction.

$$RCH_2CH_2CO\,[OH] + COAS[H] \xrightarrow[\text{thiokinase, } Mg^{++}]{ATP \;\rightarrow\; AMP + PPi} RCH_2CH_2\underset{\underset{O}{\|}}{C}.SCoA + H_2O$$

The second reaction is a dehydrogenation, removing two hydrogen atoms from α & β carbon atoms of the fatty acids. This reaction is catalysed by FAD containing enzymes, acyl CoA dehydrogenase.

$$RCH_2CH_2\underset{\underset{O}{\|}}{C}.SCoA \xrightarrow[FAD \rightarrow FADH_2]{\text{dehrdrogenase}} R-\overset{\overset{H}{|}}{C}=\underset{\underset{H}{|}}{C}-\underset{\underset{O}{\|}}{C}.SCoA$$

Three types of dehydrogenases have been identified from liver tissues. They different according to their specificity to the substrate. The first prefers to act on long chain fatty acids ($C_{14}C_{16}$, C_{18}), while other two act on medium or short chains. Next step in the oxidation involves addition of water by the enzyme enonyl hydrase & yields a hydroxy acyl CoA.

$$R-\overset{\overset{H}{|}}{C}-\underset{\underset{H}{|}}{C}.\;\;\underset{\underset{O}{\|}}{C}.SCoA + H_2O \xrightleftharpoons{\text{hydrase}} R-\underset{\underset{OH}{|}}{CH}-CH_2\underset{\underset{O}{\|}}{C}SCoA$$

β-hydroxy acyl CoA

β hydroxy group of this compound is dehydrogenated to the keto group by the enzyme hydroxy acyl dehydrogenase. The hydrogen is accepted by NAD^+.

$$R{-}\underset{\underset{OH}{|}}{CH}{-}CH_2.\underset{\underset{O}{\|}}{C}.SCoA + NAD^+ \underset{\text{genase}}{\overset{\text{dehydro}}{\rightleftharpoons}} R{-}\overset{\overset{O}{\|}}{C}{-}CH_2\underset{\underset{O}{\|}}{C}SCoA + NADH + H^+$$

β-ketoacyl CoA

This NADH is reodised through the electron transport chain & produces 3 molecules of ATP. Final step in the oxidation is thiolytic cleavage of the ketoacyl CoA & CoA

$$R\underset{\underset{O}{\|}}{C}CH_2\underset{\underset{O}{\|}}{C}SCoA + CoASH \longrightarrow R\underset{\underset{O}{\|}}{C}SCOA + CH_3\underset{\underset{O}{\|}}{C}SCoA$$

acetyl CoA

This reaction is catalysed by ketothiolase. The product is a fatty acid with two carbon atoms short. It may enter the entire process of oxidation at step 2 & further dissociation may occur. Thus, it is a process of cleaving fatty acids into fragments of 2C compound in steps. (fig. 16.1).

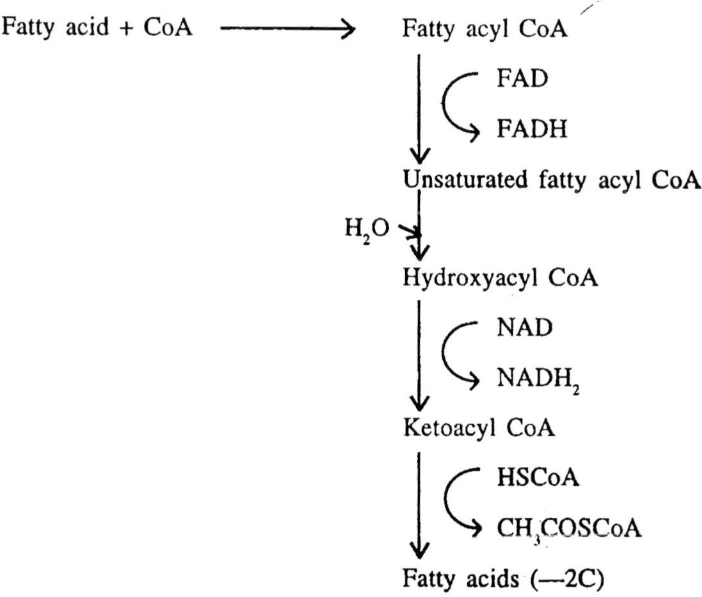

Fig. 16.1 : β oxidation

The end product of β–oxidation of fatty acid is acyl CoA. Fatty acoids with odd number of carbon atoms produce acelyl CoA & propionyl CoA. The acetyl coA is either oxidized for the production of ATP or is converted into carbohydrates. The propionyl CoA is converted to malonic semialdehyde in both plants and animals. Malonic semialdehyde may be converted to alanine by transamination with glutamic acid in animals. In plants, however, it appears that this compound is oxidized to CO_2 & H_2O.

β oxidation of fats is a good source of energy production. In each cycle of β oxidation, 5ATP molecules (3 from the oxidation of NADH & 2 from $FADH_2$) are produced. One ATP is used in the reaction of this way 35-1=34 molecules can be produced in 7 turns of palmitic acid oxidation. Besides this, 8 acyl CoA produced from palmitic acid will produce 12 × 8 = 96 ATP molecules. This will give a total as much as 130 (96 + 36) ATP per palmitic acid molecule.

2. ***Alpha oxidation***

An alternative mechanism of fatty acid oxidation was proposed by P.K. Stumpf. It is also a step wise decarboxylation of fatty acids. In the first step, fatty acid is decarboxylated & reduced to an aldehyde one carbon shorter than the original fatty acid. It is catalysed by a peroxidase.

$$RCH_2CH_2CH_2COOH \xrightarrow[\text{fatty acid Peroxidase}]{H_2O_2} RCH_2CH_2CHO+CO_2+H_2O+O_2$$

This aldehyde is then oxidized to the acid form by dehydrogenase.

$$RCH_2.CH_2.CHO \xrightarrow[\text{aldehyde dehydrogenase}]{\text{NAD} \quad \text{NADH+H}} RCH_2CH_2COOH$$

This new acid may become substrate for fatty acid peroxidase & take another turn in the β-oxidation spiral. This kind of α oxidation has been observed in the peanut cotyledons. In leaf tissues & brain microsomal preparations, a different type of α oxidation occurs, in which decarboxylation takes place after oxidation of the fatty acid.

The biological significance of α oxidation is not known. It may provide, however, fatty acids with odd number of carbon atoms. Further, the aldehyde produced during this process may be reduced so as to yield long chain alcohols.

CONVERSION OF FAT INTO CARBOHYDRATE

In some fat storing seeds, bacteria & fungi, fats are readily converted to sucrose & other complex sugars. Some of the carbon atoms of fats are diverted to the amino acids also. The process called **glyxylate cycle** operates in the special cell organells, the **glyoxysomes**. Glyoxysomes are present only in those tissues which convert fat into carbohydrate. The cycle was worked out largely by H.L. Kornberg of Oxford University in late 1950s.

To begin with the process, acetyl CoA derived from the mitochondrial fatty acid oxidation, diffuses to the glyoxysomes. There it may condense with oxalacetate to produce citrate involving the enzymes citrate synthatase. Citrate is isomerised to isocitrate by aconitase. Isocitrate may be converted to malate through Kreb's cycle intermediate producing CO_2. In glyoxysomes, however, it is cleaved primarily to succinate & glyoxylate by the enzyme isocitratase. Succinate may be diverted to the synthesis of sugars and amino acids involving some of the kreb's cycle intermediates and by the reversal of glycolytic pathway.

The other product glyoxylate, may condense with another molecule of acetyl CoA producing malate. The enzyme for this reaction is malate synthatase. Later on malate dehydrogenase oxidizes malate to oxalacetate which may go to another round of acylation producing citrate. The glyxylate cycle is, shown in fig. 16.2.

The net result of glyoxylate cycle is the conversion of 2 molecules of activated acetate to succinate, which may become a source of sugars & protein carbon atoms. Had the cycle not operated isocitrate produced in the second step (fig. 16.2), Would have been converted to malate. via succinate & fumarate, during which 2 carbons are lost as CO_2. In the glyoxylate cycle, These CO_2 producing reactions are bye passed & the 2 carbons of acetyl CoA are conserved.

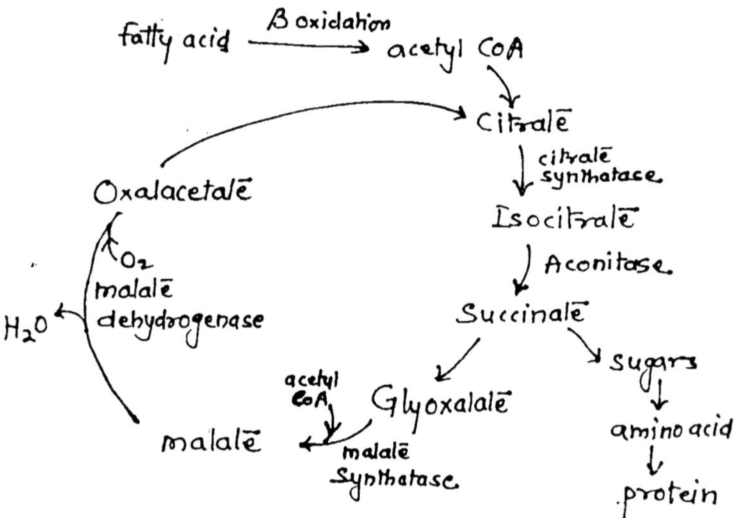

Fig. 16.2 : Glyoxylate Cycle

SYNTHASIS OF FAT

Fats are synthesized from fatty acids & glycerol by the reversal of hydrolytic process.

Starting material for the synthesis of fatty acid is acetyl CoA derived from the aerobic breakdown of carbohydrate. The Acetyl CoA is carboxylated by an enzyme acetyl CoA carboxylase. This enzyme contains biotin as a prosthetic group & requires ATP as enzyme donor for the activation of acetyl CoA.

$$CH_3CO\ SCoA + CO_2 + ATP \longrightarrow ADP + Pi + COOH\ CH_2COSCoA$$

Malonyl CoA

In the next steps, malonyl CoA is acylated several times to produce saturated fatty acids. These reactions are catalysed by a multienzyme complex: fatty acid synthatase. All Intermediate products remain attached to the enzymes in the complex form. The overall reaction for the synthesis of the palmitic acid may be summarised as:

$$CH_3COSCoA + 7\ COOHCH_2COSCoA + 14\ NADPH + 14\ H^+ \longrightarrow$$

$$CH_3CH_2\ (CH_2)_{13}\ COOH + 14\ NADP^+ + CO_2 + 8\ CoASH = 6H_2O$$

The enzyme fatty acid synthatase complex has been isolated from E.Coli, yeast, higher plants & avian liver. This enzyme complex is made up of 8 units (fig. 16.3). having atlest 7 different functions in the synthesis of fatty acids. The purified synthatase is yellow because of the presence of FMN, which can be removed by dialysis. This inactivates the enzyme also. The enzyme contains two SH group, one in the center with ACP (acyl carrier protein) and one of the periphey.

E. coli fatty acid synthatase complex has been dissociated into sub units. Experiments with these subunits suggest the following mechanism of the synthesis of palmitic acid from malonyl CoA involving this enzyme complex.

Malonyl CoA + ACP - SH $\rightleftharpoons$ malonyl - S - ACP + CoASH

Acetyl CoA + ACP - SH $\rightleftharpoons$ Acetyl - S - ACP + CoASH

Acetyl - S - ACP + Malonyl - S - ACP $\rightleftharpoons$

Acetoacetyl - S - ACP + ACP - SH + CO_2

Acetacetyl - S - ACP + NADPH + H^+ $\rightleftharpoons$

β hydroxybutryl - S - ACP + NADP

β hydroxybutyryl - S ACP $\rightleftharpoons$ Crotonyl - S - ACP + H_2O

Butryl - S - ACP + malonyl - S - ACP $\rightleftharpoons$

Ketohexonyl - S - ACP + ACP - SH + CO_2

Ketohexonyl is the reduced to hexonyl-ACP. This process of elongation is repeated until palmityl-ACP is formed & palmitate finally liberated from the enzyme complex.

The process described above yields only saturated fatty acids. Production of unsaturated fatty acids will require oxidation of saturated fatty acid.

Bloch et al. (1960) found that oxidative desaturation of palmityl to palmitoleyl was catalysed by a particulate enzyme in the yeast cells.

$$\text{Palmityl CoA} \xrightarrow{O_2\, NADPH + H^+} \text{Palmitoleyl CoA}$$

Similarly stearyl CoA was converted to Oleyl CoA by the same enzyme

$$\text{Stearyl CoA} \xrightarrow{O_2 NADPH + H^+} \text{Oleyl CoA}$$

Evidences for the production of unsaturated fatty acids by oxidative desaturation have been found in algae, higher plants & animals also. In bacteria, however, it appears that fatty acids are unsaturated when the are smaller or medium chain length. the chain grows further with the addition of Co_2 units at a time to produce desired length of fatty acid.

The other components of fat besides fatty acids is glycerol. Glycerol is synthesized from dihydroxy acetone phosphate as glycerol phosphate through glycereldehyde phospahte. The glycerol phosphate reacts with two molecules of fatty acids which are in the form of acety CoA derivatives to yield phosphatidic acid.

$$\begin{array}{l} CH_2OH \\ | \\ CHOH \\ | \\ CH_2O\ H_2PO_3 \end{array} + 2\ RCOOSCoA \longrightarrow \begin{array}{l} CH_2O\ COR \\ | \\ CHOCOR \\ | \\ CH_2OH_2H_2PO_3 \end{array} + 2\ HSCOA$$

phosphatidic acid

The phosphatidic acid (diacyl-glycerol phosphate), is dephosphorylated by an enzyme, phosphatase to yield a diglyceride, which in turn reacts with another molecule of fatty acid CoA to yield a triglyceride (triacyl glycerol).

$$\text{Phosphatidic acid} \xrightarrow[\text{Phosphatase}]{H_3PO_4} \begin{array}{l} CH_2OCOR \\ | \\ CHOCOR \\ | \\ CH_2OH \end{array}$$

Diglyceride

$$\Big\downarrow \text{RCOOSCoA}$$

$$\begin{array}{l} CH_2OCOR \\ | \\ CHOCOR + HSCoA \\ | \\ CH_2OCOR \end{array}$$

The natural fats thus synthesized may be transported & stored at specific storage site.

QUANTITATIVE DETERMINATION OF FAT BY RUSHKOVASKII METHOD

This method is convenient in that it permit several analysis to be performed simultaneously. The sample to be analysed is placed in a glass weighting bottle & dried at 65-70°C. The dry sample is carefully powdered in a porcelain morter.

PROCEDURE

Take approx. 50 mg. of dried sample in a small filter paper packet & place it in a thimble. Place the thimble in the Soxlet extraction tube. Take 200 ml. $CHCl_3$ - CH_3OH mixture (in the ratio of 2:1) in a 250 ml. round bottom flask & place in the heating unit of Soxhlet apparatus.

Arrange continuous water circulation during distillation. Extract the fat for 2 hrs. Remove the thimble, dry & weigh the defatted material.

CALCULATION

Fat content = initial wt. of dry material - final wt. of dry tissue

SEPARATION OF MACROMOLECULES (PROTEIN, FAT & NUCLEIC ACIDS) FROM A TISSUE EXTRACT

The usual techniques employed for separation of macromolecules from the tissue are based on differential or selective extraction mechanism.

SELECTIVE EXTRACTION OF LIPID

Prepare a homogenate of tissue in water in a homogeniser. They are insoluble in water but easily soluble in $CHCl_3$, ether, C_6H_6 etc. Take the homogenate in a large test tube 2-5l. & add slowly into it double the quantity of methyl alcohol in order to denature the protein on the one hand to break the lipid compound on the other.

Add into the $CHCl_3$ the same quantity as CH_3OH is added. Shake the mixture will & then pour it into a beuratte. Allow the solution to stand undisturbed for the time till separate layers are visible. Collect separate layers in different test tubes. The $CHCl_3$ layer contain all the lipids.

Chapter -17

ALCOHOLS IN BIOLOGICAL SYSTEM

These biochemicals are having one or more hydroxyl group -OH (-ol suffix). Alcohols can be considered either hydroxyl derivatives of paraffinic hydrocarbons

R—OH

or alkyl derivatives of water molecule

R—H or H—OH

they have the properties of both.

The first of the series is :-

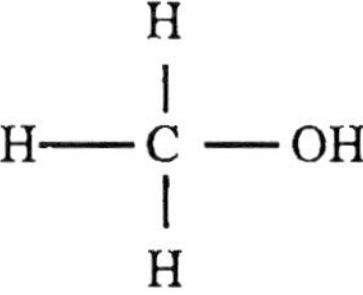

methyl alcohol or methanol, which has a low boiling point & is very soluble in water.

As the number of carbon atoms in the alkyl group increases, the less soluble & more like the parent hydrocarbons the alcohol becomes. Methyl alcohol is occasionally abbreviated MeOH.

Methyl alcohol can be synthesized but was identified with its first source the destructive distillation of hardwood, which yields wood alcohol (Destructive distillation = heat without oxygen).

ETHANOL OR ETHYL ALCOHOL

```
      H    H
      |    |
  H — C —— C —— OH
      |    |
      H    H
```

Ethyl alcohol (abbreviated Eth OH) is drinkable. Ethyl alcohol or grain alcohol is the end product of the biological fermentation.

The source of organic alcohols is the fermentation of an organic substrate (e.g. grain) in the presence of a bacterial agent (e.g. yeast).

CLASSIFICATION OF ALCOHOLS

Primary Alcohol

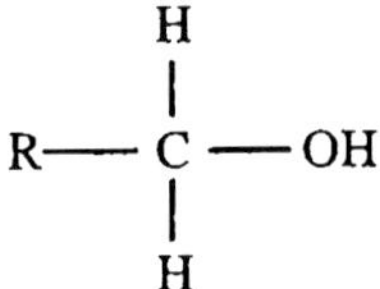

Carbonal group on the terminal carbon atom.

Secondary Alcohol

```
      R
      |
  R — C — OH
      |
      R
```

Tertiary Alocohol

```
      R
      |
  R — C — OH
      |
      R
```

$$\begin{array}{ccccccccc} & & H & & H & & H & & \\ & & | & & | & & | & & \\ H & - & C & - & C & - & C & - & OH \\ & & | & & | & & | & & \\ & & H & & H & & H & & \end{array}$$

1. Propinol

(primary alcohol)

$$\begin{array}{ccccccccc} & & H & & H & & H & & \\ & & | & & | & & | & & \\ H & - & C & - & C & - & C & - & H \\ & & | & & | & & | & & \\ & & H & & OH & & H & & \end{array}$$

2. Propinal or

Isopropyl alcohol

(Secondary alcohol)

$$\begin{array}{ccccccccccc} & & H & & H & & H & & H & & \\ & & | & & | & & | & & | & & \\ H & - & C & - & C & - & C & - & C & - & OH \\ & & | & & | & & | & & | & & \\ & & H & & H & & H & & H & & \end{array}$$

1. Butanol or butyl

alcohol

(primary alcohol)

$$\begin{array}{ccccccccccc} & & H & & H & & H & & H & & \\ & & | & & | & & | & & | & & \\ H & - & C & - & C & - & C & - & C & - & H \\ & & | & & | & & | & & | & & \\ & & H & & H & & OH & & H & & \end{array}$$

1. Butanol

(Secondary alcohol)

$$\begin{array}{ccccccccc} & & H & & H & & H & & \\ & & | & & | & & | & & \\ H & - & C & - & C & - & C & - & OH \\ & & | & & | & & | & & \\ & & H & & | & & H & & \\ & & & & | & & & & \\ & & H & - & C & - & H & & \\ & & & & | & & & & \\ & & & & H & & & & \end{array}$$

2.Methyl 1.propinal

(primary alcohol)

```
     H     H     H
     |     |     |
H—— C —— C —— C —— OH
     |     |     |
     H     |     H
           |
      H— C — H
           |
           H
```

2.methyl 1.propinal

(Tertiary alcohol)

Alcohols have a C—OH grouping or in other words, a carbonal linkage of the functional group.

All isomers of alcohols are all solvents but having different graphic formulae, they have different properties.

Now, alcohols can be synthesized, but originally were obtained through fermentation only. Different alcohols are the metabolic wastes of different organisms. *Clostridium acetobytiticum* is an obligate anaerobe which produces butyl alcohols. The yeast organisms *Saccharomyces cerevisaie* produce methyl alcohols with or without oxygen. However, yeast cells form with oxygen.

ABSOLUTE ALCOHOL

100% of that alcohol named or 200 proof. Proof = 2x the volumetric concentration of alcohol in Solution.

Denatured alcohol: Something added to alcohol to make it unpalatable, usually methyl.

Hexadecanol (16 carbon chain with terminal -OH) is used to control evaporation in reservoirs, by pouring it on the surface, it forms a one-layer film which does not volatilize. But, it is biologically attacked & must be periodically replaced.

High percentage of alcohol (95%) disinfects, but less than this is food for organisms.

Alcohol will oxidize biologically or chemically, but differ in ways of end product on whether it is a primary, secondary or tertiary alcohol.

PRIMARY ALCOHOL

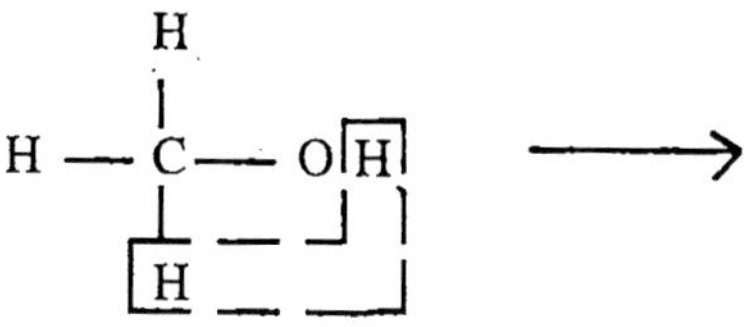

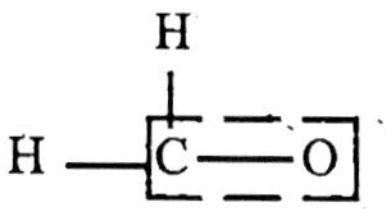

H
|
H — C — O H
|
H

→

H
|
H — C = O

carbonyl group formaldehyde

methyl alcohol

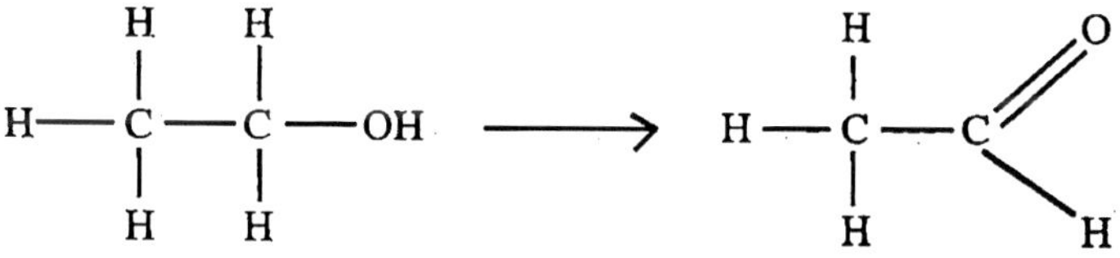

H H
| |
H — C — C — OH
| |
H H

→

H O
| //
H — C — C
| \
H H

ethyl alcohol

Acetaldehyde

(aldehydes hae an IUC suffix —al)

Secondary alcohol

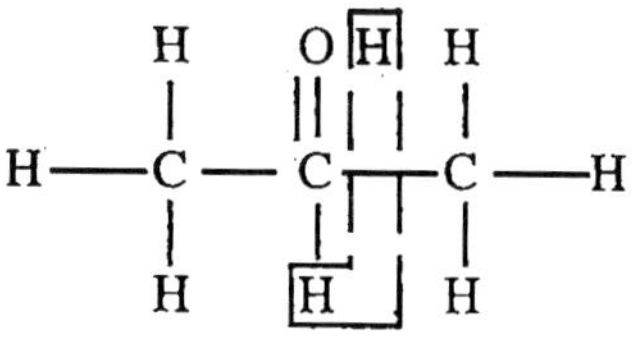

H O H H
| || | |
H — C — C — C — H
| | |
H H H

H O H
| || |
H — C — C — C — H
| |
H H

2 propinol

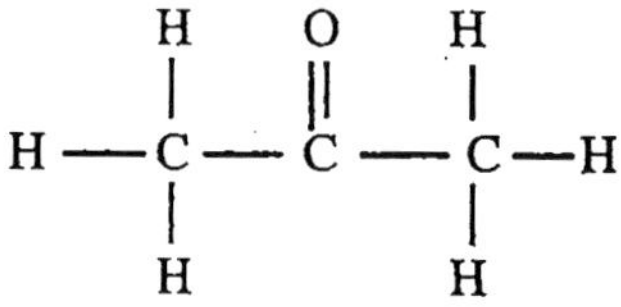

acetone

Tertiary alcohols are resistant to oxidation. They are also resistant to biodegradation. They can be made subject to degradation by acclamation or eventually building a taste for the compound in the microorganisms.

UNSATURED ALCOHOLS

```
      H     H
      |     |
H --- C === C --- OH
```

It is still uncertain that alcohol easily oxidizes to acetaldehyde.

The textile industry used to use starch for fubric sizing, which became part of that industries waste. They had to find a compound which was not BOD & now use alcohol.

POLYDYDROXY ALCOHOLS

2 or more hydroxyl groups attached to separate carbon atoms e.g. glycols, (2 hydroxyl). Two hydroxyl groups on the same carbon atom are unstable & tend to form water & an acetone. Ethylene glycol or 1.2 ethanediol: permanent anti-freeze.

```
   H H                          H  H
   | |                          |  |
H—C=C—H+HOCl  ——→           H—C—C—H + Na2HCO3  ——→
                                |  |
                               Cl OH                 glycol + NaCl
```

Preparation of glycol shows chlorine demand. A glycol mixes with water.

Butylene glycol has 6 isomers.

Glycerols

Are trihydroxy alcohols.

```
      H     H     H
      |     |     |
H --- C --- C --- C --- H
      |     |     |
      OH    OH    OH
```

1.2.3 propanetriol or glycerol tri acetate (a moistening agent)

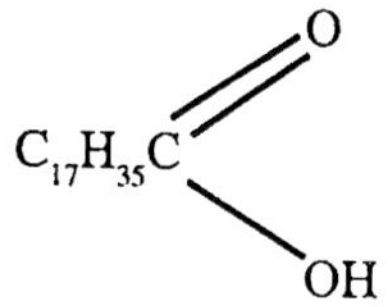

$$C_{17}H_{35}C(=O)OH$$

Stearic acid (a fatty acid)

Fatty acids have a long chain, above 12-12 carbon atoms. Stearic acid + glycerol tri-acetate is a typical structure for a fat.

Sodium salts of fatty acids yield soap.

Oxidation is not complete until the only end products are carbon dioxide and water.

Unsaturated compounds in liquid form (oils) are hydrogenated to form crystals (margarine etc.)

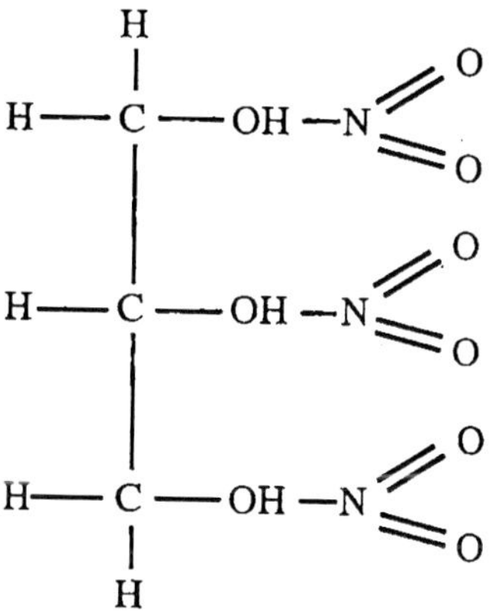

Nitroglycerin

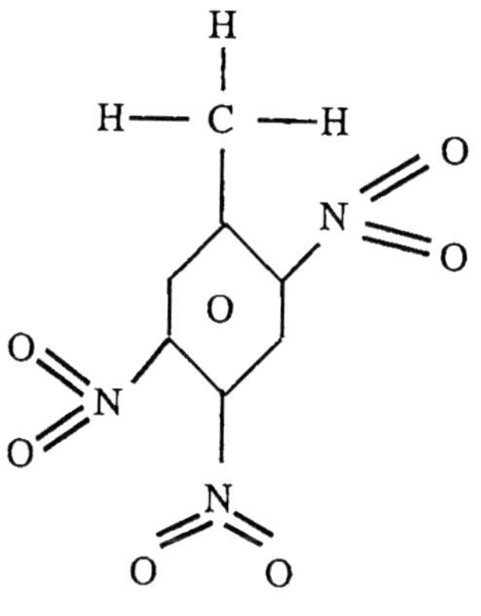

TNT (trinitro toluene)

Chapter -18

HORMONES IN BIOLOGICAL SYSTEM

Hormones are chemical substances that are produced in minute quantities by some specialised organs, producing physiological & biochemical effects on some tissue remote from the source. Most of the hormone producing organs are endocrine glands. The hormones produced by these glands are mixed in the blood stream & are transported to the target tissues. They normally stimulate some of the physiological functions, although some endocrine secretions inhibit activity which have been designated as ***chalones***.

The major hormone secreting glands are intestinal mucosa, pancreas, adreal, thyroid, parathyroid, pitutary, ovarie's & testes. A summary of the same is given in Table 18.1.

Table 18.1 : Common hormone producing organs & related hormones

Organ	*Hormones*	*Functions*
1	*2*	*3*
Gastro intestinal tract	Gastrin	Secretion of HCl by intestine
	Pancreozymin	Secretion of pancreatic juice
	Parotin	Calcification of teeth & Change in Ca & P content of the serum

1	*2*	*3*
Pancreas	Insulin	Regulation of blood sugar
	Glucagon	Regulation of blood sugar
Adrenals	Epinephrin or adrenalin	Glycogen breakdown
	Aldosterone	Mineral balance, sodium retention
	Cortisol	Metabolism, gluconeogensis
Thyroid	Thyroxin	Increase of basal metabolic rate (BMR), development
	Calcitonin	Maintenance of calcium level in blood & bones
Parathyroid	Parathormone	Maintenance of Ca & P level.
Pitutary		
Anterior lobe	Somatotropine	Growth & metabolism
	Carticotropine	Stimulation of the adrenal cortex
	Thyrotropine	Stimulation of thyroid gland
	Follicle stimulating hormones	Stimulating of maturing of gametes
	Luteinising hormones	Stimulation of production of sex hormones.
	Prolactin	Stimulation of mammary glands
Middle lobe	Melanotropin	Dilation of melanophores
Posterior lobe	Vassopressin	Renal reabsorption of water
	Oxytocin	Contraction of uterus expression of milk by lacting mammary glands
Hypothalamus	Corticotropin releasing hormones	Release of corticotropin

1	2	3
	thyrotropin releasing hormones	Release of thyrotropin
	Growth hormones	Release of somatotropin
	Gonadotropin releasing hormone	Release of luteinizing & follicle stimulating hormones
Ovary		
Follicle	Estradiol	Proliferation of the uterine mucosa & maintenance of normal size & function of other female reproductive organs
Corpus luteum	Progesterone	Converting the uterine mucosa to a secretary phase
Testis	Testosterone	Normal development of male reproductive organs
	Androsterone	appearance of secondary sexual characters

CHEMICAL STRUCTURE OF HORMONES

At least, three chemical classes of hormones can be recognized

1. Steroid hormones e.g. aldosterone, cartisol, progesterone etc.
2. Amino acid derived hormones e.g. thyroxines epinephrine etc.
3. Peptide & protein hormones e.g. thyrocalcitonin, parathormones, insulin, glucagon etc.

Steroid hormones are synthesized from cholestrol in specialized glands-adrenal cortex, ovaries, testes & placenta. They contain 18 to 21 carbon atoms & 24 to 230 hydrogen atoms. Structures of some of the steroid hormones are given below:

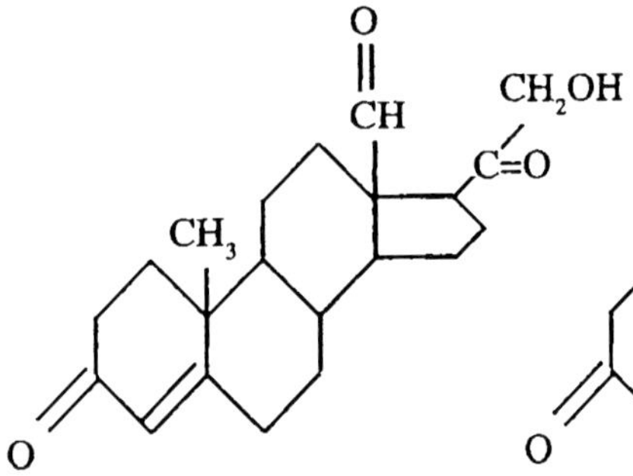

Aldosterone

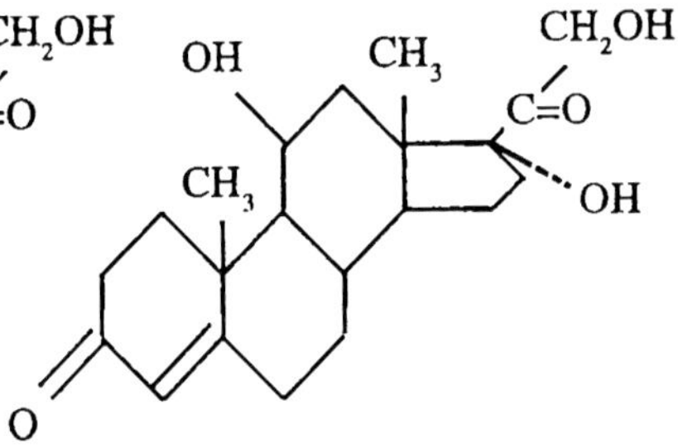

Cortisol

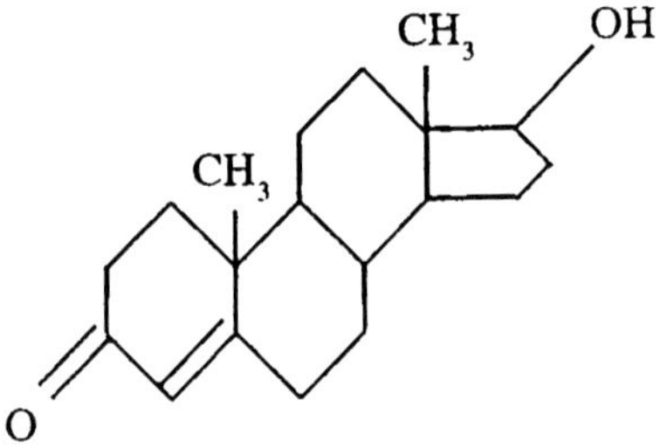

Testosterone

Other steroid hormone also have similar structure.

Harmones like thyroxine & epinephrine are derived from amino acid tyrosine. Thyroxine is produced in thyroid gland.

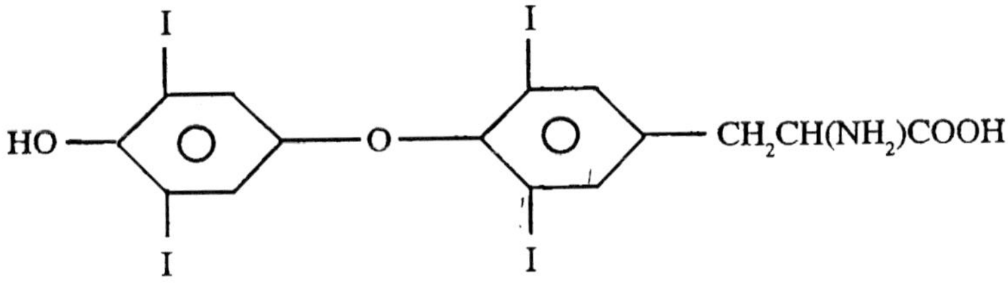

Thyroxine (tetraiodothyronine)

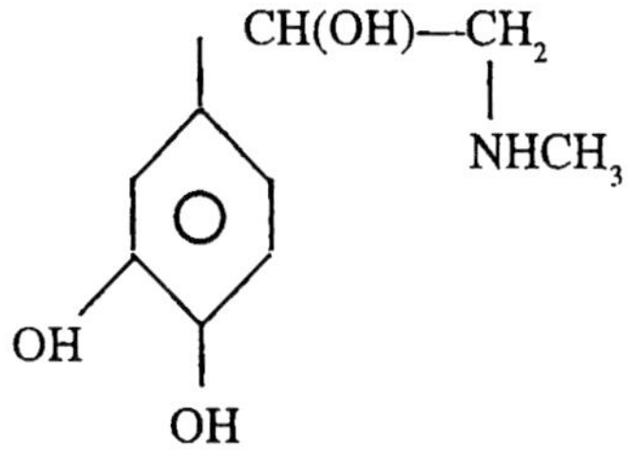

CH(OH)—CH_2

$NHCH_3$

OH

OH

Epinephrine

Peptide or protein hormones are numerous. Amino acid sequence of most of them has been worked out. Calcitonin is a single chain polypeptide containing 32 amino acids with a molecular weight of about 3600. Parathormone is a protein with 83 amino acid residues of 17 different amino acids. Similarly, vassopressin is a smaller polypeptide with only 8 amino acid residue in the following sequence:-

Cys—Try—Phe—Asn—Gln—Pro—Arg—Gly

Some protein hormones contain additional chemical groups also. Follicle stimulating hormones e.g. contains carbohydrates such as galactose, mannose, fucose, hexosamine & N-acetyl neuraminate besides amino acids. Lutenizing hormone is also a glycoprotein containing about 25% of its total weight as carbohydrate.

MECHANISM OF ACTION OF HORMONES

The hormones are circulated in the blood throughout the whole body but they act only on certain specific organs or tissues called target organs or target tissues. The cells in target tissues contain a receptor protein which binds with the hormones. Hormones may produce desired physiological effects either in this form or after being transported to its site of action inside the cell with the receptor protein.

At molecular level two types of reactions are initiated by the hormones.

1. Synthesis of a new protein

Many steroid hormones are known to induce synthesis of certain enzymes or new proteins in the target tissues. This induction is not observed in the presence of protein synthesis inhibitors such as cycloheximide, purpmycin and actinomycin D. Although hormone may induce any one of the step in protein synthesis, it is commonly held that they act at transcription level. They combine with the repressor molecule and the inhibited DNA expresses itself in the form of m-RNA molecules. These m-RNA molecules are translated to a new protein which may be an enzyme.

2. Activation of enzyme & involvement of C-AMP

Some enzymes are activated in the presence of hormones. E.W. Sutherland & T.W. Rall (1960) fouind that this activation is mediated via a special form of the adenine nucleotide, 3'5' cyclic AMP cAMP). This molecule was termed as a second messenger whereas the hormone itself was the first messenger. The hormone after reaching the receptor cell activates a membrane bound enzyme, the adenyl cyclase. This enzyme catalyses the conversion of ATP to 3'4'5'. Cycling adenosine monophosphate & inorganic pyrophosphate. Cyclic AMP activates certain rate limiting enzyme may change the metabolic rate of the tissue & the desired physiological effect is achieved. The events leading to the desired effect of a hormone through C-AMP are summarised in fig. 18.1.

The effects of several hormones such as adrenaline, glucagon, vassopressin & para thormone are known to be mediated through increase n C-AMP. The activation of enzyme by -C-AMP is also known to effect the permeability of membrane (probably Through the regulation of calcium level in the membrane) and the synthesis of certain proteins involved in the transport. The effect of aldesterone on movement of Na^+ through the kidney & increased uptake of glucose by the target organs of thyrotropin & luteinizing hormones are results of the effects of hormones on membrane perneability.

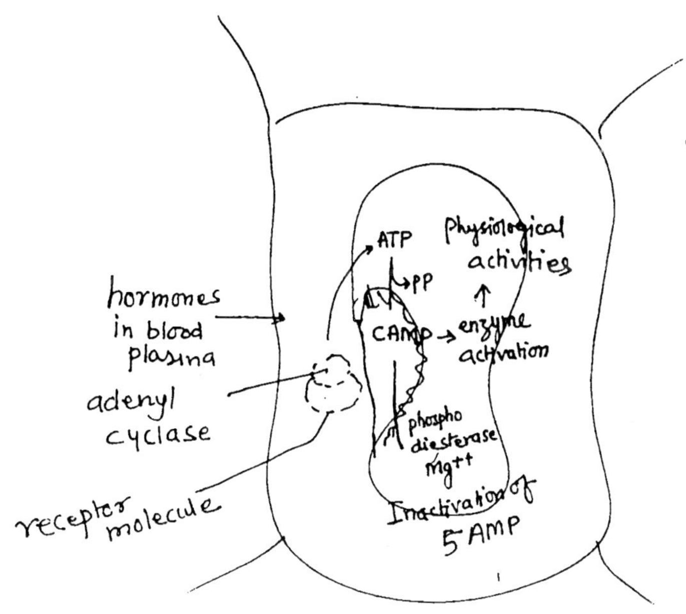

Fig. 18.1 : Cyclic AMP & hormone action

Chapter -19

ALDEHYDES, KETONES, ORGANIC ACIDS AND ESTERS IN BIOLOGICAL SYSTEM

The general formula for aldehyde and ketones are :

Aldehydes & ketones are related & have similar chemical properties and both have a carbonyl functional group.

Aldehydes have a carbonyl group attached to the terminal carbon atom and are derived from primary alcohols.

$$\mathrm{H{-}\underset{H}{\overset{H}{C}}{-}OH} \xrightarrow{\text{oxidation}} \mathrm{\overset{H}{\diagdown}\underset{H}{C}{-}O} \xrightarrow{\text{oxidation}}$$

methanol   methanal or formaldehyde

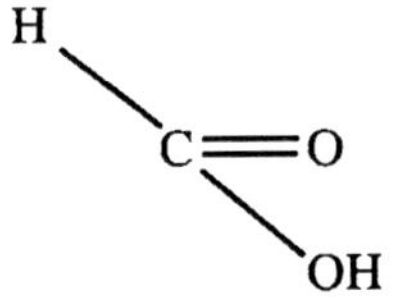

H
C=O
OH

formic acid

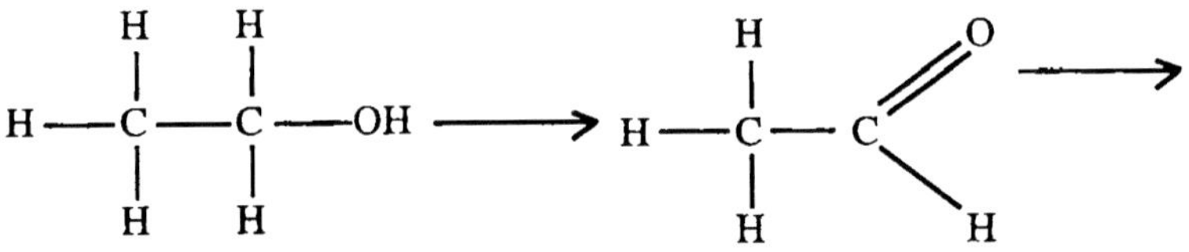

$$H-\underset{H}{\overset{H}{C}}-\underset{H}{\overset{H}{C}}-OH \longrightarrow H-\underset{H}{\overset{H}{C}}-C{\overset{O}{\diagup\!\!\diagup}}\diagdown H \longrightarrow$$

ethanol acetaldehyde

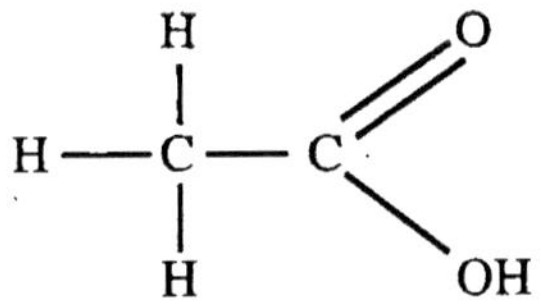

$$H-\underset{H}{\overset{H}{C}}-C(=O)OH$$

acetic acid

$$H-\underset{H}{\overset{H}{C}}-\underset{HC_3}{\overset{H}{C}}-C(=O)H$$

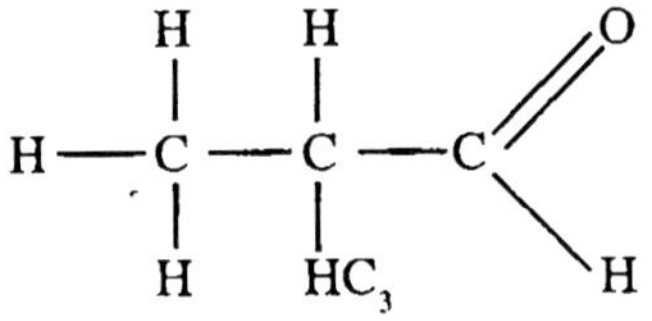

2. methyl propinal (numbering starts from functional group)

KETONES

Ketones have a carbonyl group attached to an internal carbon atom (this is also called a keto group) & are derived from the oxidation of secondary alcohols.

IUC suffix-one.

Carbohydrates contain either the aldehyde or the ketone carbonyl groups.

```
      H   O   H
      |   ||  |
  H — C — C — C — H
      |       |
      H       H
```

Propanene, propione or dimethyl ketone or acetone (solvent)

A ketone of interest of environmental scientists & technologists is the methyl-ethyl-ketone, or commonly known as MEK. It is found in the industrial waste of synthetic processes. It is also named butanone.

```
      H   O   H   H
      |   ||  |   |
  H — C — C — C — C — H
      |       |   |
      H       H   H
```

Ketones are derived by the dehydrogenation of secondary alcohols :

```
      H   O[H] H                       H   O   H
      |   |    |                       |   ||  |
  H — C — C  — C — H   ———→        H — C — C — C — H
      |   |    |                       |       |
      H  [H]   H                       H       H
```

When the compounds are low in the number of carbon atoms, they are reasonably soluble in water. Many are volatile enough to remove from aqueous solution, but they are not us volatile as ether. Both are synthesized by the petrochemical industry from crude petroleum of natural gas. Aldehydes & ketones are reactive at the carboxyl group. They are oxidized to acids or reduced to alcohols. e.g. conversion of glocol into oxalic acid.

```
       H   H                 O           O
       |   |                  \\        //
  OH — C — C — OH   ———→        C —— C          ———→
       |   |                  /        \
       H   H                 H          H
```

glycol glyoxal (a dialdehyde)

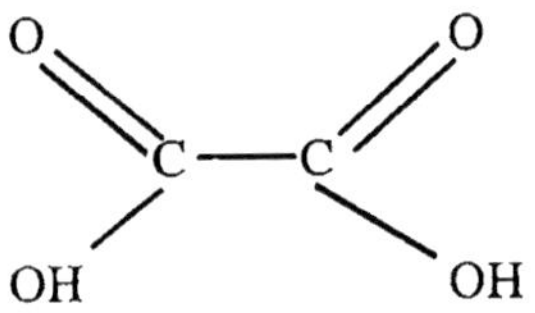

O O

C — C

OH OH

Oxalic acid

ORGANIC ACIDS

These have a carboxyl functional groups, Which is a combination of the carbonyl & the hydroxyl groups. Acids are at their highest oxidation state, further oxidation would yield carbon di oxide & water. Monocarboxylic acids have one carboxyl group. The simplest acid is formic acid.

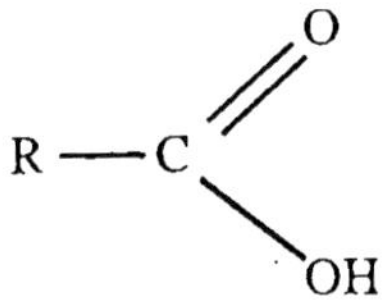

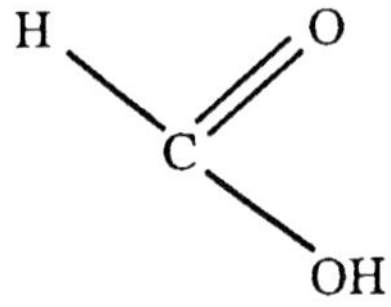

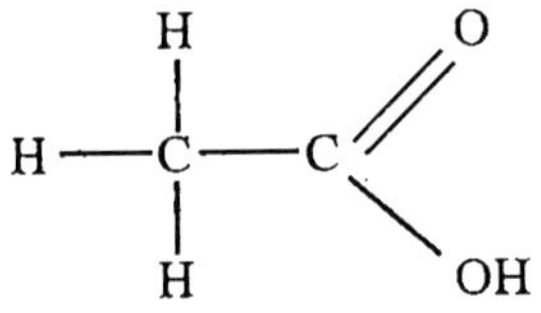

O

R — C

OH

H O

C

OH

H O

H — C — C

H OH

Acetic acid (vinegar)

Graphical formula is

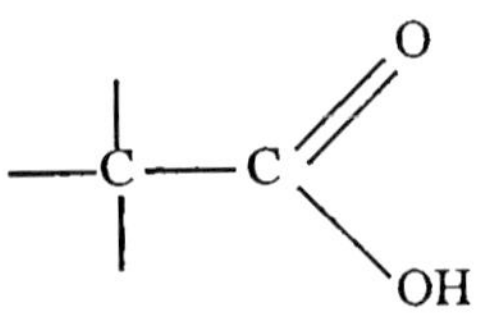

O

— C — C

OH

The general empirical for an organic acid is $C_nH_{2n+1}COOH$. The hydrolysis of fats or lipids yields fatty acids. Fatty acids refer to only long carbon chain acids. e.g. palmitic acids have 16 carbons. Volatile acids have small carbon chains & can distill at atmospheric conditions. Organic acids do have the acidic property of ionization but they are poor ionizers & are considered weak acids.

$$K^a = \frac{(AC^-)\ (H^+)}{(HAC)}$$

Where,

K_a = dissociation constant

HAC = acetic acid

Organic acids are significant to anaerobic digestion in fact they control the action of anaerobic digesters & affect stabilization of the sludge organism complex. The anaerobic digesters attack organic matters including the acids & the products of this digestion is attacked by the methane forming organisms. However, the anaerobic organisms do not function well at pH below 6.5 & the whole system will go sour from a drop in pH. Although the organic acids are relatively week, an excessive amount of them can cause a pile up-dropping pH to 3,4 or 5. Because the organisms will stop working at this level of pH, the digester will become overloaded & stuck.

To neutralize the acid like needs to be added to the digester. It takes a lot of lime to get things going again because of the low disassociation factor of the volatile acids. Determining & measuring the amount of volatile acids regularly will prevent the digester. Levels of 50-300 mg/lt. are relatively safe but levels in the thousands of mg/lt. is danger sign.

Dicarboxylic acids have a carboxyl group of each end of the alkyl radical. Monocarboxylic acids with 1-9 carbon atoms are liquid. Monocarboxylic acids with ore than 10 carbon atoms are greasy solids.

Saturated monocarboxylic acids with more than 10 carbon atoms are the fatty acids. These acids have foul odours e.g. the odour of rancid butter is produced by butyric acid. When a monocarboxylic acid is halogenated, the old method of nomenclature was to label the positions of the halogens with Greek letters starting with the carbon next to the functional group.

$$H-\underset{\underset{H}{|}}{\overset{\overset{H}{|}}{C}}-\underset{\underset{\underset{\beta}{Br}}{|}}{\overset{\overset{H}{|}}{C}}-\underset{\underset{\underset{\alpha}{Br}}{|}}{\overset{\overset{H}{|}}{C}}-C\begin{matrix}\nearrow O \\ \searrow OH\end{matrix}$$

IUC-2, 3 Dibromo Butanoic Acid

The old nomenclature is still seen in reference to proteins. Amino acids are the building blocks of proteins. One of the many varieties of amino acid is -

$$H-\underset{\underset{H}{|}}{\overset{\overset{H}{|}}{C}}-C\begin{matrix}\nearrow O \\ \searrow H\end{matrix}$$

α-amino acid-glycine

Amino acids are essential in biological processes.

Organic acids are prepared by:-

- Oxidizing aldehydes

- Using a strong oxidizing agent on ketone

$$H-\underset{\underset{H}{|}}{\overset{\overset{H}{|}}{C}}-\overset{\overset{O}{||}}{C}-\underset{\underset{H}{|}}{\overset{\overset{H}{|}}{C}}-\underset{\underset{H}{|}}{\overset{\overset{H}{|}}{C}}-H + C_2H_7 \longrightarrow C-C\begin{matrix}\nearrow O \\ \searrow OH\end{matrix}$$

Which has the effect of shortening the carbon chain.

- a reaction which adds to the carbon chain

$$C - C - OH + KCN \longrightarrow \underset{|}{\overset{|}{C}} - \underset{|}{\overset{|}{C}} - C \equiv N$$

a nitrile (alkyl cyanide)

By this method, chemists can synthesize organic acids.

Reactions which use the ionizing properties of organic acids, are important in organic synthesis. The reaction yields sodium acetate which reacts as an acid but is used as a salt.

$$-C-C(=O)-O + NaOH \longrightarrow -C-C(=O)-ONa + H_2O$$

Sodium salts of long carbon chain fatty acids yield natural soap with its soluble detergency properties. Minerals such as calcium makes hard water which in reaction with the soap makes the insoluble deposits which form the ring in the bath tub.

$$Ca^{++} + \text{Sod. palmitate} \longrightarrow \text{Calcium palmitate}$$

OTHER REACTION OF ORGANIC ACIDS

- Acid plus alcohol in the presence of dehydrating agent yield an ester.

$$-C-C-OH + HO-C(=O)-C- \xrightarrow{\text{dehydrate}} -C-C(=O)-O-C_2H_5$$

- Reduce acids to alcohols.
- Halogenate acids.

$$Cl_3C-C(=O)-OH$$

Trichloroacetic Acid

ESTERS

May be organic or inorganic. Organic esters are derive from a reaction of organic acid & alcohol.

— an inorganic reaction —

$$C_{12}H_{35}OH + H_2SO_4 \longrightarrow C_{12}H_{25}{-}O-\overset{\overset{\large O}{||}}{\underset{\underset{\large O}{||}}{S}}-O^{-}H^{+}$$

Lauryl alcohal — Sulfuric acid (inorganic acid) — Lauryl sulfate (inorganic ester)

By adding sodium one gets sodium lauryl sulphate,

$$C_{12}H_{25}-O-\overset{\overset{\large O}{||}}{\underset{\underset{\large O}{||}}{S}}-O.Na$$

an ionic, synthetic detergent. Another synthetic detergent or syndet is liner alkyl sulphate or LAS.

Organic esters have the general graphic formula :

$$R-\overset{\overset{\large O}{||}}{C}-O-R$$

Organic esters are widely distributed in nature in plants & animals. Esters give the fragrant scent to flowers & the bouquet to wine.

The acetate radical is:

$$-C-\overset{\large O}{\overset{//}{C}}-O-$$

An ester can be returned to its respective acid & alcohol by hydrolysis.

$$-\overset{|}{\underset{|}{C}}-\overset{O}{\overset{||}{C}}-O-\overset{|}{\underset{|}{C}}-\overset{|}{\underset{|}{C}}-$$

$$CH_3COOH + C_2H_5OH \longrightarrow C_2H_5COOCH_3 + H_2O$$

acetic acid ethyl alcohol ethyl acetate

Waxes are organic esters made up of 2 long carbon chain compounds with an ester linkage. The ester linkage is :

$$-C\begin{matrix}\nearrow\!\!\!\!\!= O \\ \searrow O\end{matrix}$$

Palmitic acid (C_{16}) + a similar alcohol → Wax

Glycereldehydes (fats, lipids, oils) are esters of glycerol long carbon chian acids. Glycerol esters in solid form are lards & tallows & in liquid forms are oils such as cotton & corn oils.

The general structurl formula of a fat or oil is

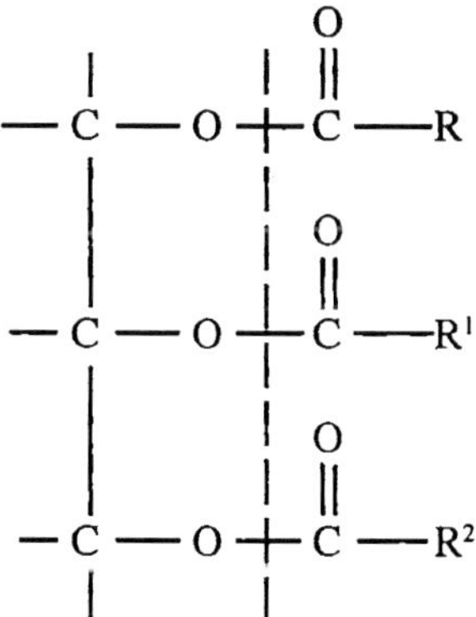

hydrolysis will divide this to alcohol & fatty acid.

The radicals can all be the same or different. e.g. if all three radicals are $C_{15}H_{3L}$ the compound will be glycerol tri palmitate or if—

R = $H_{33}C_{17}$ (unsaturated oleic acid)
R_1 = $H_{31}C_{15}$ (palmitic acid)
R_2 = $H_{35}C_{17}$ (stearic acid)

} glycerol oleopalmito stearate

Saponification

the alkaline hydrolysis of a fat to form the tri hydroxy alcohol & the corresponding fatty acid.

Physical properties of fats & oils :-

— A saturated alkyl chain of 10 or more carbons = solids.

— less than 10 carbons or a longer unsaturated chains = liquids.

The significance of using oils to prepare loeomargarine & shortening is that hydrogenation breaks that double bond.

$$-C=C- \longrightarrow -C-C-$$

The iodine number indicates the degree of unsaturation. A known amount of iodine is added to a known concentration of a compound which will break its double bond & join with the iodine. The residual iodine is measured for the degree of unsaturation organic acids.

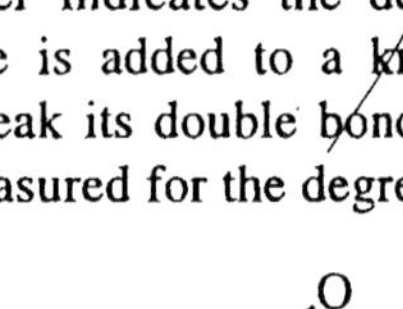

$$R-C(=O)-OH$$

will ionize a bit $C(=O)-O^-H^+$

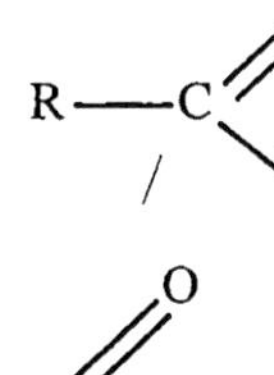

The simplest dicarboxylic acid is :

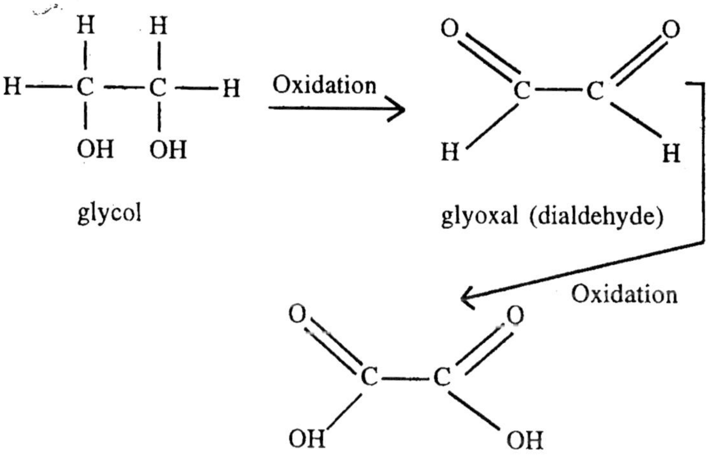

```
     H     H                           O             O
     |     |                            \\          //
H —— C ——— C —— H   Oxidation            C ——— C
     |     |       ————————→            /       \
     OH    OH                          H         H

     glycol                         glyoxal (dialdehyde)
```

Oxidation

```
   O           O
    \\        //
      C ——— C
     /       \
   OH         OH
```

Oxalic acid
(Important to organism metabolism)

The general formula of dicarboxylic acid is

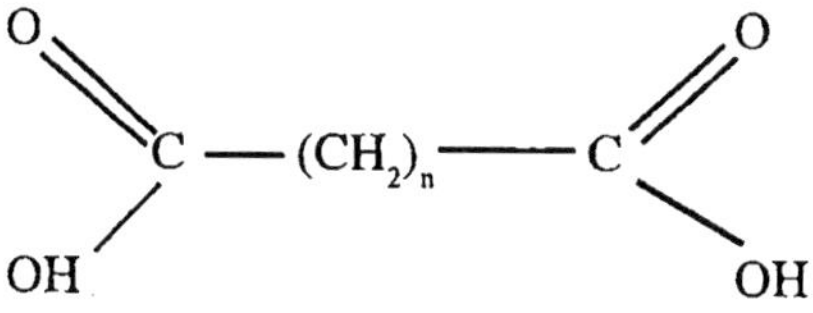

```
 O                          O
  \\                       //
    C —— (CH2)n ——— C
   /                       \
 OH                         OH
```

Aliphatics — straight or branched carbon chain compounds.

Aromatics — rings of 6 carbon atoms with alternate double bonds.

The parent compound of aromatics is benzene, an organic solvent.

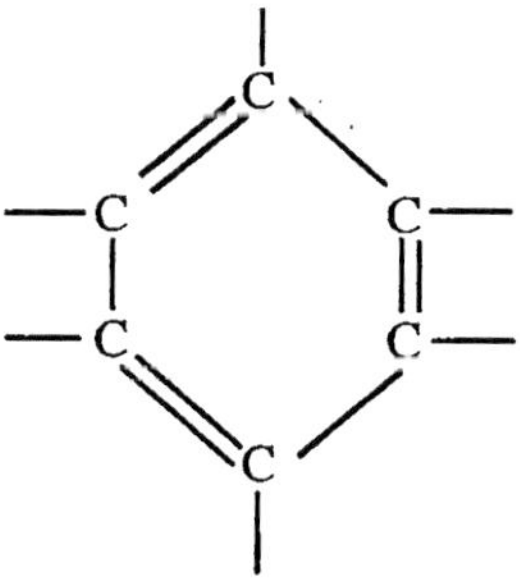

```
        |
        C
      //  \
 — C        C —
    |       ||
 — C        C —
     \\   /
        C
        |
```

Not to confused with aromatics are cycle-aliphatics, which are often present in petroleum—

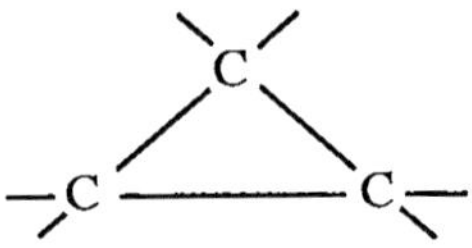

Cycloproprane

The aromatic ring of simple benzene is occassionally written—

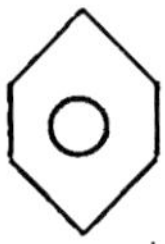

to indicate the resonating double bonds.

Some monosubstituted benzene rings are :-

CI

H

C—H

H

OH

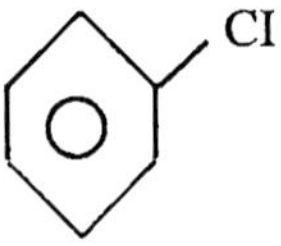

monochlorobenzene

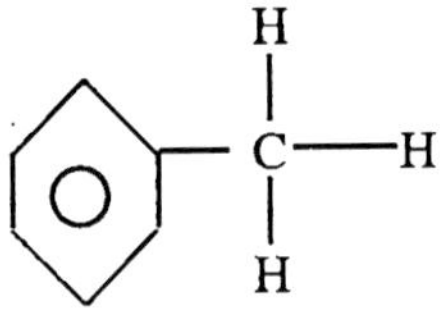

toluene (methyl benzene)

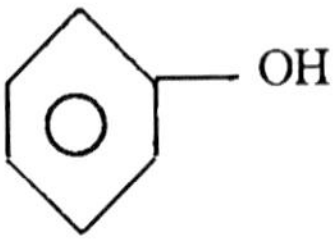

Phenol (monohydroxy benzene)

The phenol coefficient is the measure of antiseptic properties by response of an organism to phenol.

Other monosubstituted benzene rings :

R

O

C

OH

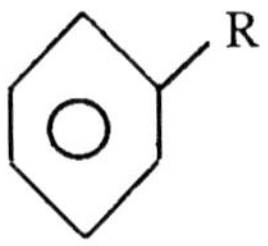

Phenyl group

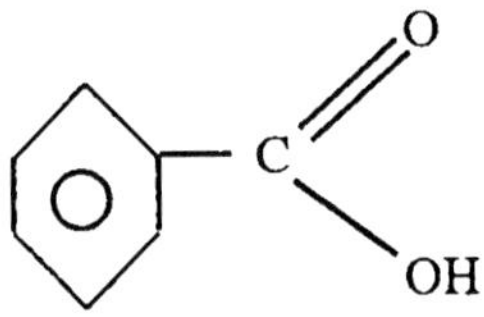

Benzoic acid

Disubstituted benzene rings are numbered to give the smallest numbers. Also, adjacent positions yield a different compound than the same substitutions in alternate positions.

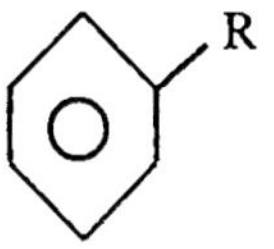

R

1.2. dichlorobenzene or othro dichloro-benzene or O-dichlorobenzene

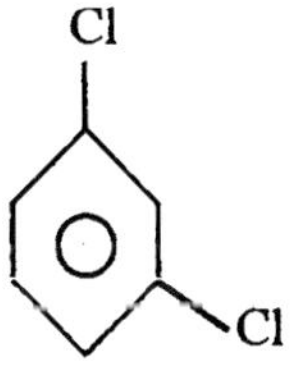

Cl

Cl

1.3. dichlorobenzene or metadichlorobenzene or m-dichlorobenzene

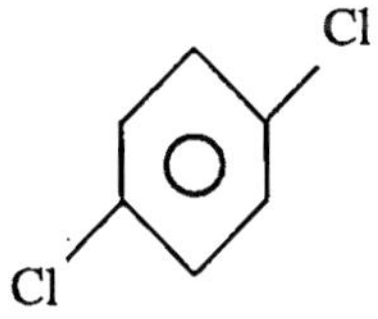

Cl

Cl

1.4 dichlorobenzene or para dichlorobenzene or P-dichlorobenzene

Ortho - adjacent carbons
Meta - Free carbon between
Para - Attached to opposite carbons.

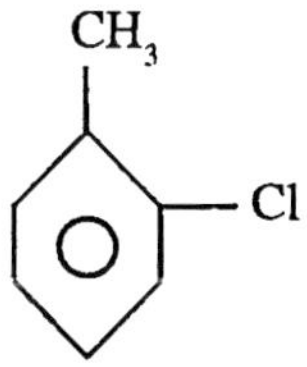

CH_3

Cl

Ortho chloro toulene or 2 chloro toluene

Polysubstituted benzene rings

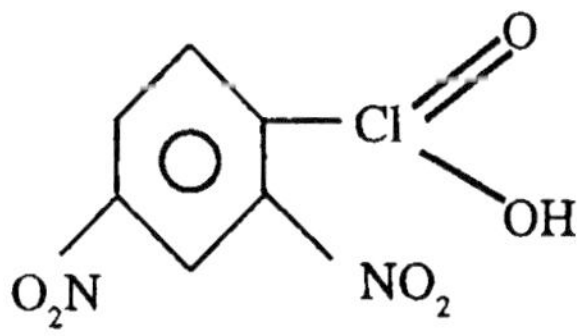

O

Cl

OH

O_2N NO_2

2.4 dinitrobenzoic acid

Cl

Cl Cl

1.3.5 trichlorobenzene

Polyring Structure

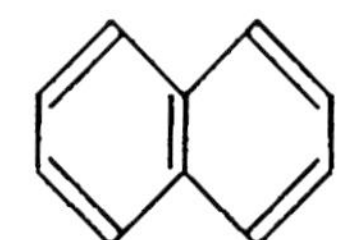

Napthalene

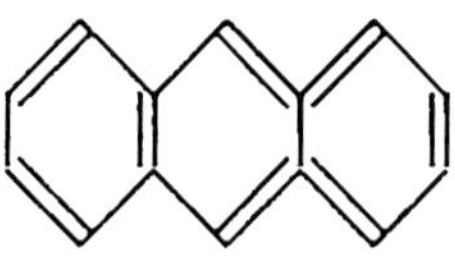

Anthracene

Phenanthrolene

Chapter -20

ETHERS AND SECONDARY METABOLITES IN BIOLOGICAL SYSTEM

The graphic structure of an ether,

R—O—R

Shows why it is considered a dialkyl derivative of water. Ethers of low molecular weight are extremely volatile. Ethers are used as solvent or anaesthetics. The common anaesthatic ether is diethyl ether. Ethers are derived from monohydric alcohols in the presence of a dehydrating agent e.g.

$$\begin{array}{ccccccc} & & H & & H & & \\ & & | & & | & & \\ H & — & C & — & C & — & OH \\ & & | & & | & & \\ & & H & & H & & \end{array} \quad + \quad \begin{array}{ccccccc} & & H & & H & & \\ & & | & & | & & \\ HO & — & C & — & C & — & H \\ & & | & & | & & \\ & & H & & H & & \end{array} \longrightarrow$$

ethyl alcohol

$$\begin{array}{ccccccccccc} & & H & & H & & & & H & & H & & \\ & & | & & | & & & & | & & | & & \\ H & — & C & — & C & — & O & — & C & — & C & — & H \\ & & | & & | & & & & | & & | & & \\ & & H & & H & & & & H & & H & & \end{array}$$

ethyl ether

The general empirical of ethers

$$C_nH_{2n+2}O$$

is the same as that for monohydric alcohols.

Diethyl ether & n-butyl alcohol have the same empirical formulae & are therefore considered *isomers.*

Simple ethers

Where both alkyl radicals are the same.

Mixed ethers

where the 2 alkyl radicals are different e.g.

```
     H         H    H
     |         |    |
H----C----O----C----C----H
     |         |    |
     H         H    H
```

Ethers are poorly soluble in water. Chemical industries using ether solvents have ethers in their industrial waste where they cause problems for engineers & scientists. Because ethers are poorly soluble, they tend to float on the top in sedimentary tanks. Ethers plus oxygen can form very explosive peroxides e.g.

```
     H    H              H    H
     |    |              |    |
H----C----C----O----O----C----C----H
     |    |              |    |
     H    H              H    H
```

Ethers are considered to be resistant to biodegradation, but there is little good data on this because of the volatility.

Because ethers tend to form explosive peroxides, they are often found stored in explosion proof refrigerated containers.

Ethers may be formed with unsaturated linkages e.g. divinyl ether

```
     H    H         H    H
     |    |         |    |
H----C====C----O----C====C----H
```

SECONDARY METABOLITES

ALKALOIDS

The alkaloid chemistry began within the announcement of isolation of morphin from opium in 1805 by F.W. Serturner. The term alkaloid (meaning alkali like) was first proposed by a pharmacist W. Meissner in 1819. Today more than 3000 alkaloids are known to occur in the more than 4000 species of plants.

Alkaloids are mostly toxic substances having primary effect on central nervous system. They are heterocylic ring compound containing nitrogen & have basic properties. Most of them contain oxygen also. In most cases, they are of limited distribution in plant kingdom & have pharmacological activity. Alkaloids are mostly found in vascular plants. They rarely occur in cryptogames (exception-ergot alkaloids), gymnosperms or monocotyledons.

They are abundantly present in dicotyledonous family, like Apocyanaceae, Papavaraceae, Papilionatae, Rananculaceae, Rubiaceace, Rutaceae, Solanaceae, etc. Pappararaceae is an usual family that all its member contain alkaloids. Alkaloids have been isolated from the roots, leaves, bark of stem or seeds of the plants. In the leaves of some species, it constitutes 0.5 to 8% of the dry weight. They are stored in the vacuoles or in the cytoplasm in the solid form.

CLASSIFICATION & STRUCTURE OF ALKALOIDS

The two commonly used systems classify alkaloids either according to plant genera in which they occur or on the basis of their chemical structure. In the second system, the name of the alkaloid is given according to the chemical nucleus in the structure present. Different chemical groups of alkaloids & some example of the groups are listed below:

1. ***Phenylethyl Amino Group :***

Tyramine

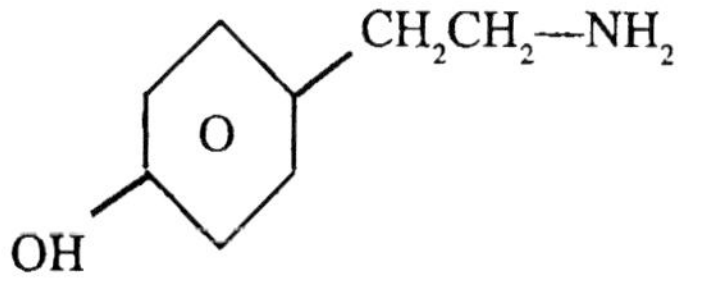

$CH_2CH_2—NH_2$

O

OH

Hordeninc

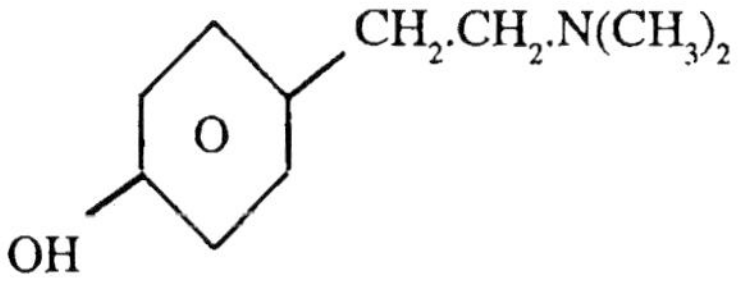

$CH_2.CH_2.N(CH_3)_2$

O

OH

2. ***Pyrollidine Group :***

Hygrine

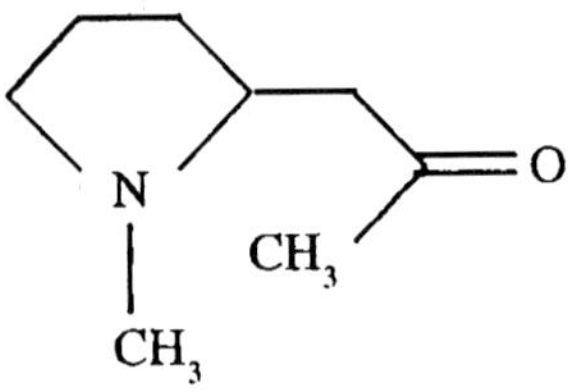

N

CH_3

CH_3

O

Stachydrine

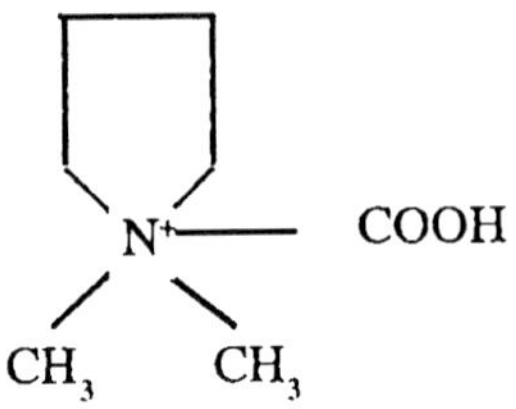

N^+

COOH

CH_3 CH_3

3. ***Pyridine Group :***

Trigonelline

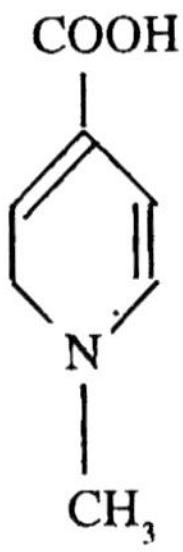

COOH

N

CH_3

Ricinine

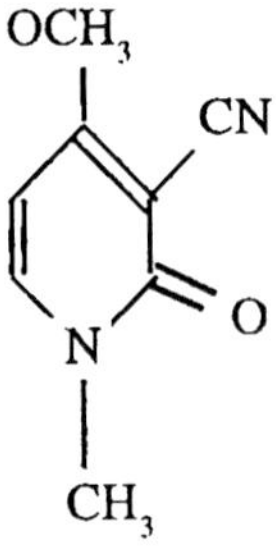

OCH_3

CN

O

N

CH_3

Coniine

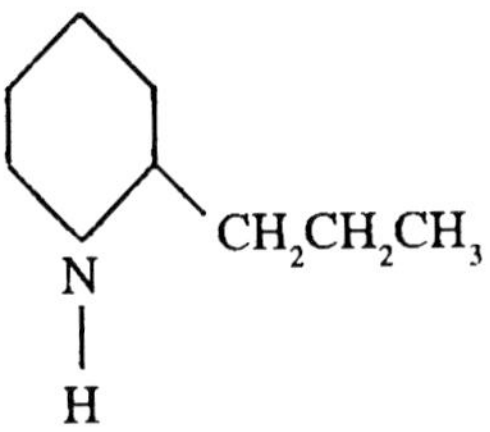

N

H

$CH_2CH_2CH_3$

Piperine

O O

O

N

CO $(CH = CH)_2$

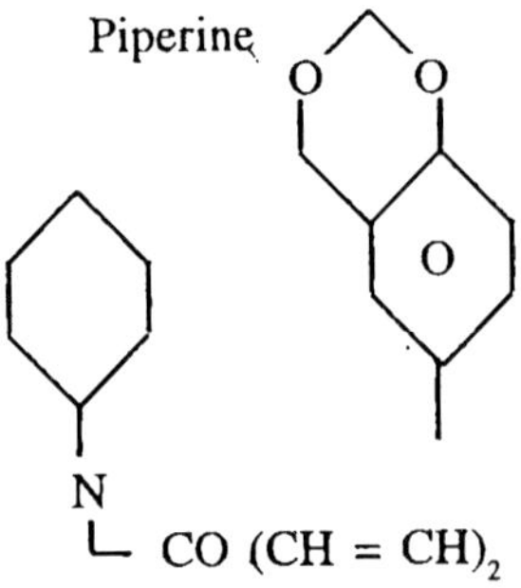

4. ***Pyrollidine-pyridine (tropane) group :***

Nicotine

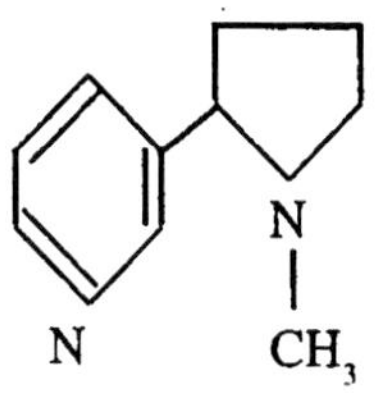

Hyocyamine

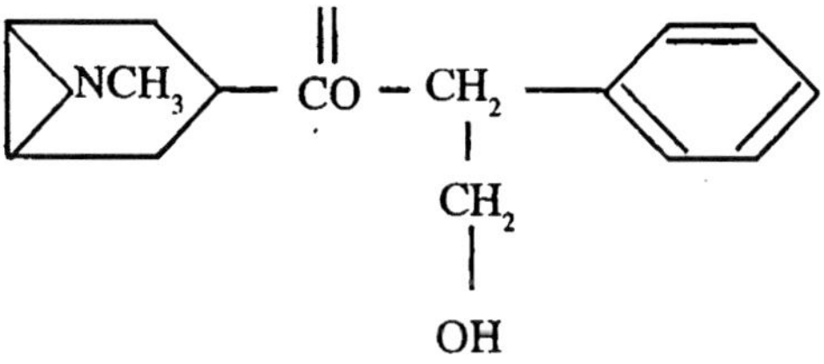

N N CH_3

NCH_3 CO – CH_2 CH_2 OH

5. ***Quinoline Group :***

Strychnine

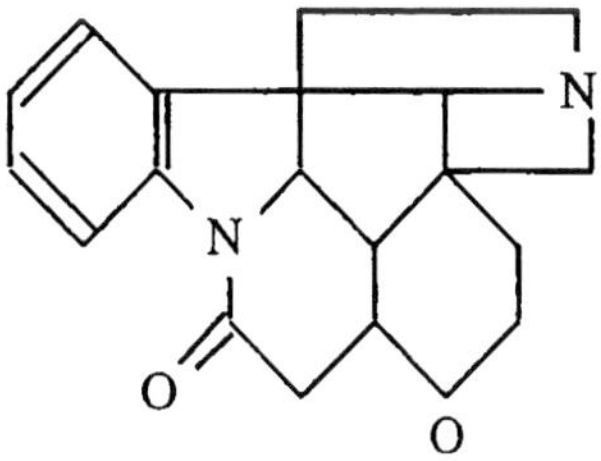

Cinchonine

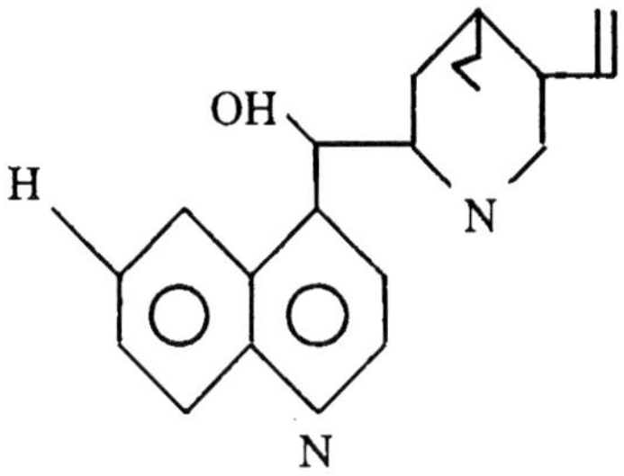

N N O O

OH H N N

6. *Phenanthrene Group :*

Morphine

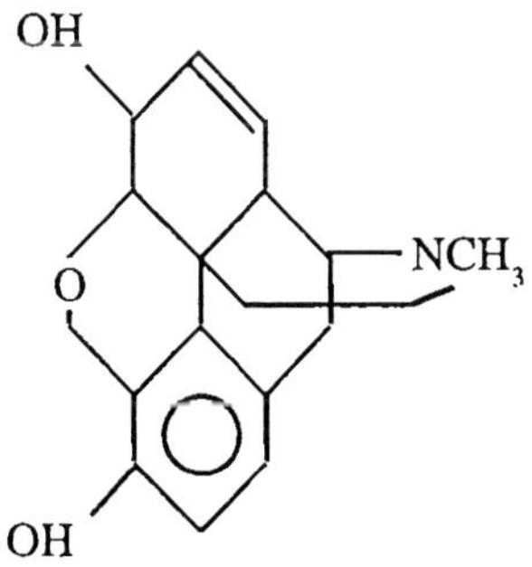

Thebaine

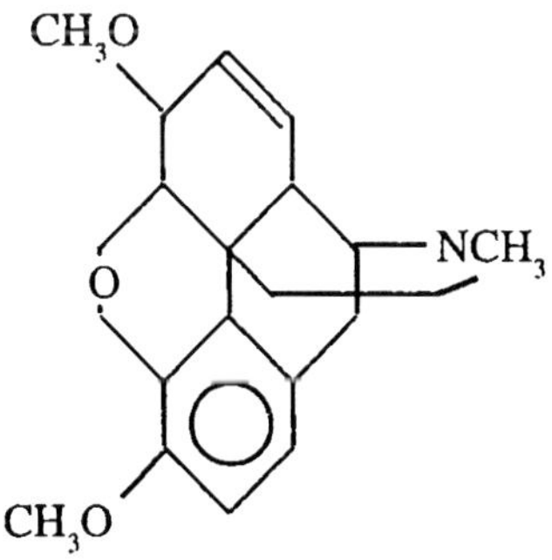

OH O NCH_3 OH

CH_3O O NCH_3 CH_3O

GENERAL PROPERTIES

The alkaloids are usually colourless, crystalline & nonvolatile solids. They are slightly soluble in water but soluble in ethanol, ether and chloroform. Some are colourless liquids & soluble in water e.g. conline & nicotine and a few are coloured as barberine is yellow. The alkaloids form insoluble precipitates with phosphotungaustic acids, phosphomolyblic acid, picric acid, potassium mercuric iodide etc. Many of these precipitates have definite crystalline shape & are used for the identification of the alkaloids.

Because of their physiological effects on human systems, alkaloids have been used in the pharmacy since time immemorial. Conniine is principle alkaloid of hemock (Conium masculatum) which was responsible for the death of Socrates in 339 B.C. when he drank the cup of poison-hemlock oil.

BIOSYNTHESIS

Alkaloids are synthesized from various amino acids or their derivatives mostly in the roots from where they are translocated to leaves & other parts. The amino acid precursors of various alkaloids are glutamate, orinithine. Besides this, the amide asparagine also serves as a precursor for the synthesis of pyridine alkaloids. The synthesis of various alkaloids does not follow a common pathway, some pathways of alkaloid biosynthesis are as follow:

Phenyl-ethyl-amine alkaloids : The alkaloids tyramine, hordenine and lycorine etc. Are synthesized from amino acids phenylanine & tyrosine. Methyl group of in those alkaloids is derived from methionine. The pathway is summarized in the fig. 20.1.

TROPANE GROUP ALKALOIDS

Tropane group alkaloids contain one pyrollidine and one pyridine ring. Various amino acids serve as precursors of these alkaloids. Pyrollidine is derived from amino acids, ornithine, proline or glutamate & pyridine from aspirate & glycerol. The two groups then fuse to form tropane ring (fig. 20.2).

QUINOLINE ALKALOIDS

The quinine group of alkaloids contains one quinoline ring. The amino acid precursor for these alkaloids is tryptophan. Tryptophan condenses with the terpenoid residue to form a corynanthenine type

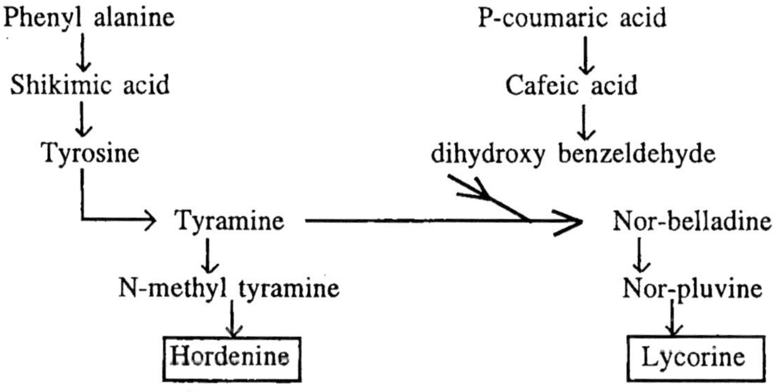

Fig. 20.1 : Bisynthesis of phenyl ethyl amine alkaloids

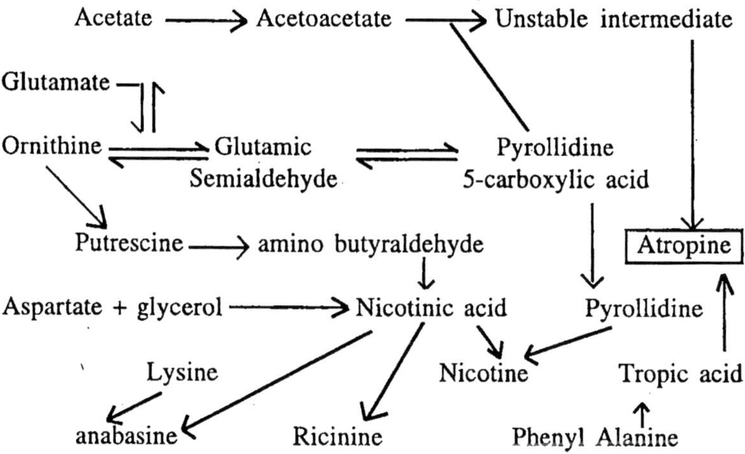

Fig. 20.2 : Biosynthesis of trapane alkaloids

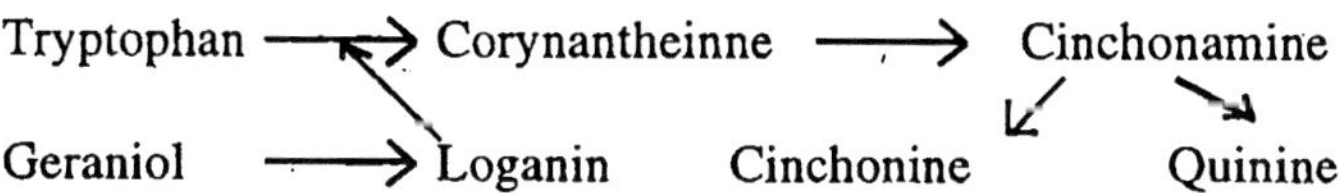

Fig. 20.3 : Biosynthesis of quinoline alkaloids

ring which is converted to cinchonamine & then to cinchonine or quinine (fig. 20.3).

PHENANTHRENE ALKALOIDS

Alkaloids such as narcotine, papaverine, morphone & the baine are synthesized from amino acid, tyrosine. Methyl group is donated by the methionine in this process (Fig. 20.4).

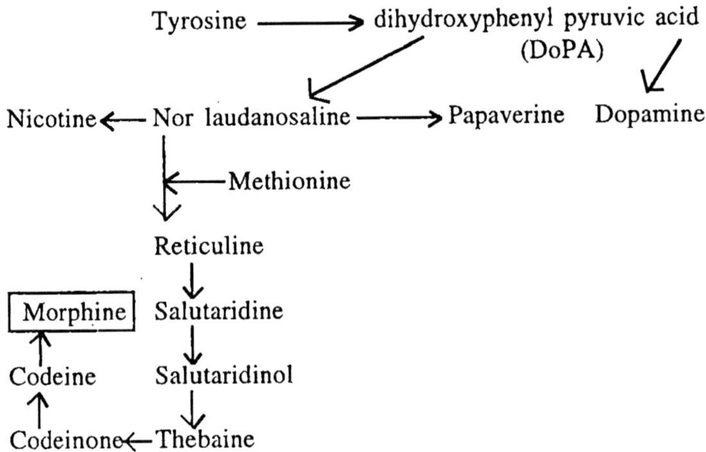

Fig. 20.4 : Biosynthesis of phenanthrene alkaloids

BIOLOGICAL SIGNIFICANCE

Many suggestions have been made regarding the function of alkaloids. The alkaloids may afford protection to the plants against fungi, insects and animals. It is also suggested that they are excretory products of plants & excess of ammonia is excreted by this method. That they may work as a nitrogen reserve has also been suggested, but there is little evidence that they are utilized in the conditions of nitrogen deficiency. Alkaloids have chelating properties & in this way they behave like growth regulators. There is some evidence that morphine plays an essential role in the formulation of viable seeds in poppy. Lupin alkaloids inhibit seed germination & some may remove the inhibitory affects of tannins. Inhibition of enzyme activity by alkaloids is also known.

Chapter -21

VITAMINS IN BIOLOGICAL SYSTEM

According to Hofmeister ***Vitamins are substance which are indispensable for the growth & maintenance of the animal organism; which occur both in animals & plants & are present only in small amount in food.*** History of the discovery of vitamins goes as early as 16th century when Jacques cartier & other sea voyageurs noticed that small quantities of some unknown compound(s) were essential component of the diet. They also found that scurvy could be treated by an infusion of pine needles. Funk in 1911, isolated from rice polishing & yeast cells, a crystalline substance that was effective in curing polyneuritis in pigeon. He named this substance, '**Vitamin**' since it was essential to life & chemically appeared to be an amine. The spelling since then has been changed to vitamin. Originally it was believed that there is only one vitamin but soon a list of vitamins had to be started.

Vitamins differ from the hormones in that they are supplied with the diet whereas hormones are produced by glands in the body. Most vitamins & hormones, however, are involved in affecting the enzyme activity directly or indirectly. Several vitamins are incorporated as coenzymes or prosthetic groups of enzymes in living beings.

VITAMIN A

Vitamin A is a complex primary alcohol with empirical formula $C_2H_{29}OH$ & exists in two forms-Vitamin A_1 or retinol and Vitamin A_2 or dehydro retionol.

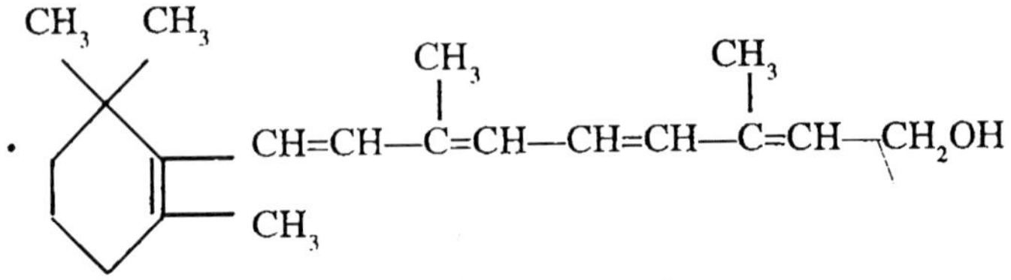

CH_3 CH_3 CH_3 CH_3

CH=CH—C=CH—CH=CH—C=CH—CH_2OH

CH_3

Vitamin A_1 (retinol)

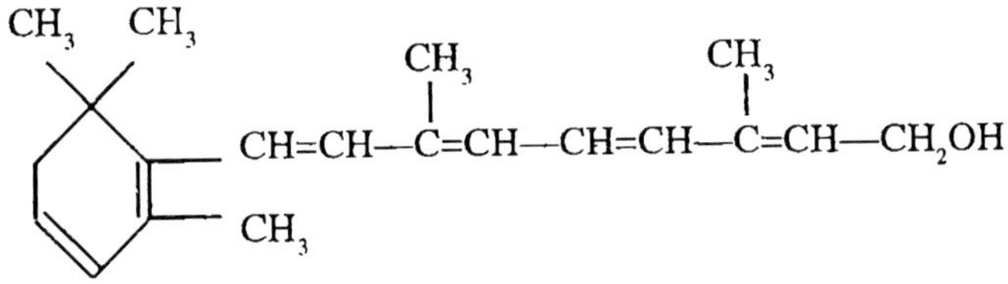

CH_3 CH_3 CH_3 CH_3

CH=CH—C=CH—CH=CH—C=CH—CH_2OH

CH_3

Vitamin A_2 (dehydroretinol)

In dehydroretinol, a double bond is created between carbon 3 & 4 of the ring.

Vitamin A occurs in cod liver oil, butter, eggs, cheese, carrot & green vegetables. In plant materials, it occurs as β-carotence (provitamin A) which splits up to yield retinol. Its deficiency in animals causes retarded growth & an eye disease called xerophthalmia.

Vitamin A combines with certain structural proteins in the cell to stabilise them. These proteins are probably involved in the formation of epithelial cells. Vitamin A is also involved in the biosynthesis of mucopolysaccharide at an enzymatic level. Its role in electron transport system is also suggested.

VITAMIN B_1 (THIAMINE)

Thiamine is a white crystalline compounds readily soluble in water & slightly soluble in ethanol. In dry conditions, it is stable upto 100°C in the presence of moisture, it is destroyed at high temperature. Thiamine occurs mostly as its hydrochloride.

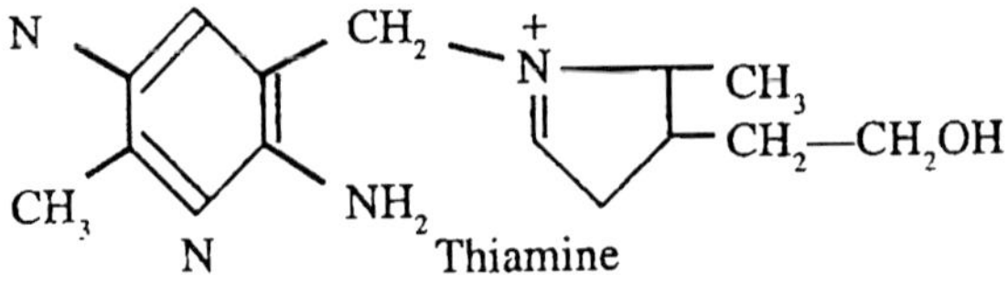

N CH_2 — N^+ CH_3

CH_2—CH_2OH

CH_3 N NH_2

Thiamine

Cereals, legumes, beef, pork, liver, nuts and yeast are good sources of thiamine. The deficiency of this vitamin causes reduction in growth of young animals, polyneuritis in birds & rats & beriberi in human beings.

Thiamin affects the physiology through carbohydrate metabolism. Thiamine pyrophosphate is a coenzyme of decarboxylases & aldehyde transferases. Thus, it is closely associated with the decarboxylation of two keto acid, pyruvate & α-ketoglutarate.

VITAMIN B_2 (RIBOFLAVIN)

Riboflavin is an orange yellow, water soluble crystalline compound. It is relatively heat stable but is destroyed by light. In nature, it occurs in bound forms as flavin adenine nucleotides or as flavoproteins.

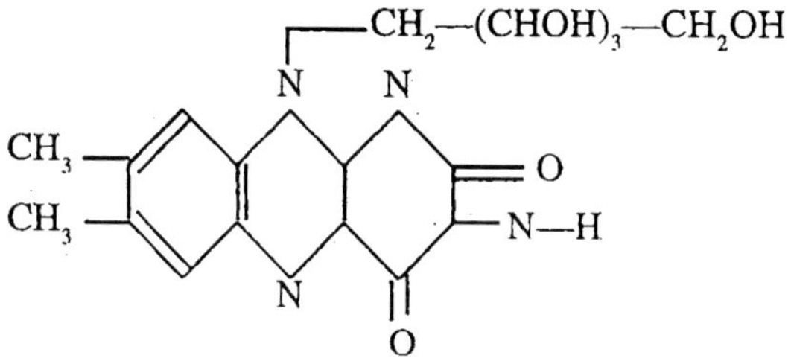

CH_2—$(CHOH)_3$—CH_2OH

N N

CH_3 O

CH_3 N—H

N

O

Riboflavin

Milk and leaf vegetables are good source of riboflavin. Deficiency of this vitamin impairs growth & causes skin diseases. Dermatitis & cataract in human beings are also associated with riboflavin deficiency.

Riboflavin is involved in the respiratory metabolism. Role of riboflavin compounds as coenzymes has been discussed elsewhere in this volume.

NICOTINAMIDE

Nicotinamide occurs as white needle like crystals. It is stable in air & moderate heat. It may exist in nature either in or in amide form.

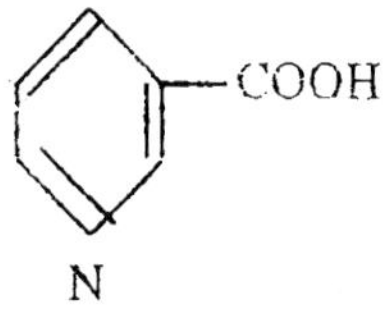

COOH CONH$_2$

N N

Nicotinic acid 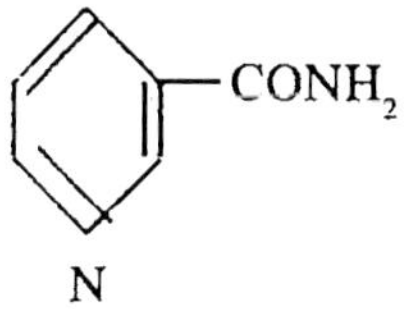 Nicotinamide

Nicotinamide occurs in liver, meat, fish & eggs. Its deficiency cause a disease called pellagra, which is a special kind of dermatitis, involving borrowing of skin.

Nicotinamide functions as a constituent of pyridine nucleotides (NAD^+, $NADP^+$), which have an important role as hydrogen transferring coenzymes.

VITAMIN B_6 (PYRIDOXINE)

Vitamin B_6 exists either in its aldehyde (pyridoxal) or amine (pyridoxamine) form, in addition to alcohalic pyridoxine form. It is also a water soluble vitamin, sensitive to heat & ultraviolet radiations.

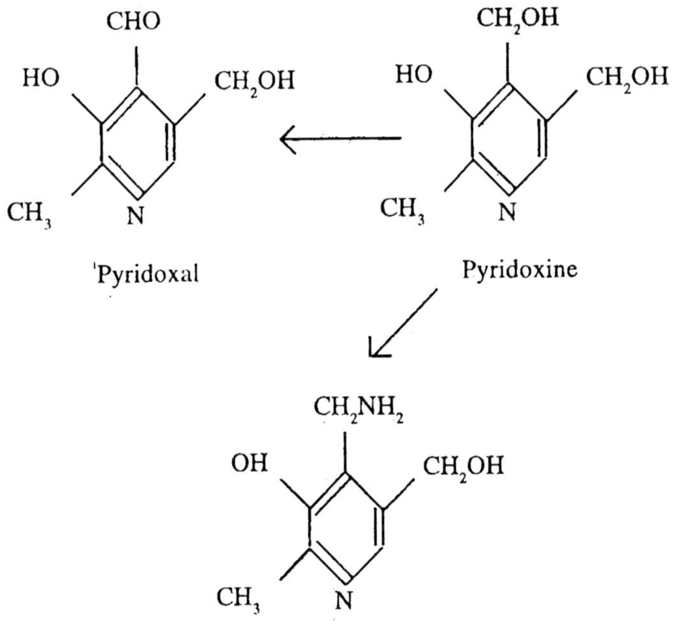

CHO, HO, CH_2OH, CH_3, N

Pyridoxal

CH_2OH, HO, CH_2OH, CH_3, N

Pyridoxine

CH_2NH_2, OH, CH_2OH, CH_3, N

Pyridoxamine

The good sources of vitamin B_6 are egg yolk, meat, milk, fish, cereals, legumes & cabbage. Symptoms of pyridoxine deficiency in adult human beings are not normally observed. In children, however, epileptic convulsions due to deficiency have been observed. At biochemical level, pyridoxine is involved in the metabolism of amino acids. It acts as a coenzyme for decarboxylases, reassembles & transferases. It is also necessary for the conversion of tryptophan to nicotinic acid.

VITAMIN B_{12} (COBALAMIN)

Cobalamin is a deep red, crystalline solid containig a cobalt atom & cyanide group. It is also soluble in wter, alcohol & acetone but not in choloroform. The structure of the cobalamin is complex. Its empirical formula is $C_{63}H_{90}N_{14}PCo$.

The chief source of cobalamin is liver. Although it is also present in milk, meat, egg & fish to certain extent. It is not present in vegetarian diet. Its deficiency courses pernicious anemia in man.

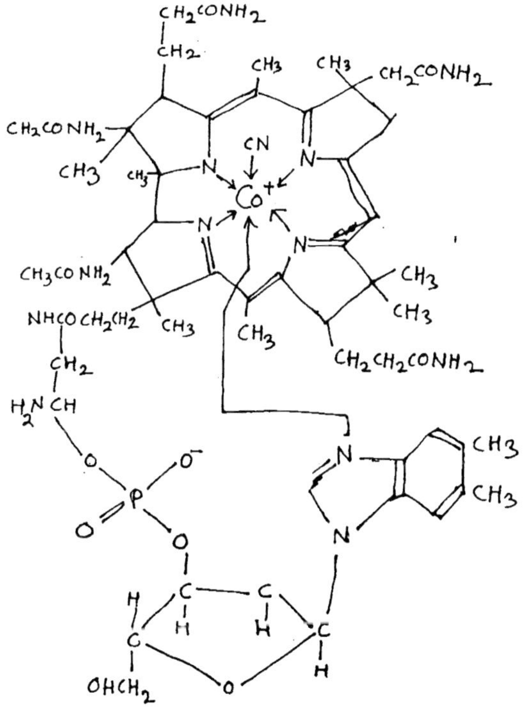

Cobalamin

But this deficiency is usually due to poor absorption of the vitamin rather than its inadequate intake.

the mechanism of action of cobalamin is not clearly understood, although it is known to influence several metabolic processes. It affects the synthesis of nucleic acids, metabolism of glycine, serine, methionine & choline. It also works as a coenzyme (commonly called coenzyme B_{12}) in the enzymes transferring methyl groups. Its role in transmethylation reaction involving methylated tetrahydrofolic acid has been summarised in fig. 21.1.

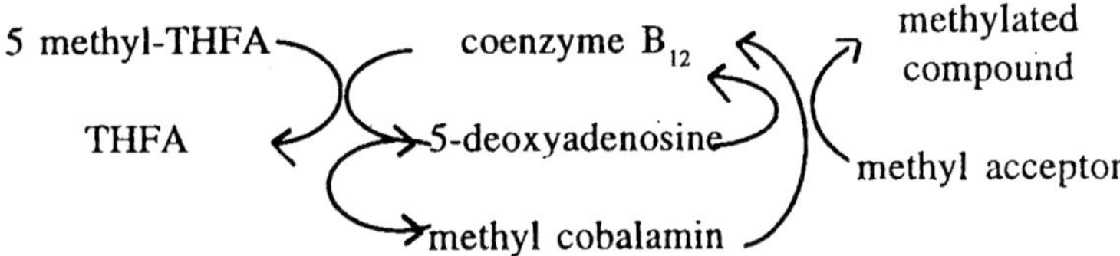

Fig. 21.1 : Role of cobalamin in transmethylation reaction.

It works as a coenzyme in several other reaction also, such as reduction of formate to methyl groups, anaerobic degradation of lysine, reduction of disulphife to sulphyhydryl groups & activation of amino acids for protein syntrhesis.

VITAMIC C (ASCORBIC ACID)

Ascorbic acid is soluble in water but insoluble in fats and oils. It is one of the important biochemical redox systems, since it can be oxidized reversibily to dehydroascorbic acid. Chemically, it is a hexose derivative with empirical formula $C_6H_8O_6$.

Citrus fruits & tomatoes are the best sources of ascorbic acid. Other good sources are green peppers, parsley, spinach, cabbage, potatoes, fresh peas & lettuce. It is not considered as a vitamin for most mammals because it is synthesized in their body according to the requirement. Deficiency of ascorbic acid in human being causes a disease called scurvy, which is probably the first disease reported to be associated with a food deficiency. In fact, the name ascorbic acid is derived from its antiscurvy properties.

```
O ═ CH ──┐                          O ═ C ──┐
    |    |                              |   |
HO ─ C   |                          O ═ C   |
    ‖    |                              |   |
HO ─ C   O          -2H             O ═ C   O
    |    |        ⇌                     |   |
H ── C ──┘          +2H             H ── C  |
    |                                   |   |
HO ─ C ── H                             C ──┘
    |                                   |
H ── C ── OH                        H ── C ── OH
    |                                   |
    H                                   H
```

L-ascorbic acid — Dehydro L- ascorbic acid

At physiological level, the major function of ascorbic acid appears in the formation of collagen & intercellular cement substances. This is affected through the enzyme proline hydroxylase, which catalyses the conversion of proline to hydroxy proline . Because of its oxidation-reduction properties, it is also believed to be associated with electron transport system in the mitochondria probably at cytochrome C level. An NAD^+ oxidase isolated from microsomal fraction of the cell, also requires ascorbic acid for activity.

VITAMIN D (CALCIFEROLS)

Calciferols are white crystalline substances soluble in fat solvents only. They are resistant to oxidant & moderate heat. Chemically the calciferols are steroids & exist in two important forms- cholecalciferol & ergocalciferol.

Fish liver oil is the best source of vitamin D. Its deficiency in children causes rickets, which affects the development of bones. Its mechanism of action is through calcium & phosphorus metabolism. Vitamin D facilitates the transport & absorption of calcium from the intestine. This is mediated through protein synthesis, since addition of actinomycin D (an inhibitor of m-RNA synthesis) inhibits vit. D. action in experimental materials.

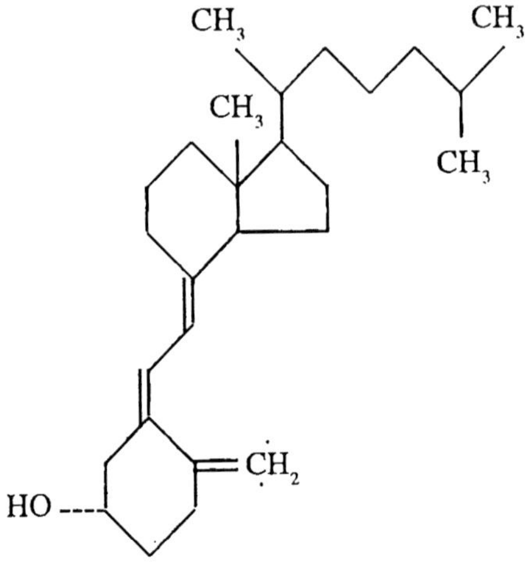

Cholecalciferol

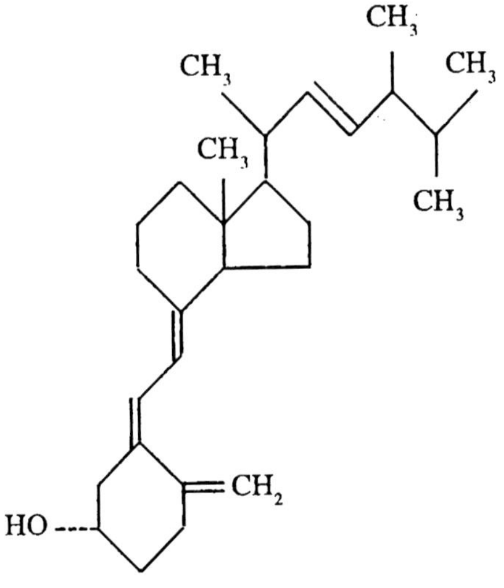

Ergocalciferol

VITAMIN E (TOCOPHEROLS)

Several types of tocopherols are active as vitamin E. They are methyl derivatives of the parent compound tocopherols.

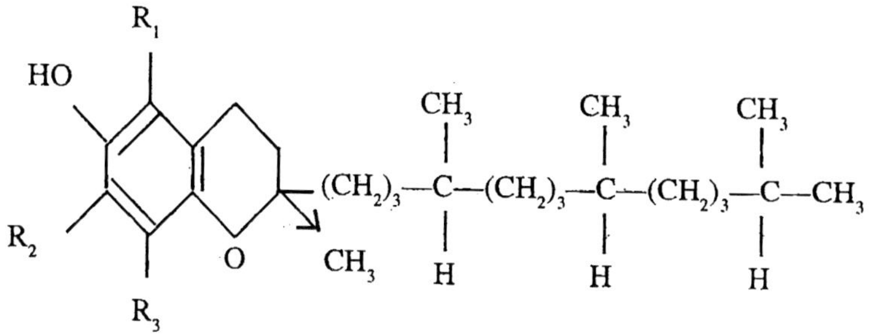

R_1, HO, R_2, O, R_3, CH_3

$(CH_2)_3-\underset{H}{\overset{CH_3}{C}}-(CH_2)_3-\underset{H}{\overset{CH_3}{C}}-(CH_2)_3-\underset{H}{\overset{CH_3}{C}}-CH_3$

Tocopherols

In α tocopherol R_1, R_2 & R_3 are CH_3, in .b.-tocopherol R_1 & R_3 are CH_3 & R_2 is H, and in λ - tocopherol R_2 & R_3 are CH_3 & R_1 is H. Tocopherols are light yellow fat soluble oils. They are widely distributed in plant & animal tissues. Good sources are cotton oil, maize oil, peanut oil & wheat germ oils.

In man, a deficiency disease of vitamin E is not known. In rats, however, deficiency of tocopherol results in damage to the reproductive system in both male and females. α-topopherol can be oxidized easily to a quinone (tocoquinone). Because of this property, it is used as an antioxidant in industries. It also appears that its primary action in biological system is to protect certain intercellular enzymes of respiration & lipid configuration of cell membranes. It exerts some influence on rate of DNA synthesis also.

VITAMIN K (QUINONES)

Two types of vitamin K are found in nature : Vitamin K_1 (Phylloquinone) & vitamin K_2 (Menaquinones), Phylloquinones is fat soluble & stable to heat & reducing agents.

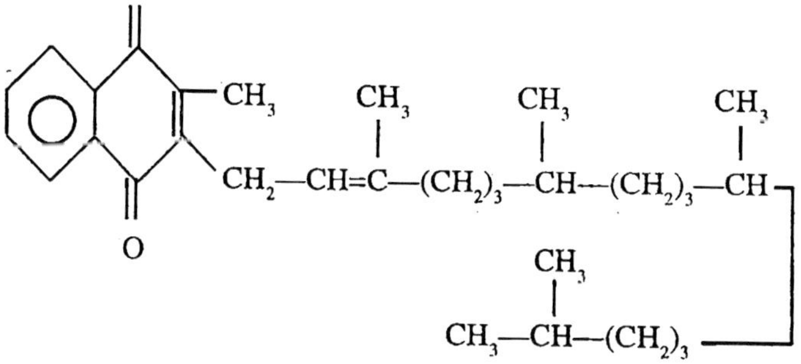

O, O, CH_3

$CH_2-CH=\overset{CH_3}{C}-(CH_2)_3-\overset{CH_3}{CH}-(CH_2)_3-\overset{CH_3}{CH}-$

$CH_3-\overset{CH_3}{CH}-(CH_2)_3-$

Phylloquinone (Vitamin K_1)

Sources of vitamin K are alfalfa, cabbage, cauliflower, spinach & other green vegetables. The deficiency of the vitamin causes profuse bleeding from minor wounds & disturbances in blood clotting. This is the result of lack of prothrombin in the blood. Administration of vitamin K induces prothrombin synthesis in the parenchymal cells of liver. Synthesis of blood clotting proteins under the influence of this Vitamin involves formation of new RNA. This vitamin also works as component of the election transport system, as coenzyme Q.

Chapter -22

MEMBRANES IN BIOLOGICAL SYSTEM

Cytoplasm of the living cells is bound by a well organised cell membrane. The membrane which is known as plasma membrane or plasmalemma is protected by an outer covering of the cell wall in plant cells. In animal cells, however the plasma membrane is outer most covering of the cell. The membranes of somatic cells do not lie right against, one another but are separated by a space filled up with some amorphous substances. In some plant tissues, on the other hand, cells are joined by fibrous cytoplasm passing through the pores in plasma membranes of the adjacent cells. These fibrous bridges are called ***plasmodesmata***.

Many of the cell organelles are also bound by a membrane. The membrane around vacuoles is called ***tonoplast***. All types of membrances are similar in their essential chemical components & physical organisation.

CHEMICAL COMPOSITION OF THE MEMBRANE

A typical cell membrane is composed of lipids, proteins & some carbohydrates, nucleic acids & enzymes are also associated with the membrane.

Lipids account for about one quarter to half of the dry weight of membrane. The important types of lipids found in membranes are cholesterol, phospholipids & galactolipids. Choline substituted phospholipids (lecithins) make about half of the total phospholipids in the membranes. In other membrane phospholipids, phosphate group is attached to inositol, ethanolamine or serine. Phosphatidic acid is also found in membranes. Small amount of galactolipids have been measured in mitochondrial membranes.

The membrane proteins, generally called stromatin is firbours in nature & has high m.w. Ultracentrifugation analysis of proteins from red blood cell membranes has revealed that there are at least two types of proteins, sedimenting at 5S to 10S. The amino acid composition of these proteins is similar to that of acting in muscles.

In membranes of the liver and red blood cells, hexoses, hexosamine, fucose & sialic acid carbohydrates are found in association with protein molecules. In Gram negative bacteria, the outer membrane (bacterial cells) is made up of lipopolysaccharides and liporoteins. In Gram positive bacteria, technic acid is also found in the outer membranes. The inner membranes of bacteria do not have barbohydrate. In between the two membranes, however, there is a layer of mucopeptide (fig. 22.1).

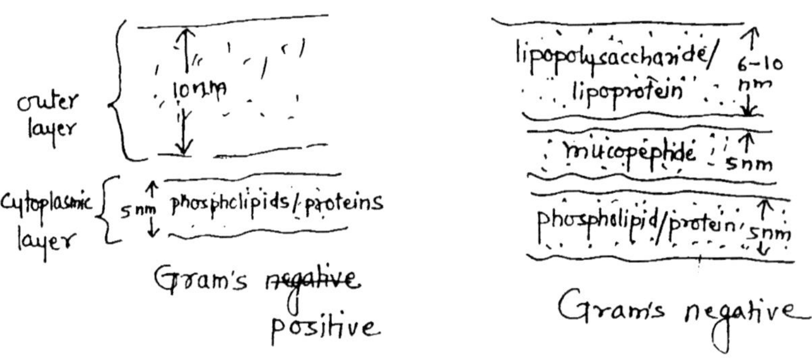

Fig. 22.1 : Envelope of Gram's positive and negative bacteria - diagrammatic representation

More than 30 enzymes have been detected in the biological membrane. The most common ones are ATPase, alkaline phosphatase & nucelases. ATPase has an important role in the active transport of ions across membrane. It is activated by Mg^{++}. The activity appears to be dependant upon membrane lipids, because when the lipids are extracted from the membrane, it is inactivated, outer membrane of the mitochondria house the enzyme involved in the activation, oxidation

& elongation of fatty acids, & many coenzymes of Kreb's cycle. The enzymes of electron transport chain appear to be located on inner membrane.

MOLECULAR STRUCTURE OF THE MEMBRANES

Knowledge of the molecular structure of membranes came out of the study of different physico-chemical properties of membranes. J.F. Danielli (1934) proposed that plasma membrane contained two layers of lipids & two layers of protein adhering to both lipid surfaces. The most satisfactory model was advanced, however, only after the introduction of electron microscopy & refined histochemical techniques. A study of myelin sheath has revealed that the membrane is about 7.5 nm wide. The membrane is divisible into three layers - a central (3.5 nm) & two peripheral (2.0 nm each) layers. The central layer is made up of lipid molecules, oriented in such a way that their hydrophobic fatty acid tails soluble glycerol ends towards the periphery. The two peripheral layers are made up of protein. A hypothetical structure of membrance in presented in Fig. 22.2.

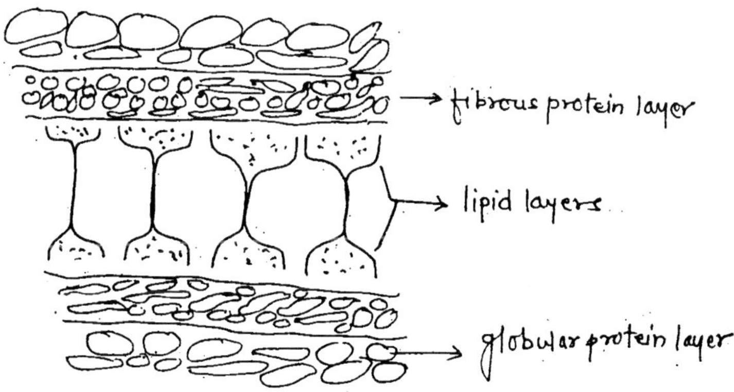

Fig. 22.2 : Membrance structure - lipid bilayer model

This three layered structure of membrane has been called, the unit membrane. A survey of various animal & plant tissues has demonstrated that unit membrane is characteristic of other structures, such as golgi bodies, glyoxysomes, lysosomes & endoplasmic reticulum. The membranes of mitochondria & nucleus, however are double. The nuclear membrane is made up of two unit membranes, each 8 nm wide. The space between the two membranes is about 15 nm. These membranes are connected with the endoplasmic reticulum & possibly with the plasmolemma through structure, it has been postulated that there are two types of proteins- intrinsic & extrinsic, in a unit membrane. The intrinsic proteins are dispersed within the lipid layer while extrinsic proteins form the outer covering (fig. 22.3).

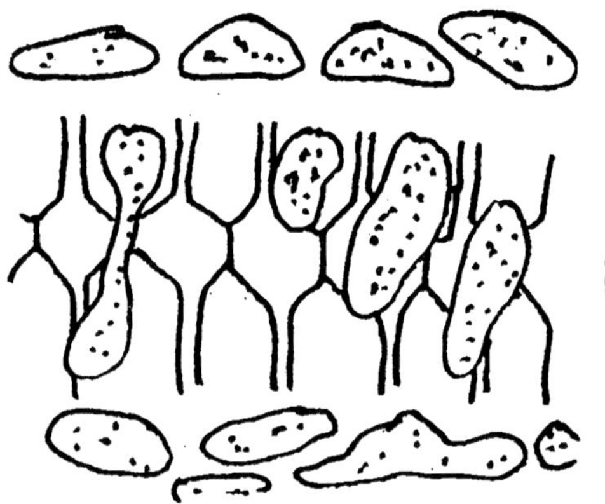

Fig. 22.3 : Membrance structure : current model

FUNCTIONS OF THE MEMBRANE

Membranes, being the outermost covering of the cell or cell organelles, provide protection to them. The physical separation of the organelles is also possible only due to the presence of the membranes. This separation is important for the regulated functioning of the cell. Other biological functions of membranes are as follows :

Active transport

One of the important function of biological membranes is exchange of ions or solutes against concentration gradient. The membranes of kindney cell for example, will allow the pumping out of Na^+ in a medium which already has higher concentration of Na^+. Transport against concentration gradient (active transport) requires expenditure of energy. This energy is derived from respiration because certain poisons of respiration inhibit active transport. The work done by the cell against concentration gradient can be calculated from Nernst equation as expressed below.

$$\text{Work done (Mv)} = 2.3\ RT \log 10\ \frac{C_2}{C_1}$$

Where,

R = the gas constant.

T = absolute temperature.

C_1 = molal conc. of ions in concentrated solution, and

C_2 = molal conc. of ions in dilute solution.

If a cell is pumping out Na^+ in a medium which already has higher concentration of Na^+ than the cell, then C_1 will be the concentration of Na^+ in the medium & C_2 will be concentration inside the cell.

How the electrolytes are carried away against concentration gradient? A carrier hypothesis has been proposed to explain the mechanism. The carrier molecule picks up the substances to be transported on one side of the membrane & carries it to the other side. The carrier may be a protein, a lipid or an enzyme. The enzymatic nature of the carrier, however, appears to be more convincing. Several indirect experiments have lead to believe that ATPase & permease participate as carrier molecules. These enzymes are activated by ATP. For the movement of Na^+ & K^+ across mitochondrial membrane, it has been postulated that ATP combines with ATPase to form a ATP-ATPase complex. This complex binds with internal Na^+ & transfers it to the outside of membrane. The same complex, binds with K^+ outside & release it inside. The membrane shows biological specificity in uptake of electrolytes. Mitochondria from the rate liver, for example, can take in citrate extracted from liver cells only & not from heart cells. This specificity may be lying in carrier molecules.

The carrier hypothesis accounts far most of the known facts about active transport.

ENDOCYTOSIS

Another activity of the plasma membrane is ingestion of solid or liquid material inside the cell. The process of ingestion of solid material is referred to as ***phagocytosis***. Phagocytosis serves as a mode of nutrition in most of the protozoans. Although, it is observed in some metazoans also but its role in those organisms is mostly to defend against foreign organisms.

During phagocytosis the foreign particles adhere to the protoplasm first. The protoplasm on two sides of the particle is then extended. These extensions grow slowly to surround the particle completely. Later on the membranes of the extensions fuse together & the particle is ingested by the cell (fig. 22.4.)

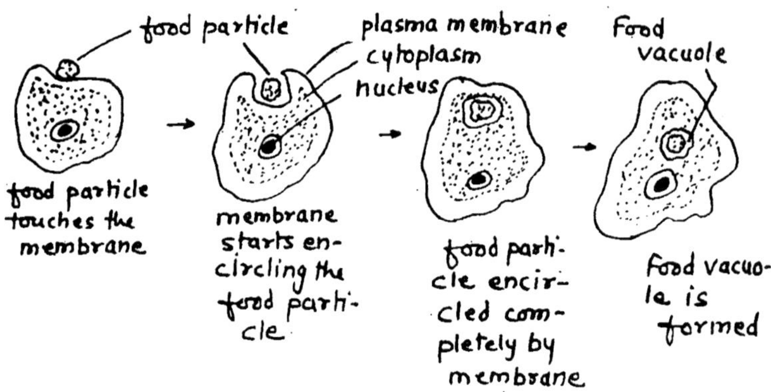

Fig. 22.4 : Phagocytosis

The process of intake of fluid vesicles by living cells has been called pinocylosis (Greek pinenin = to drink). the mechanism is similar to phagocytosis. Vigorous protoplasmic movement around the fluid visicle encircles & engulfs it. These vesicles (now vacuoles) are surrounded by the membranes as in case of food vacule formed periphery of the cell are later on transported to other portions of the cell. The material engulfed by endocytosis are digested by the enzymes present in membrane or added by lysosomes to the ingested particles.

Energy is required for the operation of endocytosis. Increase in oxygen consumption, glucose uptake & glycogen breakdown are observed during phagocytosis by leukocytes. In cultured cells, endocytosis is activated by the addition of ATP and inhibited by respiratory & other metabolic poisions.

MECHANISM OF ATP SYNTHESIS

The most important function of mitochondrial & chloropolastic membranes may be considered as the synthesis of ATP. During electron transport in the oxidation of NADH in mitochondria or in transfer of electrons from water to $NADP^+$ in illuminated chloroplast energy is liberated according to the difference in redox potentials of electron donating & electron receiving components. A part of this energy is used up in the synthesis of high energy ATP molecules form ADP and iP. This synthesis occurs on the inner surfaces of the membranes.

Three hypothesis have been proposed regarding the mechanism of ATP synthesis.

Chemical Hypothesis

The chemical hypothesis was proposed originally by E.C. Slater (1953) explain mainly the substrate level phosphorylation. According to this hypothesis, some definite chemical compounds participate in phosphorylation. The energy of electron transfer from one substrate to the other is picked up by these compounds. Later on these energized compounds activate inorganic phosphate which combines with ADP to form ATP. The sequence of events may be written as follows -

$$AH_2 + B + I \rightleftharpoons A \sim I + BH_2$$

$$A \sim I + X \rightleftharpoons A + X \sim I$$

$$X \sim I + iP \rightleftharpoons X \sim P + I$$

$$X \sim P + ADP \rightleftharpoons X + ATP$$

Where AH_2 and B are two substrates & X & I are hypothetical intermediates.

Confirmational hypothesis

The confirmational hypothesis of ATP synthesis was proposed originally by P.D. Boyer (1967). According to this hypothesis, the energy of electron transport is picked up by a macromolecule or a portion of the membrane, bringing about a conformational change in the macromolecule or membrane. The activated with changed configuration later on, relaxes releasing the energy for ATP synthesis (fig. 22.5.)

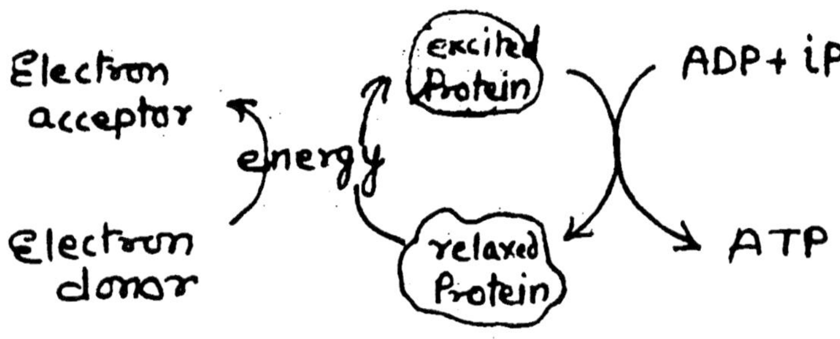

Fig. 22.5 : ATP synthesis - involvement structural protein

Chemo-Osmatic Hypothesis

Peter Mitchell (1961, 1968) suggested that electron transport could be coupled to phosphorylation of ADP by a chemo-osmatic process. This hypothesis is equally applicable to both photophosphorylation in the chloroplast & oxidative phosphorylation in the mitochondria.

This hypothesis requires the involvement of enzyme. ATPase and a well organised membrane. In the reversible reaction -

$$ADP + iP \xrightleftharpoons{ATPase} ATP + H_2O$$

Removal of water from active centre of ATPase will favour synthesis of ATP according to law of mass action. This assumption is the fundamental basis of Mitchellis -Chemo-osmotic hypothesis. In the later modifications of the hypothesis, involvement of chemical intermediates X & I has also been suggested. The detailed mechanism is outlined in fig. 22.6.

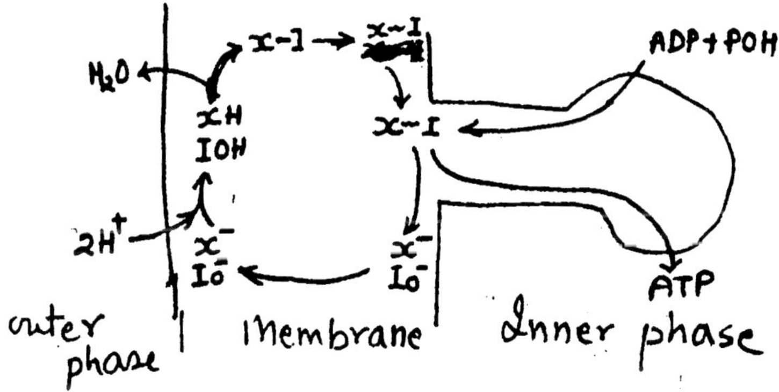

Fig. 22.6 : Chemiosmotic hypothesis - Mitchell's modification

The electron transport in the membrane has two types of effects related to ATP synthesis :

(1) Ionization of water & separation of N^+ and OH^- and

(2) formation of high energy X ~ I complex from two unknown chemical intermediates X & I. Positively charged hydrogen ions combine with the complex X^- I O^- on the outer surface of the membrane & H_2O is liberated. Later on X & I from a complex which is converted to a high energy X ~ I by the energy liberated during electron transport. This moves towards the inner side of the membrane where it combines with ADP i and OH^- to yield ATP & released H^+ on the inner side of the membrane. The complex XIO moves again to the outer side of the membrane to combine with the hydrogen ion & the process continues further, generating one ATP molecule in each cycle on the inner surface of the membrane. Under conditions, where ATP synthesis does not occur the ejection of protons (H^+) is balanced by the uptake of Ca^{++}. For every pair of protons one Ca^{++} is absorbed inside.

Chemo-osmatic hypothesis also requires the involvement of few chemical intermediates as in case of chemical hypothesis. In chemo-osmotic hypothesis, however, these chemical intermediates do not have to enter into combination with an electron carrier in the respiratory chin. Further, the propounders of chemo-osmatic hypothesis nature of X & I than those of chemical hypothesis, who predict that these are discrete covalent compounds.

Chemo-osmotic hypothesis appears to accommodate most of the experimental observation. The generation of pH gradient across membrane, as predicted by this hypothesis, is frequently observed during phosphorylation. The pH of the medium containing actively phosphorylating isolated mitochondria decreases. This decrease can be stopped by adding antimycin, an uncoupler of phosphorylation. Further, synthesis of ATP in isolated mitochondria from the rat liver or beet roots is stimulated by lowering the pH of the medium. The stimulation does not occur in the presence of dinitrophenol or oliogmycin.

Although most of the experimental observations swing in favour of this hypothesis it is difficult to accept it as an universal model of ATP synthesis. More experiments have to be performed to elaborate each step of the model & hypothesis further tested.

Chapter -23

STRUCTURE & PROPERTIES OF THE MOLECULES

Hydrogen, Oxygen, Carbon & Nitrogen atoms constitutes about 99% of the total atoms in most of the living systems. Calcium, phosphorus, magnesium, potassium, sodium, sulfur, iron & chlorine are also found in appreciable amounts. Several other atoms are also found in small amounts. These atoms, however, do not participate as such in biochemical reactions. They aggregate & join through various bonds to form small (micro) or big (macro) molecules. Thus, water molecules are made up of two hydrogen & one made up of two hydrogen & one oxygen atoms & glucose molecule contains & carbon, 12 hydrogen % 6 oxygen atoms.

ELECTRONIC CONFIGURATION OF AN ATOM

Although atoms are conceived to be the smallest particles of objects in elementary studies, they are further divisible into sub atomic particles. An atom is made up of a central nucleus containing protons & neutrons and one or more orbits of electrons spinning around the nucleus (fig. 23.1). Electrons are negatively charged, protons are positively charged and neutrons have no charge. The magnitude of positive charge on electrons inmost of the atoms. The number of electrons and protons is always the same in an atom to keep the atom electrically neutral. An atom which lacks an electron has a charge of -1. These opposite & equal charges between protons & electrons provide a definite structure of an atom. Particles with similar charges in continuous spin while all the protons & neutrons are aggregated in the centre. Electrons are always arranged indifferent orbits. The first orbit (k shell) does not contain more than two electrons. An atom containing complete K shell only, is resistant to chemical reactions, as shown by the inert nature of the helium. When the number of electrons exceed over two, rest of the electrons are arranged in second (L) and subsequent shells. The L shall contains upto the maximum of 8 electrons as in neon & third (M) shell also contains 8 electrons as in argon. All these elements which have closed outermost shell are stable and do not form molecules under normal conditions. Elements with incomplete outer shell are chemically active. In a chemical reaction these elements reach in such a way that their outmost shell is saturated with electrons & the element then becomes stable.

Fig. 23.1 : Structure of a carbon atom

A neutron has nearly the same mass as the proton. The mass of an electron is negligible as compared to the mass of a proton or neutron. Thus about 99.98% of the mass of an atom is concentrated & the properties of the lithium atoms with 3 or 4 neutrons differ slightly from each other.

The number of protons in an atom is known as the ***atomic number***. Thus the atomic number of hydrogen is 1, lithium 3, calcium 20, Uranium 92 & so on. As seen with lithium, the number of neutrons in an element is also not constant. e.g. hydrogen may not have any neutron or may have 1 neutron or 2 neutrons in its nucleus. Atoms having the same number of protons but different number of neutron are ***isotopes*** of the same element.

ASSOCIATION OF ATOMS INTO MOLECULES

Chemical bonds

Association of atoms into molecules involves different types of interactions. The type of interaction or binding depends upon the electronic configuration of the atoms involved. Some of these bonds are discussed in following paragraphs :

Ionic bonds

In early years of twentieth century, the ideas was propounded that electrons can be transferred from one atom to the other so that both could achieve the inert gas configuration. This formed the basis for understanding ionic bonds in molecules. Sodium atom has 11 electrons arranged in 3 orbits of 2, 8 & 1 electrons. The 10 electrons making up the first & second orbits shield 11 protons in the nucleus from attracting the 11th electron of the third orbit. This electron can be easily lost to achieve a stable neon like shell. Fluorine on the other hand, lacks in one electron of having a neon like closed shell. It can accept the electrons from Na & become stable In this transfer Na achieves a positive charge whereas F gets negative charge. The two atoms with opposite charge attract each other & a molecule is formed when both charge combine. Thus, a molecule formation in this case involves the transfer of electrons, ionizaiton of the atoms & then holding the atoms together by a attraction of oppositely charged ions. The number of charges of an ion (an element which has lost or gained electrons) is called the electro valency. Sodium carrying a positive charge has an electrovalence of +1 and chlorine with a negative charge has -1. Some elements may have two or more electro-valencies like iron having either two (ferrous) ro three (ferric).

The ionic bonds are important in molecules having crystalline structures & in ionic compounds dissolved in solution.

Covalent bonds

The most common type of bond in organic molecules is covalent bond. In covalent bonds, the electrons are get transferred but they are shared between atoms in such a way that participating atoms achieve inert configuration. The simple example of covalent bond is of atoms of a hydrogen molecule. In this case, two hydrogen atoms are held together by continuous sharing of two electrons. Bonding in other gas molecules such as N_2, O_2, O_3 and so on & in most organic compounds involve covalent type of bonding. In methane e.g. the 4 outer electrons in c atoms are shared by 4 different hydrozen atoms.

$$\cdot\dot{\underset{\cdot}{C}}\cdot \; + \; 4H \longrightarrow H:\overset{H}{\underset{H}{\overset{\cdot\cdot}{\underset{\cdot\cdot}{C}}}}:H$$

At one time the carbon atom can accommodate 4 electrons of hydrogen atoms to form a stable atom (with 2 & 8 electrons in the outer orbit) of neon type. On the other hand, H atoms may share the outer 4 electrons of C to form helium type stable atom (with 2 electrons). In longer molecules, a carbon atom can be linked covalently to 1, 2, 3 or 4 carbon atoms. Thousands of carbon atoms may join together in this way to form a large chain of carbon compounds. The number of covalent bonds an atom can form is called the ***covalence***. Thus, carbon has covalence of 4.

Secondary bonds

Secondary bonds are weaker that the primary ionic & covalent bonds. But these are most important in joining the biochemical compounds. Generally, weak bonds do not bind atoms to another but serve as links between different molecules or different part of the same molecule. These bonds can be broken & formed easily for they involve very little energy. The three dimensional structures of important biological molecules such as protein and nucleic acid involve several weak secondary bonds. Two important secondary bonds are hydrogen bonds & vander Waal bonds.

1. ***Hydrogen bonds*** :

   Hydrogen bonds are formed by electro static interaction between the molecules. It does not involve transfer or sharing of electrons between atoms. When a hydrogen atom carrying somewhat acidic charge in bonds O-H, F-H, N-H & so on approaches an atom with basic charge such as oxygen in water, alcohol or ether or N in amines, here an association occurs. This bonding is called as the ***hydrogen bond***. Similar bonds may be formed between two water molecules when the positively charged hydrogen atoms of one water molecule come in contact with a water molecue containing negatively charged oxygen atom. In water, each hydrogen bond is very short lived, lasting only for about 10^{-11} seconds. Hydrogen bonds are quite stable when they occur in multiples. The energy of hydrogen bond ranges between 3 & 7 K cal./mol. & thus weaker than the covalent bond which bears energy from 50 to 100 K cal./mole. Hydrogen bond always has a specific length and destination. In since, the length of hydrogen bond is 1.77 A.

The compounds which can form hydrogen bonds with water molecules are soluble in water. Glucose & pyruvate e.g. have large number of groups that contain = O or -OH to form hydrogen bonds. When such groups are inserted in water lattice, the molecular architect of water is broken & hydrogen bonds between glucose & water or pyruvate & water are formed. benzene on the other hand, can not from hydrogen bonds with water molecules & it is immiscible with water.

Wander Waal's bonds

Wander Wall bonds are the weaked bonds & act only between molecules that are brought close together. The bonds results from the fluctuating charged caused by the nearness of the molecules. However, when the molecules are too close, repulsive forces operate because of overlapping of the outer electron shells of the atoms involved. Thus, wander wall bonds operate only at a specific distance.

Wander Wall bonds are very weak secondary bonds having energy of about 1 K cal. per mole. They are formed between all types of molecules polar as well as nonpolar. In biological systems they are very important bonds. The compounds are held together by several such bonds in physiological conditions. When needed, they can be broken with little energy.

PHYSICAL PROPERTIES OF MOLECULES

Diffusion & Osmosis

Movement of molecules of the particles from higher to the lower concentration is called ***diffusion***. If, e.g. few drops of a concentrated solution of copper sulphate are dropped carefully over distilled water in a glass vessel, the blue copper sulphate slowly disperses in the water till the entire fluid becomes coloured. Likewise, gases also diffuse from higher pressure to a lower pressure. When the cork of the ammonia bottle is opened in a room, the presence of ammonia can be smelled in all parts of the room gradually.

If a porous partition is placed between a concentrated & a dilute solution, only particles of smaller size can move from one compartment to the other through the partition. There may be a partition which will allow the diffusion of only one component the solution, generally the solvent. Such porous partitions or membranes are called semipermeable membranes. Diffusion of solvent through a semipermeable membrane is called ***osmosis***. If two solution of unequal concentrations are separated through a semipermeable membrane, the solven articles will flow from their higher concentration to lower concentration. e.g. if a strong solution of sugar is contained in a bag made up of semisolution or pure water, the water will osmose inside the bag. The osmosis occurs till the equilibrium is achieved. At this stage, the pressure developed by water inside will prevent further entry of water molecules. The pressure which has to be applied to membrane is called osmotic pressure. The osmotic pressure of a solution is directly proportional to the concentration of the solute. The osmotic pressure of a concentrated solution will be higher than that of the dilute solution with the same solute. In other words, if the entry of a solvent in a solution has to be prevented more external pressure has to be applied in case of concentrated solution than dilute one. If the solutions have equal osmotic pressure they are said to be iso-osmatic.

Osmotic pressure depends upon the number of solute particles contained in a given volume of the solvent & not upon chemical composition of the solute. Therefore, the molecules which ionize, exhibits greater osmotic pressure that non-ionizing ones. Such properties of molecules are called ***colligative properties***. The osmotic pressure can be measured with mercury manometer. It can also be calculated from the concentration of solute according to the following equation

$$O.P. = C_iRT$$

Where,

C_i = molar concentration

R = gas constant

T = absolute temperature

For osmosis to occur, solute particles should not pass through the membrane. Most of the biological membranes, however, allow the passage of selective solute particles also, although the process is descried as osmosis. Under electron microscope, it has been observed that many such membranes contain some pore like structure, which may allow solute particles. They are known as ***plasmodesmata.***

DONNAN EQUILIBRIUM

Aqueous solutions of most proteins are colloidal in nature & they do not pass through the biological membranes. If a solution of sodium proteinate containing positively charged sodium ions & negatively charged protein ions is placed in a cellophate bage & immersed in a solution of NaCl, only movement of Na^+ & Cl^- across the membrane occurs. Let us assume that initial concentration of Na^+ inside the bag is C_1, that of protein also C_1 & of Cl^- is zero. The concentration of Na^+ & Cl^- outside, at this time, are C_2 each (fig. 23.2 a & b). After sometime, if x concentration of Cl^- has entered the bag, they will be accompanied by the same amount of Na^+ also so that the neutrality of charge is maintained. At that time the concentration of Na^+ inside the membrane will be C_1 + X & that of Cl^- inside will be. The concentration of Na^+ outside will be C_2- & that of Cl^- outside also will be C_2-.

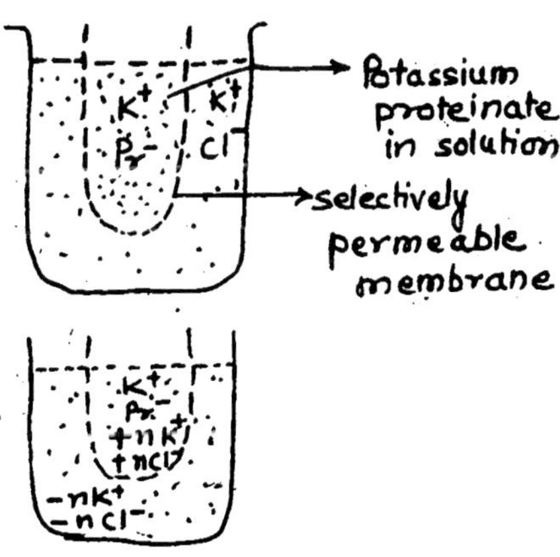

Fig. 23.2a & b : Establishment of Donnan Equilibrium

F.C. Donnan (1927) found that at equilibrium the product of concentrations of diffusible ions inside the bag equals to the product of the ions outside. In this particular case therefore,

$$(Na^+) \text{ in } (Cl^-) \text{ in} = (Na^+) \text{ out } (Cl^-) \text{ out}$$

or

$$(C_1 + x)\ (x) = (C_2 - x)\ (x)$$

or

$$C_1x + x^2 = C_2^2 - 2C_2x = x^2$$

or

$$x = \frac{C_2^2}{C_1 + 2C_2}$$

If the initial concentrations of Cl (C_1) ana Na^+ (C_1 and C_2) are known the amounts of ions entering inside the bag (x) can be calculated according to this equation. The equilibrium achieved in such a system is called ***Donnan equilibrium or Gibbs-Donnan equilibrium***. This kind of equilibrium is achieved only when at least one component is in colloidal form. If these are several aolloidal & non-colloial ions the distribution of ions becomes more complex, but they always follow the rule of Donnan equilibrium & their individual concentration diffusing across the membrane may be calculated.

Donnan equilibrium is of great importance in biological system. In animal tissue, different kinds of cells immersed in the same body fluid differ in their chemical composition. The colloidal & non colloidal components of the cell determine the diffusion of ions across the cell membrane.

VISCOSITY

Viscosity is the property of matters which renders resistance in their flow. It is opposite to fluidity which expresses the tendency to flow. All liquids do not flow with equal rapidity. Liquids such as glycerine, honey, oil & coaltar etc. flow slowly while water, alcohol, ether etc. flow rapidly, viscosity of a liquid is due to internal friction of the molecular layers. Unit of viscosity is ***poise***. viscosity is measured as coefficient of viscosity & normally expressed relative to water, whose viscosity taken as unit at 20°C (1.005 centipoise). the viscosity coefficient of castor oil is about 1000 times more than the water & that of blood serum is about 1.5 to 2.0. The viscosity is dependent upon temperature. With increase in temperature, the viscosity decrease.

Several methods have been employed to determined viscosity within the cell. The rate of layering of various inclusion in a cell during centrifugation has been used in several cases. If the viscosity in relation to only one inclusion is known, the same for others may be calculated.

ACIDITY, ALKALINITY & HYDROGEN ION CONCENTRATION

Acidity, alkalinity of a compound depends upon the kind of ions produced in the solution. Acids from hydrogen ions (H^+) in the solution & hydrogen ion in an acid may be displaced by a solution & hydrogen ion in an acid may be displaced by a metal. An alkali or base forms hydroxyl ion (OH^-) in solution & it reacts with acid to form the salt & water. many salts also produce H^+ or OH^- in aqueous solution & depending upon the relative concentration of H^+ or OH^- they are acidic, basic or neutral. The strength of an acid or base in determined by the degree of its ionization in aqueous solution. Acids like HCl or H_2SO_4 which ionize completely to produce H^+ are stronger acids. Acetic acid & many other organic acids are examples of weak acids. Similarly NaOH or KOH, which dissociate readily to produce OH^-, are strong alkali whereas solutions of organic amine are slightly alkaline. When an acid is mixed with a base, both H^+ & OH^- are produced in the medium. The combine to form neutral water molecules. If the concentration of H^+ is equal to the concentration of OH^-, mixture is neither acidic nor basic. This phenomenon is called ***neutralization***.

An aqueous solution containing more than 1 x 10^{-7} mol/litre OH^- (or less than 1 x 10^{-7} mol/litre OH^+) is alkaline. Thus by the knowledge of concentration of any one of these ions (H^+ or OH^-) the nature of the solution may be determined. The acidic or basic nature of a solution, however, is normally expressed in terms of hydrogen ion concentration. e.g. H^+ conc. of 1 N HCl is about 1 whereas that of 1 N NaOH is about 10^{-14} moles/litre.

pH SCALE

For expressing the hydrogen ion concentration of a solution logarithmic scale was used by a Danish biochemist Sorensen. This scale was called ***pH scale*** where P stands for power potential & H for hydrogen ion concentration. As defined by him the ***pH of a solution is the negative logarithm of the concentration (in moles/litre) of hydrogen ions whithin it contains.***

Thus :

$$pH = -\log (H^+)$$

or

$$= \log \frac{1}{(H^+)}$$

Pure water is only slightly dissociated producing equal amount of H^+ & OH^- according to equation

$$HOH \rightleftharpoons H^+ + OH^-$$

At all time it-exists in dissociated and undissociated stages. The ionization product for water (H^+ or OH^-) has been determined to be 10^{-14} at 25°C. since the amount of $H^+ = OH^-$, the value of any one of these ions will be 10^{-14} or 10^{-7} moles/litre. The pH value for water will be therefore log $1/10^{-7}$ or 7. It is the neutral pH. Solution with less than 7 pH are acidic & those with more than 7 pH are basic. The pH of 1N NaoH containing 10^{-14} moles/litre H^+ will be 1. Generally in biological systems pH values normally ranges between 5 to 8.

BUFFERS

A buffer solution is that which tends to maintain it pH when small amounts of acid or base are added to that. It contains a hydrogen ion donor & a hydrogen ion acceptor form of weak acids & weak bases. A buffer system is most effective when the concentration of H^- donor & H^+ acceptor is equal.

Carbonic acid bicarbonate is a common buffering system in blood plasms. The weak carbonic acid dissociates into H^+ and HCO_3^- as follows

$$H_2CO_3 \rightleftharpoons H^+ + HCO_3^-$$

When a small amount of HCl is added to this system, H^+ ions produced from the acid combine with HCO_3^- ions to form H_2CO_3. If a small amount of NaOH is added, the OH^- produced reacts with H^+ to form water molecules. Thus this system soaks the H^+ or OH^- produced from strong acid or base & tend to maintain the original pH. similarly weak bases & their salts also work as buffer systems.

Buffer systems in the organisms help in carrying on most of the biochemical reaction in a narrow pH range of 6 to 8. The blood for example, maintains its constant pH of about 7.4. despite the fact that it carries a large number & variety of chemicals. Buffer systems provide protection of cells & tissues against sudden change in pH.

ISOMERISM & OPTICAL ACTIVITY

Isomerism is the property of certain organic molecules which renders great variability to some important biological molecules. Different isomers of a compound have same empirical formula but they differ at least one of their physical or chemical property. This difference is either due to difference in the functional groups or in the position of the same functional group. Thus, isomerism may be either positional or geometric type. This is discussed in detail elsewhere. To refresh the same, description in nut shell is as follows :

1. ***Positional Isomerism***

   Positional isomerism is found when one or more groups assume different positions on a carbon skeleton. e.g. dihydroxyacetone & glycereldehyde are positional isomers where position of side groups differ.

2. ***Geometric isomerism***

   When a molecule has at least one assymetric carbon atom, the geometric isomerism arises. A assymetric carbon atom is an atom is an atom to which are attached four groups or atoms will assume different sides. The compounds will thus differ only in the spatial arrangement of the groups. Such isomers are also called stereoisomes or enantiomerphs. Maleic acid and fumaric acid may be cited as examples of geometric isomers which differ only in the position of -COOH & -OH groups on C atom.

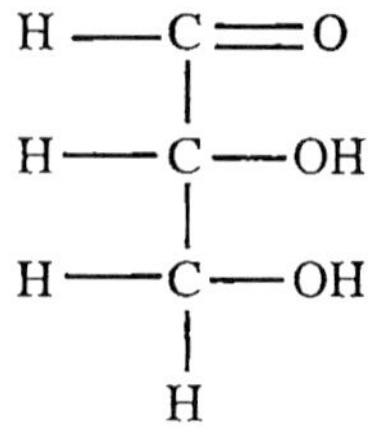

```
H —— C ═══ O
      |
H —— C —— OH
      |
H —— C —— OH
      |
      H
```

Glucereldehyde

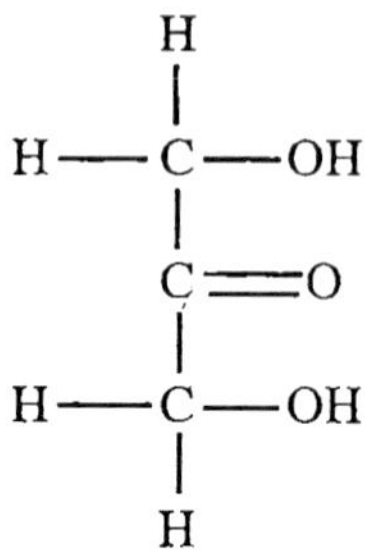

```
        H
        |
H —— C —— OH
        |
        C ═══ O
        |
H —— C —— OH
        |
        H
```

Dihydroxyacetone

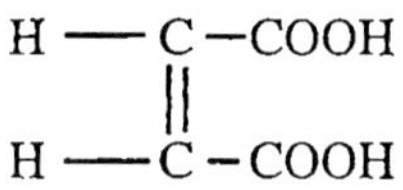

```
H —— C – COOH
      ||
H —— C – COOH
```

Maleic acid

```
HOOC – C —— H
        ||
     H —— C – COOH
```

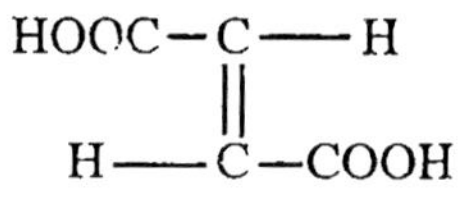

Fumaric acid

Because of asymmetric carbon atom, stereoisomers have the property of turning of plane of polarized light. Polarized light is the light which vibrates in only one plane. This can be achieved by passing ordinary light through a polarimeter. Every optically active compound has a corresponding isomer that roates the plane of polarized light in exactly opposite directions.

Chapter -24

ENERGETICS IN BIOLOGICAL SYSTEM

Energy in the living systems is stored in the form of energy rich or high energy compounds, which are usually phosphate containing molecules. Whenever, the energy is required for metabolic processes, these high energy compounds are broken and energy is released in utilised. The building of these compounds itself is the result of various metabolic activities, oxidation of NADH in the mitochondria being the main reaction.

The theory of biochemical energetics is almost the same as of physico chemical. The molecules involved in bioenergetics, however, are big and complex and the environment in which they play is also complicated. For proper understanding of energy exchanged in biological systems, some basic principles of physicoenergetics will be described.

ENERGY & ITS FORMS

Energy is defined as the capacity to do work. This capacity may be bound in the molecules. The bound form of energy is called potential energy. Potential energy is always measure in terms of work that could be performed if the energy were released. It may be electrical, chemical, photicor light, atomic or positional. In the biological world, potential energy comes from the sun, which is fixed up in the form of chemical energy of carbohydrate molecules during photosynthesis in green plants. Other organisms depend directly or indirectly upon this process (photosynthesis) for the energy. The molecules in motion have capacity to perform work. The energy which a molecule or a body has be virtue of its motion is known as ***kinetic energy***. Kinetic energy depends upon the mass of the body & its velocity of motion according to the equation -

$$KE = \frac{Mv^2}{2}$$

Where,

KE = kinetic energy

m = mass of the body

v = Velocity of the body.

The most important type of energy in biological system is the free energy. It is a energy liberated or consumed in a reaction which can be used in or supplied to another system. Change in free energy of a reaction is sum of the change in the free energies of the products & the reactions :-

$$A + B \rightleftharpoons C + D \pm \text{free energy}$$

Such reactions, which release energy are exergonic as happens in the process of respiration. On the other hand, processes which require energy, such as photosynthesis are endergonic :

In the reversible reaction

$$A + B \rightleftharpoons C + D$$

the equilibrium constants of the products and the reactants according to the equation :

$$K = \frac{(C)\ (D)}{(A)\ (B)}$$

The active concentrations, however, will be related to the energy contents of the products & the reactants. The change in the free energy (ΔF) can be calculated as follows :

$$\Delta F = -RT \ln K$$

Where,

R = gas constant

T = absolute temperature

ln = natural logarithm to the base 10

K = equilibrium constant

Standard free energy changes & equilibrium constants for many reactions have been calculated. A negative value of ΔF means that energy is given off during the reaction e.g. oxidation of glucose has ΔF as — 686 K cal/mole at 25°C. These reactions are exergonic as defined earlier. The endergonic reactions on the other hand, has positive ΔF values. Ionization of acetic acid e.g. has ΔF as + 6.3 K cal/mol. If the value of ΔF is zero a state of equilibrium exist & no net chemical change occurs.

FIRST LAW OF THERMODYNAMICS

Law of conservation of energy

A energy change in a system follows certain basic rules. According to the first law of thermodynamics, ***the energy can neither be created not destroyed but it can changes its form***. It means than in an isolated system, there is no change in the total energy since energy may change form, in calculating the total energy, three quantities have to be known, the work that is done, the heat that is exchanged & the energy that is stored in the system.

In these terms, the first law of thermodynamics may be expressed by the following equation -

$$\Delta E = q - w$$

Where,

ΔE = change in the internal energy of the system,

q = heat absored by the system and,

w = work done.

The value of q & w may be positive or negative, depending upon the nature of reaction. It heat is absorbed e.g. q has a positive value. If heat is given out q has negative value. Likewise if system perform work, the value of w is positive but if the surroundings perform work on the system, the value of the w is negative.

The change in total internal energy does depend upon the pathway but on the two states transformation. e.g. if a molecule a is convertible into b either by path I or by path D the total energy is ΔE_1 & in the other case ΔE_2 then

$$\Delta E_1 = \Delta E_2$$

The equation for first law of thermodynamics inthis case may also be written as

$$\Delta E = E_b - E_a$$

Where,

E_a = energy level of a &

E_b = energy level of b.

Many experiments have been performed to show that energy is conserved in one form or their other conserved in one form or the other during a reaction & first law of thermodynamics is universally accepted. The discovery of Elbert Einstein that the mass is convertible in to energy, however, necessitated the change in definition of first law of thermodynamics. Mass may be converted into energy according to the following equation,

$$E = mc^2$$

Where,

E = energy produced in ergs

m = matter

c = velocity of lighy in cm/sec.

The reverse of this process also occurs that energy can be converted to matter. Therefore, for calculating energy change in a system its mass should also be taken into account. It we include these criteria, then the first law of thermodynamics should be stated as matter and energy can neither be created nor destroyed although they can be interconverted.

Matter to energy or energy to matter conversion, however, are of little importance in biological systems.

SECOND LAW OF THERMODYNAMICS

All the natural processes have a tendency to go to the equilibrium. The second law of thermodynamics is related to the equilibrium state of process. Two types of statements have been given for the second law-

It is impossible by a cyclic process to take heat from a reservoir & convert it to work without at the same time transferring heat from a hot to a cold resevoir

or

It is impossible to transfer heat from a cold to a hot reservoir without at the same time converting a certain amount of work into heat.

Thus, while first law of thermodynamics tell us about the conservation of matter & energy, the second law tells about direction of flow of energy. System try to reach at equilibrium stage from unequilibrium stage from unequilibrium stage. In this kind of change, the system loose their capacity to work slowly. If a system is at equilibrium, no process occurs spontaneously & there is nothing to produce work (Fig. 24.1). As an example of this statement, production of hydro-electric may be cited. Still water can not do any work in that state. But when it is flown from high level to a low level, energy is obtained. Functioning of refrigerator may be another example of this law. Spontaneous flow of energy is from high temperature to low temperature & the reverse is possible only when the work is expanded.

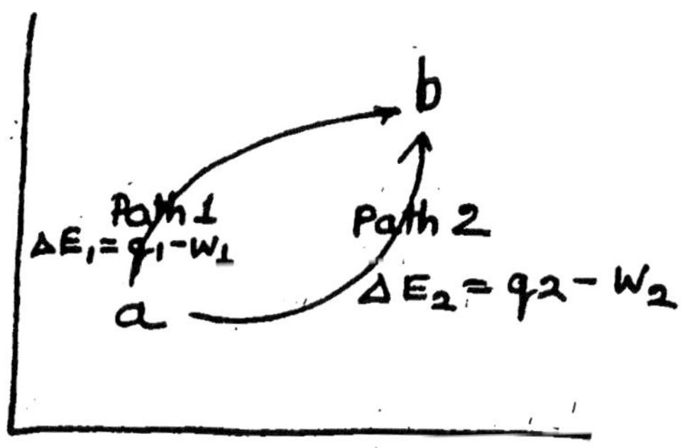

Fig. 24.1 : Pathways for energy conversion of a into b.

It is apparent that a machine like refrigerator is capable of absorbing heat from the surrounding to proceed to a high energy state & to perform work. If we recall the first law of thermodynamics

$$E = q - w$$

the value of q depends upon two factors intensity factor & capacity factor. The intensity factor is the force or the resistance which it is overcome when a body performs work or expends energy. Thus flow of energy will depend more upon the temperature of the system & the surrounding.

$\frac{q}{T}$ of the system + $\frac{q}{T}$ of the surrounding = a positive value

i.e. $\Sigma (q/T) = > 0$

If $\frac{q}{T} = 0$ then the system is at equilibrium.

Therefore, the ratio q/t is a better expression than q-w, because it tells us the state of the system. On the other hand, q may be positive, negative or zero and the reaction will proceed. This new ratio (q/T) is called ***entropy*** & is denoted by the symbol ΔS. It is related to the other parameter of energy as follows -

$$\Delta F = \Delta H - T\Delta S$$

Where,

ΔF = change in free energy

ΔH = change in heat content

T = Absolute temperature

ΔS = entropy

The entropy can be calculated from the knowledge of the increase in heat content & temperature of a particular system according to the equation -

$$\Delta S = \frac{q}{T}$$

Suppose, the entropy change in melting of 1 gm of ice will be 80/273 or 0.293 cal per degree per gram where 80 is the latent heat of the melting of ice & 273 is absolute temperature.

Entropy in physical & biological world is always changing. Whenever there is a spontaneous change in the process the entropy increases. It is also accompanied by the increase in disorder of the system. Let us consider e.g. flow of heat from hot end of a conductor to the cold end. Before the transfer takes place, there if certain amount of order in the conductor. Since hot molecules are at one end & cold molecules at the other. When the spontaenous variation in the concentration of hot & cold molecules in all parts of the conductor. The process of spontaneous flow occurs till the equilibrium is achieved i.e. all parts of the conductor have molecules of similar heat. At equilibrium, the entropy change is zero.

ENERGY RICH COMPOUNDS & ENERGY COUPLING

The importance of high energy phosphate bonds was realized by Fritz Lipmann for the first time. He observed that ΔE of ATP hydrolysis was large and negative and this energy could be used for driving other endergonic reactions. Subsequently, other compounds were also discovered to posses high energy, which was liberated during their hydrolysis. Normally the compound which have ΔE -1 to -5 k cal/mole are low energy compounds, while those having -5 to -15 K cal/mole are high energy compounds. Energy liberated during the hydrolysis of high energy compounds exceeds the ΔE required for formation of most of the biochemical compounds.

In a high energy compound, the bond which is called high energy bond is that bond, which is broken during the reaction to liberate energy. A wriggle bond (~) is used to symbolise these bonds, which involve free energy change of more that 5 K cal., when broken. Thus a molecule of ATP can be written as A P ~ P ~ P where A represents the nucleoside adencsince. Some low and highs energy compounds are listed in Table 24.1.

The high energy content of the molecules is simply the function of their molecular structure. Hydrolysis of high energy compounds releases products which may undergo spontaneous secondary reactions to form stable products. Phosphate compounds are most suitable energy rich compounds because they & their hydrolic products are

Table 24.1 : Some High & Low energy biological compounds

General Structure	*Example*	*ΔE Kcal/mol.*
1	*2*	*3*
$R-\overset{O}{\overset{\Vert}{C}}-NH_2$ (amide)	glutamine	— 3.4
$R-\overset{O}{\overset{\Vert}{C}}-OR$ (esters)	ethyl acetate	— 1.8
> C—O—P (sugar Phosphate)	glucose-6-phosphate	— 2.9
$RO-\overset{O}{\overset{\Vert}{\underset{OH}{\underset{\vert}{P}}}}-O\sim P$ (pyrophosphate)	adenosine diphosphate	—8.0 to 12.0
$R-\overset{NH}{\overset{\Vert}{C}}-\underset{H}{\underset{\vert}{N}}\sim P$	creatine phosphate arginine phosphate	—10.5
$R-\underset{CH_2}{\underset{\Vert}{C}}-O\sim P$ (enol phosphate)	Phosphoenol pyruvate	— 12.8
$R-\overset{O}{\overset{\Vert}{C}}-O\sim P$ (acyl phosphate)	acetyl phosphate	—10.1

stable in water. Other compounds like acetic anhydride also have higher ΔE but they are not suitable as high energy compounds because they are easily hydrolysed in water. To function as a high energy compound for a longer period, the energy rich compound must be stable in water, the biological medium. These compounds are hydrolysed only in the presence of suitable enzymes.

The free energy of high energy compounds in determineded from equilibrium constant of hydrolysis according to the equation,

4E = –2.303 RT log 10K

Measurements have indicated ΔE for hydrolysis of ATP to ADP from 7.6 to 7.8 K cal/mol. of pH 7.0 and 37°C. The free energy of hydrolysis of ADP to AMP is also about 8.0 K cal/mol. but that of AMP to adenosine is only 2.2 K cal. per moles. The electrostatic repulsion of similar charges (phosphate) in ATP & ADP may have added to the release of extra energy in these molecules-such situation does not exist in AMP & hence it is a low energy compound.

In biological systems, ATP serves as a source of energy for most of the endergonic reactions. ATP itself is syntherised in many exergonic process. Thus an endergonic reaction is coupled with an exergonic reaction through ATP. The net effect of this coupling is the transfer of kinetic energy from one system to the other. This coupling is facilitated by several enzymes. Coupling of reaction for efficient transfer of energy is a common biological feature. Let us consider a specific example of conversion of phospho-enol-pyruvate to pyruvate. This is an energy releasing process, energy liberated can be captured inthesynthesis of ATP from ADP. This ATP can be hydrolysed to provide energy for other reactions. e.g. formation of glucose-6-phosphate from glucose & phosphoric acid is an energy requiring process with ΔE = -4 K cal/mol. Hydrolysis of ATP provides energy for the formation of glucose-6-phosphate bonds & the extra energy of high phosphate bond is liberated as such. This series of coupled reaction, may be summarised as follows -

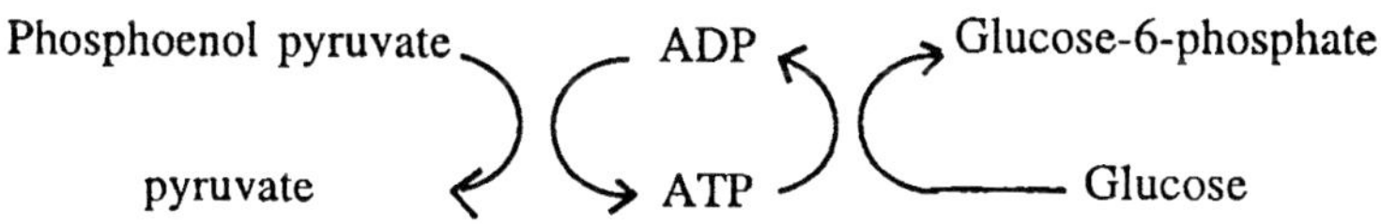

By means of energy coupling a complex compound is synthesized apparently in contradiction to the second law of thermodynamics. But the algebraic sum of the components of a coupled reaction is negative e.g. the sum of the energy liberated is more than the same of the energy consumed. In biological system, the hydrolysis of energy rich compounds is always accompanied by an endergonic reaction so that the energy liberated in the farmer is not wasted. This kind of control is enforced by numerous enzymes occurring in the cell. The high energy compounds dissociate only in the presence of enzymes.

REDOX POTENTIALS

During oxidation of NADH in the mitochondria, to produce ATP, electrons are transferred from one substrate to the other. This flow occurs according to the relative affinities of the substrate to gain or lose electrons. The process of electron transfer is accompanied with the oxidation-reduction of the system. A compound losing electron is oxidized while a compound gaining it is reduced. e.g. ferric ion is oxidised to ferrous ion by losing one electron & vice versa.

$$Fe^{3+} \underset{\text{reduction}}{\overset{\text{Oxidation}}{\rightleftharpoons}} Fe^{2+} + e$$

The quantitative measure of the affinity of a compound to lose or gain electrons is the redox potential (reduction-oxidation potential).

A redox system can be compared to a dry electric cell. In a cell, the electrons are transferred through a wire from one electrode to the other. This generates an electric current. The capacity to gain or lose electrons in such a system is electrode potential. It can be measured through a standard hydrogen electrode, which has an arbitrary electrode potential of zero at 1 N concentration & 1 atmospheric pressure. The electrode potential of a reducing oxidizing system can also be measure in a similar way. Electrode potential in this case will be the redox potential.

Measurement of redox potential of an organic compound

Redox potential of an organic compound can be measured in the laboratory by using a standard platinum electrode as shown in fig. 24.2. Platinum, being an inert material does not dissolve in solution or form iron. One compartment of the apparatus contains reducing system in buffer. Nitrogen gas is passed through the systems simultaneously. The known amount of oxidant is added through a funnel & the mixture is stirred continuously. Potential of the mixture is measured with a potentiometer through platinum & hydrogen or calomel electrodes. Hydrogen electrode is dipped in a different compartment containing KCl solution & connected to the first compartment through a salt bridge. Salt bridge is a KCl-gelatin solution.

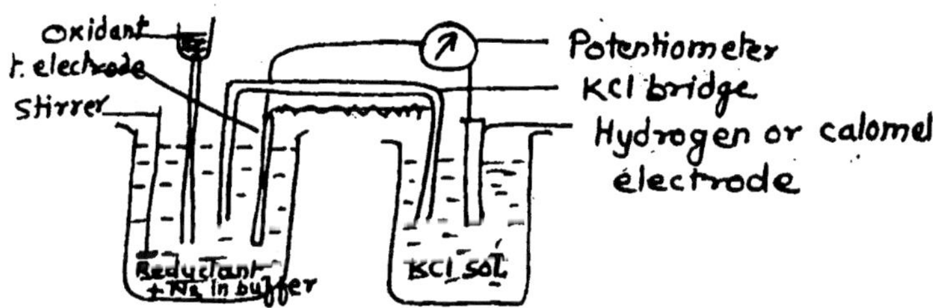

F. 24.2 : Redox potential measurement of an organic compound

Redox potential can be determined indirectly using Peter's equation -

$$\text{Redox potential (E)} = EO + \frac{RT}{nF} \ln \frac{(\text{Oxidant})}{(\text{reductant})}$$

Where E_0 is the redox potential of a mixture containing equimolal concentrations of the oxidant & the reductant.

R is gas constant in joules per mole per degree, T is the absolute temperature, F is Faraday number (96,500 coulombs), n is the number of electrons per gm. equivalent of reactant & in is the logarithm to the base e.

Peter's equation may be simplified under normal conditions of temperature 30°C (absolute 303°C), valency change of 2 & converting in to log 10 as follows,

$$E = E_0 + 0.030 \log_{10} \frac{(\text{Oxidant}}{(\text{reductant})}$$

When the concentration of oxidant is equal to the concentration of reductant the factor $\log_{10}$ (oxidant)/(reductant) becomes O and $E = E_0$.

E_0 values for some of the biological redox systems are given in Table 24.2.

Table 24.2 : Standard redox potential of some biological systems at 25°C

System	E_o
Ferredoxin (red-oxin)	– 0.432
H_2=(2H$^+$ + e)	–0.414
NAD(H)	–0.320
NADP (H)	–0.316
FAD (H_2)	–0.120
FMN (H_2)	–0.120
Cytochrome b (F^{++}/Fe^{+++})	–0.040
Cytochrome C (Fe^{++}/Fe^{+++})	+0.260
Cytochrome A (Fe^{++}/Fe^{+++})	+0.290
Cytochrome A (Fe^{++}/Fe^{+++})	+0.365
H_2O (1/2 O_2 + H^+ +e)	+0.815

Effect of pH on redox potential

Many oxidation reduction reactions involve not only electron transfer but also hydrogen transfer. e.g. the oxidation of hydrogquinone to quinone releases H^+ also

$$\underset{\text{Hydroquinone}}{H_2Q} \underset{\text{reductant}}{\overset{\text{oxidation}}{\rightleftharpoons}} \underset{\text{Quinone}}{Q} + 2H^+ + 2e$$

In these reactions, hydrogen ion concentration (pH) will effect the oxidation -reduction process according to the law of mass action. In the redox system of Fe^{+3}/Fe^{+2}, however, where hydrogen transfer is not involved, pH has little effect on the system.

For calculation of redox potential of hydrogen involving systems, Peter's equation is modified as follows -

$$E = E_0 + \frac{RT}{nF} \ln \frac{(\text{Oxidant})}{(\text{reductant})} + RT \ln (H^+)$$

If the later half of equation is not taken into account, it must be stated that at what pH, E_0 has been measured. In that case E_0 is written as $E_{0'}$, symbolising standard redox potential at a known pH.

REDOX SYSTEMS & FREE ENERGY

As described earlier, during electron transfer in a redox system, electricity is generated which can be used to perform work. The difference in the redox potential of two systems determines the degree of electron flow or in other words the change in free energy. The change in free energy in a redox system is calculated according to the following equation-

$$\Delta E = -nF\ \Delta E_0$$

Where

n = number of electron transferred

F = Farudary number (96,500 Coulombs)

E_0 = difference in redox potential.

In a dry cell with zinc & copper electrodes, 2 electrones are transferred from zinc to copper. The redox values of Zn/Zn^{++} & Cu/Cu^{++} systems are -0.76 & + 0.34 respectively. Therefore, change in energy for the reaction between zinc & copper can be calculated.

E = -2F (0.34) - (-0.76)

= -2 (96.5 K cal) (1.10)

= -212 K cal/molecule

The most important energy releasing biological redox system is the oxidation of NADH in the mitochondria. During this process, the electrons are transferred from NADH to molecular oxygen through cytochromes according to their redox potentials. NADH has a strong affinity to lose electrons, while O_2 can accept it. Each individual step in the process is associated with its own ΔE, which can be calculated from the difference in redox potential (fig. 24.3). When the difference in E_0 in large, the energy generated is used in the chemical synthesis of ATP from ADP +iP. The free energy required for the synthesis of one ATP molecule is about 29K cal/mole. Three ATP molecules are synthesized during complete oxidation of NADH at three steps which have large E_0 to generated enough energy for chemical reaction. If we take the whole series as one reaction ΔE from NADH to molecular oxygen in 1.13. The free energy with this value of ΔE will be 2x96.5x1.13 = 208 K cal according to the equation.

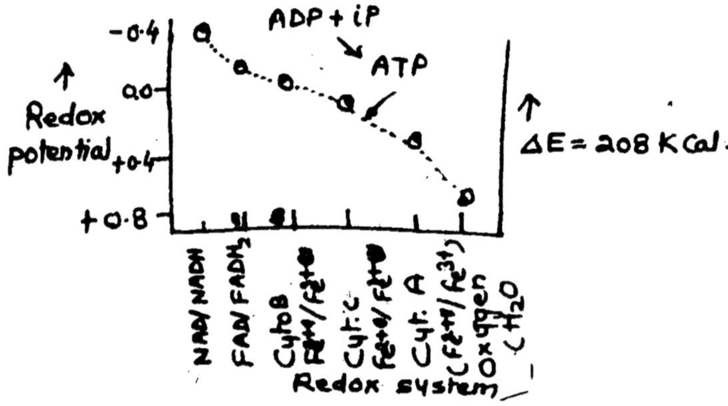

Fig. 24.3 : ATP generation & redox system in ET system.

Synthesis of three molecules of ATP will require 29x3=87 K cal. of energy. Thus, about 42% of the total energy liberated during this process is conserved in the chemical synthesis of ATP. This is quite an efficient system as far as conservation of energy is concerned. Rest of the energy probably escapes as heat.

Chapter -25

REGULATION OF BIOCHEMICAL REACTIONS IN BIOLOGICAL SYSTEM

A normal cell is a complex chemical factory, designed to make the most efficient use of available resources. The two important biochemical functions of a cell are to provide chemical substrates for various reactions & energy to derive reactions. These two functions are performed with a remarkable precision even under diverse environmental conditions of the metabolism, normally do not accumulate in the cell except when an inhibitor is present or there is a genetic block by mutation. The energy produced at any moment is also slightly above the level required for driving the biochemical reactions. The demand for a particular substrate or a particular level of energy, however, may not be uniform during entire life history of an organism. The physiological requirements of a pregnant woman, for example will be different that those of an adolescent child. Similar a plant growing in a desert soil will have different physiology from growing near a pond. Therefore, at certain stages of development or under certain environmental conditions, some biochemical reactions have to be accelerated or suppressed so as to achieve a specific physiology. Besides, biochemical reaction are well regulated to achieve co-ordination among different functions also.

RATIONALE OF THE CONTROL

At cellular level, biochemical reactions are regulated with the following objectives :

1. Unnecessary expenditure of energy in the reaction is avoided.
2. The cell sap may have limited capacity to dissolve the metabolites. By regulating the concentration of metabolites, the sap is always maintained below saturation level.
3. Some intermediate compounds to be used in other metabolism are conserved.

DETERMINATION OF RATE LIMITING STEP

The overall activity of a series of biochemical reactions is normally controlled by one or two steps in the series. The step or the reaction controlling the activity of the series is called the rate limiting step or pace maker reaction. e.g. in a series involving reduction of anitrate & incorporation of amino nitrogen into amino acids & proteins, the reduction of nitrate to nitrite is accelerated, the syntheses of various amino acids and proteins are also accelerated. To identify the rate limiting step in a series, following principles are applied :

IDENTIFICATION OF THE EQUILIBRIUM OF DIFFERENT STEPS

Let us consider a hypothetical example of a reaction converting substrate A to a product F through intermediate compounds B, C, D, & E.

$$A \rightleftharpoons B \rightleftharpoons C \rightleftharpoons D \rightleftharpoons E \rightleftharpoons F$$

If these steps are reversible, the concentration of all the intermediates at equilibrium will be almost equal. The measurements of equilibrium constants of different steps will give an idea of the relative concentrations of different intermediates. The step which is far from equilibrium is normally the rate limiting step.

In a study of apparent equilibrium constants of various steps in glycolysis, it was observed that the constant for conversion of fructose-6-phosphate to fructose-1-6-disphosphate, was far from equilibrium. It has been shown by various other techniques also that this step is rate limiting in glycolysis.

DETERMINATION OF ENZYME ACTIVITY

Each step in a series of biochemical reactions is catalysed by an enzyme. Determination of the activities of different enzymes therefore, may indicate the rate limiting step. Normally, the crude homogenates of the cell are prepared & various activities of various enzymes assayed under varying conditions of substrate concentration, pH & temperature. The enzyme which has the lowest rate of activity may be the pace-maker. e.g. in a study of activities of enzymes of glycolysis in rat liver homogenates, the activity of phosphofructokinase was found to be the lowest in other tissues such as kidney, skeletal muscles & heart also, the level of their enzyme was the lowest. This indicated that phosphofructokinase is the rate limiting enzyme in glycolysis.

The critical enzymes are isolated & purified and their optimum activities are determined.

APPLICATION OF CROSS OVER PRINCIPLE

Application of cross over principle to a series of reactions is probably the most definile technique for determining the pace maker reaction. In the hypothetical example of conversion of A to F, if the first step is rate limiting, exogenous supply of A will have no effect on the production of F. The limiting step is bypassed by supplying B. If a step beyond the synthesis of B is limiting, supply of B will have no effect on the final product. This technique has been applied in determining the rate limiting step in many reactions, where detains of pathway are known.

MECHANISM OF CONTROL

Mass action in a reversible reaction

In a reversible process, the accumulation of products, normally inhibits their own synthesis. e.g. in the reaction-

$$A+B \rightleftharpoons C+D$$

accumulation of C and D prevents further conversion of A & B. If C & D are removed from the scence, however, A & B react to form more C & D. This kind of control is responsible for maintaining equilibrium conditions in many biochemical reactions.

Formation of creatine phosphate (Lohman's reaction) is an example of this type of control.

$$\text{ATP + Creatine} \rightleftharpoons \text{Creatine Phosphate + ADP}$$

The accumulation of creatine phosphate occurs till ATP is in excess. When ATP is limiting then creative phosphate combines with ADP to form ATP.

Substrate sharing

Sometimes, two or more reactions use the same metabolite or cofactor. The availability of metabolites in those cases, determines the rate of reactions. e.g. α-keloglutarate is an intermediate metabolite of Kreb's cycle as well as of flutamate synthesis. In the presence of ammonia, glutamate is syntesized from α Ketoglutarate. Therefore, when ammonia is in adequate supply, Kreb's cycle is suppressed. An example of the competition for same cofactor is that of reduced ferredoxin, which is required for the reduction of both CO_2 (through NADPH) & for nitrite. It has been observed by many investigators that when NO_2 is being assimilated either by isolated chloroplasts or by detached leaves, apparent rate of photosynthesis decreases. Ferredoxin, generated in the light is consumed in the reduction of nitrite & very little reduction of CO_2 occurs.

REGULATION OF ENZYME ACTIVITY

Study of many biochemical reactions have shown that the pace of reaction is maintained by the additives of the enzymes. Enzymes are rather unstable molecules & are synthesized & degraded simultaneously. Their activities may be regulated either through their syntheses or by modifying the existing enzyme molecules. The activities of enzyme molecules are regulated by several ways.

Allosteric regulation

Allosteric regulation is a fine mechanism of controlling a reaction through the enzyme activity. Some enzymes (e.g. allosteric enzymes) shown sigmoidal curve between the substrate concentration & the activity. The activity of these enzymes is modified by several metabolites. The effect of different concentration of activator or inhibitor on these enzymes is also signoid. These effector molecules have a structure different that the substrate molecules. In most of the cases, allosteric inhibitors are the end products of the reaction, inhibiting the first enzyme in the series. Thus, this kind of inhibition is called ***feed back inhibition***, end product inhibition or retroinhibition. The allosteric activators are normally one of the substrates or cofactors of the enzyme. The effect of the allosteric inhibitor or activator on the enzyme is reversible, when they are withdrawn, the enzyme resumes the original activity.

Inhibition of thereonine deaminase by isoleucine is an example of allosteric inhibition. Threonine deaminase deaminates threonine to α ketobutyrate. The final product of the reaction is isoleucine occurs, conversion of threonine to α-ketobutyrate & consequently formation of further intermediates in the biosynthess of isokeucine is stopped when the isoleucine is used up, threonine deaminase is reactivated & reactions for the biosynthesis of isoleucine start again.

Activation of glycogen synthatase by glucose-6-phosphate is an example of allosteric activation. Another example of allosteric regulation (of both inhibitory and activating type) is observed during Pasteur effect. Pasteur effect is the inhibiton of glycolysis and fermentation by Oxygen. The molecular basis of this effect is the allosteric inhibition of enzyme phosphofructokniase by ATP and citrate and its activation by AMP. Like many others, this kind of regulation is also of adaptive significance. As the level of AMP increases due to increased use of ATP in the cell, glycolysis is increased by the ATP. When ATP level exceeds normal requirement of the cell, inhibition of glycolysis occurs through the same enzyme phosphofructokinase and ATP synthesis is shut off.

Regarding the mechanism of allosteric regulation, it is proposed that allosteric enzyme have two active centres, one for the substrate and the other for effector. These two sites apparently lay on two subunits. Binding of a effector molecule to one type of subunit changes the structure of the enzyme molecule in such a way that binding of the substrate to the other submit is effected. To explain the mechanism, an example of allosteric regulation of aspartate transcarboxylase may be cited. Aspartate transcarboxylase contains two types of subunits. The two type of subunits may be split apart by treatment with mercurials, with one type retaining the ability to bind with the substrate, where as the other to recognize the inhibitor. When these two species of subunits are together (active enzyme molecule), binding the inhibitor (CTP) to one type of subunit changes the structure of other subunits in such a way that the binding of the substrate is inhibited. When the subunits containing binding sites for the inhibitor is removed, enzyme is not affected by CTP. Further, it gives a typical Michaelis-Menten curve with the substrate concentration. Similarly, the binding of activator may change the molecular structure in such a way that the binding of substrate is facilitated.

ISOZYME FORMATION

Another phenomenon that controls cellular metabolism is the formation of isozymes (isoenzymess). Isozymes are different physical foms of the same enzyme performing the same general function at different rates. They differ to some extent in their amino acid composition also, so that they may be separated by electrophoresis. Lactic dehydrogenase is a classic example of isozyme formation. It catalyses the oxidation of lactate to pyruvate with the help of NAD.

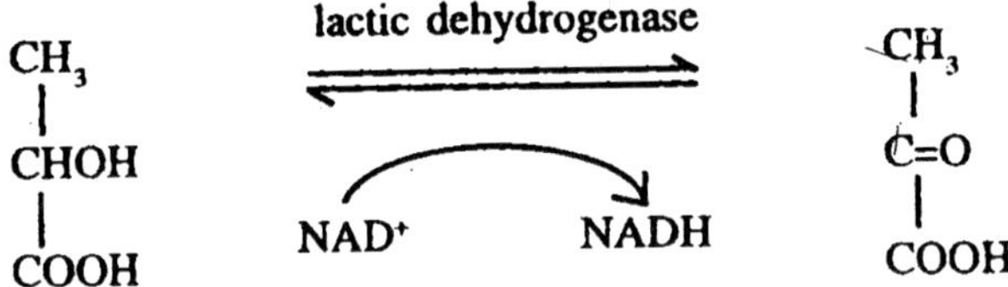

The enzyme is a tetrameter consisting of two distinct types (H & M types) of subunits. Depending upon the relative amounts of two types of subunits, lactate dehydrogenase forms 5 isoenzymes as follows:

LD_1 — HHHH

LD_2 — HHHM

LD_3 — HHMM

LD_4 — HMMM

LD_5 — MMMM

The molecular weight of enzymes is 13,500 but when it is treated with urea or guanidine hydrochloride, it dissociates into two subunits each having a m.w. of about 35,000. The regulation of different isozymes is different. LD_1 (HHHH) type of lactate dehydrogenases is found in the heart muscles. This species is most active as at low pyruvate concentration & is inhibited by high concentration & of pyruvate. LD_5 (MMMM) type of enzyme is found in skeletal muscle cells & it remains active at high pyruvate concentrations.

Another example of isozyme formation is that of aspartokinase, which catalyses the reaction between aspartic acid & ATP to form aspartyl phosphate. Amino acids Lysine, methionine & threonine are final product of the reaction. The enzyme aspartokinase exist in 3 forms - aspartokinase I, II, & III.Aspartokinase I is inhibited by threonine & III by lysine. Aspartokinase II is insensitive to any of these amino acids. Thus, when any once of these amino acids accumulate, the synthesis of the other is affected very little.

MULTIENZYME SYSTEM

Some enzymes exist notes individuals but as aggregates of several enzymes & coenzymes, so as to channel the metabolites in pathway efficiently. En a aggregate each component is arranged in such a way that the product of one enzyme becomes the substrate for the other & so on.

An example of enzyme aggregation is that of pyruvic acid dehydrogenase in E.coli. This complex consists of three enzymes - pyruvate decarboxylase, dihydrolipoic dehydrogenase & lipoyl reductase transacetylase. The co-enzyme associated with the complex are thiamine pyrophosphate (TPP) & flavin adenine dinucleotide (FAD). The step wise reaction catalysed by this complex may be written as follows :

Pyruvate + thiamine pyrophosphate $\longrightarrow$
α-hydroxyethyl tiamine pyrophosphate + CO_2

α-hydroxyethyl thiamine + lipoate $\longrightarrow$
Thiamine pyrophosphate + acetyl dihydrolipoate

Acetyl dihydroipoate + CoASH $\longrightarrow$
acetyl CoA + dihydrolipoate

Dihydrolipoate + NAD^+/FAD^+ $\longrightarrow$
lipoate + $NADH/FADH_2$

REGULATION OF ADENYLATE ENERGY CHARGE

The importance of adenosine phosphate in metabolic process has been well recognized for living systems. The adenylate energy charge is the measure of total pool of adenosine phosphates in the form of ATP, ADP & AMD. Atkinson (1969) defines adenylate energy charge as follows -

$$\text{Adenylate energy charge} = \frac{(\text{ATP}) + 1/2\ (\text{ADP})}{(\text{AMP}) + (\text{ADP}) + (\text{ATP})}$$

In most systems, an increase in adenylate energy charge in the physiological range causes stimulation of regulatory enzyems. Although it is a well known phenomenon in animals & microorganism, some instances have been recorded in plants also. Recently konotz et al (1977) has shown that the adenylate energy charge affects the activity of pyrophospho mevalonate decarboxylase, which is the key enzyme in the biosynthesis of keurene from emvalonate. An increase in enzyme activity is observed between adenylate energy charge of 0.8 & 1.0

ENZYME INDUCTION & ENZYME REPRESSION

In bacterial & certain other microbes, it has been observed that enzyme are synthesized according to the need of the organism. In higher plants and animals, also some examples of this kind are observed. The enzyme nitrate reductase in the plants appears only when the substrate (NO_3) is provided. This involves de novo synthesis of the enzyme because in the presence of protein synthesis inhibitor, cycloheximide, the enzyme activity is low even if there is plenty of nitrate. Ammonium salts & many amino acids inhibit this substrate induction of nitrate reductase, especially in bacteria & fungi. In maize roots, it has been demonstrated that the inhibition of nitrate reductase by an amino acid analogue, canavanine is through repression.

Induction & repression of enzymes are governed by genes. The mechanism has been described elsewhere in this book. The process of induction & repression of enzymes is a slow method of metabolic regulation as compared to the regulation by allosteric or other kind of enzyme manipulation. The process normally requires several minutes to become effective. This lag period is the time required for expression of different genes in the operon. The inducer or repressor acts only when it combines the product of the regulator gene. A hypothetical graph showing time kinetics of enzyme activity.

Many hereditary metabolic disorders in human being are the result of enzyme induction & repression. Altered activities of certain genes may change the production of some key enzymes & thus the level of metabolites. When the operator part of a gene cluser (operon) is inactivated, probably by mutation, the products of structural genes go on accumulating irrespective of the need of the cell.

COMPARTMENTATION OF THE METABOLITES

The cell is divided into numerous specialized cell compartments each performing a specific metabolic functions. Vacuoles, mitochondria and chloroplasts, are bigger compartments of the cell. In addition to these, many smaller compartments such as glyoxysomes, peroxisomes, lysosomes and golgi bodies also exist in a cell. These compartments, which are segregated from others by membranes have all the essential enzymes for carrying out the specific function. This serve as a control on the metabolism of the organism. Thus synthesis of fatty acids (occurring on endoplasmic reticulum) is separated from its degradation (in glyoxysomes & mitochondria). Golgi bodies are concerned with the cell wall synthesis & lysosomes produce only some destructive hydrolytic enzymes.

Compartmentation may regulate biochemical reactions in several ways. Separation of substrate & enzyme in different compartments may inhibit some of the reactions. They are mixed up together only when required, either changed permeability of the membrances isolating them or by activated diffusion. The same metabolite may also be distributed among different compartments. It has been well recognized that there are two pools of amino acids in the ecytoplasm. The amino acids derived from the degradation of proteins are accumulated in the storage pool, whereas those destined to be incorporated into new proteins are stored in metabolic pool. The amino acids of the storage pool are largely catabolized to other carbon compound & CO_2.

Two pools for a simpler compound, nitrate, have been recognized it he plant cells. An interesting case of nitrate compartmentalisation has been observed in maize seedlings.

INDEX

C

D

E

F

G

H

I

J

K

L

M

R

S

T

W

Z